The Biology of Echinostomes

Bernard Fried • Rafael Toledo
Editors

The Biology of Echinostomes

From the Molecule to the Community

Springer

Editors
Bernard Fried
Department of Biology
111 Kunkel Hall
Lafayette College
Easton, PA 18042
USA

Rafael Toledo
Departamento de Biología
Celulary Parasitología
Facultad de Farmacia
Universidad de Valencia
Av. Vicent Andrés Estellés s/n
46100 Burjassot, Valencia
Spain

ISBN 978-1-4419-1867-3 e-ISBN 978-0-387-09577-6
DOI: 10.1007/978-0-387-09577-6

Printed on acid-free paper

springer.com

Preface

Echinostomes are parasitic intestinal trematodes that infect a wide range of vertebrate host species, including humans, in their adult stage, and also parasitize numerous invertebrate and cold-blooded vertebrate hosts in their larval stages. Echinostomes have been studied for many years in relation to their role as basic research models in biodiversity and the systematics of helminths, particularly since the systematics of echinostomes are problematic because of interspecific homogeneity of characteristics to distinguish species. Apart from these aspects, echinostomes have been used extensively as experimental models in parasitology. In a pragmatic sense, echinostomes offer many advantages in experimental parasitology, i.e., easy maintenance of their life cycles in the laboratory, wide distribution of larval and adult stages, and a broad spectrum of intermediate and definitive hosts. For these reasons, echinostomes have contributed significantly to numerous developments in many areas studied by parasitologists and experimental biologists.

In this context, the application of novel techniques is moving the echinostomes into the frontline of parasitology in several areas. In recent years, a number of findings of importance for present and future developments in parasitology have been made using echinostomes as experimental models. Of particular interest are advances in fields such as immunobiology of snails and rodent hosts against parasitic infections, effects of echinostomes and echinostome-like trematodes on natural populations of amphibians, metabolic profiling of trematodes, assessment of anthelminthic drugs, and proteomic studies of intestinal helminth infections. Extensive coverage of the aforementioned topics is included in this book.

A book which presents echinostomes as experimental models should be well received by research workers and advanced students. This book reviews the recent literature on echinostomes, mainly from 1998 to 2007, supplementing the literature covered in the coedited book by B. Fried and T.K. Graczyk in 2000 on Echinostomes as Experimental Models for Biological Research (Kluwer, Dordrecht). In this new book several chapters cover the research and literature on such echinostome-like trematodes as Ribeiroia spp. Emphasis is placed on recent advances and gaps in knowledge that must be filled to determine the importance of this group of digeneans as experimental models. This is critical to gain a full understanding of the potential role of echinostomes in the field of experimental parasitology.

The list of chapters includes some basic subject matter as well as some new topics. All chapters are covered from a modern point of view, considering matters such as the applications of novel techniques and analysis of data in the context of host-parasite interactions. In summary, the main goal of this book is to present the echinostomes in the context of modern parasitology and to show applications of new methodologies and concepts to a group of trematodes that may be useful to obtain information of great value in both parasitology and general biology.

Easton, PA Bernard Fried
Burjassot, Valencia Rafael Toledo

Contents

1 **Echinostomes: Systematics and Life Cycles** 1
José Guillermo Esteban and Carla Muñoz-Antoli

2 **Echinostomes and Snails: Exploring Complex Interactions** 35
Christine Coustau, Benjamin Gourbal, Guillaume Mitta, and Coen Adema

3 **Echinostomes in the Second Intermediate Host** 61
Shamus P. Keeler and Jane E. Huffman

4 **Echinostomes in the Definitive Host: A Model for the Study of Host–Parasite Relationships** 89
Rafael Toledo

5 **Maintenance, Cultivation, and Excystation of Echinostomes: 2000–2007** 111
Bernard Fried and Robert C. Peoples

6 **Echinostomes in the Wild** 129
Arnaldo Maldonado Jr. and Reinalda M. Lanfredi

7 **Echinostomes in Humans** 147
Jong-Yil Chai

8 **Immunology and Pathology of Echinostome Infections in the Definitive Host** 185
Rafael Toledo

9 **Echinostomes: Genomics and Proteomics** 207
Antonio Marcilla

10 Studies of Echinostomes Using Chromatography and Atomic Spectrometry 229
Joseph Sherma and Bernard Fried

11 Effects of Environmental Change on Helminth Infections in Amphibians: Exploring the Emergence of *Ribeiroia* and *Echinostoma* Infections in North America 249
Pieter T. J. Johnson and Valerie J. McKenzie

12 Concurrent Infections of Echinostomes with Helminths and Protozoans 281
Gregory S. Noland and Thaddeus K. Graczyk

13 Use of *Echinostoma* spp. in Studies on Chemotherapy and Metabolic Profiling 295
Jasmina Saric, Jia V. Li, Yulan Wang, Elaine Holmes, Jürg Utzinger, and Jennifer Keiser

Index 325

Contributors

Coen Adema
University of New Mexico, Albuquerque, NM, USA

Jong-Yil Chai
Department of Parasitology and Tropical Medicine, Seoul National University College of Medicine and Institute of Endemic Diseases, Seoul National University Medical Research Center, Seoul 110-799, Korea

Christine Coustau
U547 Inserm, Institut Pasteur de Lille, Lille, France

José Guillermo Esteban
Departamento de Biología Celular y Parasitología, Facultad de Farmacia, Universidad de Valencia, Av. Vicent Andrés Estellés s/n, 46100 Burjassot, Valencia, Spain

Bernard Fried
Department of Biology, 111 Kunkel Hall, Lafayette College, Easton, PA 18042, USA

Benjamin Gourbal
UMR 5244, CNRS Université de Persignan, Via Domicia, Perpignan Cedex, France

Thaddeus K. Graczyk
Department of Environmental Health Sciences, Division of Environmental Health Engineering, Johns Hopkins Bloomberg School of Public Health, Baltimore, MD 21205, USA

Johns Hopkins
Center for Water and Health, Johns Hopkins Bloomberg School of Public Health, Baltimore, MD 21205, USA and
Department of Molecular Microbiology and Immunology, Johns Hopkins Bloomberg School of Public Health, Baltimore, MD 21205, USA

Elaine Holmes
Department of Biomolecular Medicine, Division of Surgery, Oncology, Reproductive Biology and Anaesthetics (SORA), Faculty of Medicine, Imperial College London, Sir Alexander Fleming Building, South Kensington, London SW7 2AZ, UK

Jane E. Huffman
Applied DNA Sciences, Northeast Wildlife DNA Laboratory, Fish & Wildlife Microbiology Laboratory, East Stroudsburg University, East Stroudsburg, PA 18301, USA

Pieter T.J. Johnson
Ecology and Evolutionary Biology Department, Ramaley N122, CB 334, University of Colorado, Boulder, CO 80309-0334, USA

Shamus P. Keeler
Southeastern Cooperative Wildlife Disease Study (SCWDS), Department of Population Health, College of Veterinary Medicine, University of Georgia, Athens, GA 30602, USA

Jennifer Keiser
Department of Medical Parasitology and Infection Biology, Swiss Tropical Institute, 4002 Basel, Switzerland

Reinalda M. Lanfredi
Laboratório de Biologia de Helmintos Otto Wuchere, Programa de Biologia Celular e Parasitologia, Instituto de Biofísica Carlos Chagas Filho, Universidade Federal do Rio de Janeiro, 21941-902 Rio de Janeiro, RJ, Brazil and
Laboratório de Biologia Celular, Programa de Pós-graduação em Biologia de Agentes Infecciosos e Parasitários, Universidade Federal do Pará, Belém, Pará, Brazil

Jia V. Li
Department of Public Health and Epidemiology, Swiss Tropical Institute, 4002 Basel, Switzerland and
Department of Biomolecular Medicine, Division of Surgery, Oncology, Reproductive Biology and Anaesthetics (SORA), Faculty of Medicine, Imperial College London, Sir Alexander Fleming Building, South Kensington, London SW7 2AZ, UK

Arnaldo Maldonado Jr.
Laboratório de Biologia e Parasitologia de Mamíferos Silvestres Reservatórios, Instituto Oswaldo Cruz, Fundação Oswaldo Cruz, 21045-900 Rio de Janeiro, RJ, Brazil

Antonio Marcilla
Departamento de Biología Celular y Parasitología, Facultad de Farmacia, Universidad de Valencia, Av. Vicent Andrés Estellés s/n, 46100 Burjassot, Valencia, Spain

Valerie J. McKenzie
Ecology and Evolutionary Biology Department, Ramaley N122, CB 334, University of Colorado, Boulder, CO 80309-0334, USA

Guillaume Mitta
UMR 5244, CNRS Université de Persignan, Via Domicia, Perpignan Cedex, France

Carla Muñoz-Antoli
Departamento de Biología Celular y Parasitología, Facultad de Farmacia, Universidad de Valencia, Av. Vicent Andrés Estellés s/n, 46100 Burjassot, Valencia, Spain

Gregory S. Noland
Division of Infectious Diseases and Global Pediatrics Program, Department of Pediatrics, University of Minnesota, Minneapolis, MN 55455, USA

Robert C. Peoples
Department of Biology, 111 Kunkel Hall, Lafayette College, Easton, PA 18042, USA

Jasmina Saric
Department of Public Health and Epidemiology, Swiss Tropical Institute, 4002 Basel, Switzerland and
Department of Biomolecular Medicine, Division of Surgery, Oncology, Reproductive Biology and Anaesthetics (SORA), Faculty of Medicine, Imperial College London, Sir Alexander Fleming Building, South Kensington, London SW7 2AZ, UK

Joseph Sherma
Department of Chemistry, 326 Hugel Science Center, Lafayette College, Easton, PA 18042, USA

Rafael Toledo
Departamento de Biología Celulary Parasitología, Facultad de Farmacia, Universidad de Valencia, Av. Vicent Andrés Estellés s/n, 46100 Burjassot, Valencia, Spain

Jürg Utzinger
Department of Public Health and Epidemiology, Swiss Tropical Institute, 4002 Basel, Switzerland

Yulan Wang
Department of Public Health and Epidemiology, Swiss Tropical Institute, 4002 Basel, Switzerland

Chapter 1
Echinostomes: Systematics and Life Cycles

José Guillermo Esteban and Carla Muñoz-Antoli

Contents

1.1 Echinostomes: Systematics and Life Cycles 2
1.1.1 Introduction 2
1.1.2 Systematic Studies 2
1.1.3 An Overview and Recent Advances in the Biology of Echinostomes and Echinostome-Like Digeneans 8
1.1.4 Concluding Remarks 25
References 26

Abstract This chapter provides a review of the most significant literature in the last decade on the systematics and biology of echinostomes and echinostome-like digeneans. This review is primarily concerned with members of the genus *Echinostoma*, although members of other genera (*Echinoparyphium, Echinochasmus, Himasthla, Hypoderaeum, Petasiger, Euparyphium, Stephanoprora, Isthmiophora,* and *Acanthoparyphium*) and echinostome-related genera (*Parorchis, Philophthalmus* and *Ribeiroia*) are also considered. The literature on molecular systematics and morphometrics of these trematodes is reviewed. Specific mention is made of the life cycle patterns of echinostome and echinostome-like digeneans along with an overview of recent advances on different topics in the biology of these trematodes; the review covers various aspects of the different stages of these organisms, i.e., free-living stages (miracidia and cercariae), and parasitic stages in the invertebrate hosts (sporocysts, rediae, and metacercariae) and vertebrate hosts (adults).

J.G. Esteban (✉) and Carla Muñoz-Antoli
Departamento de Biología Celular y Parasitología, Facultad de Farmacia, Universidad de Valencia, Av. Vicent Andrés Estellés s/n, 46100, Burjassot-Valencia, Spain

B. Fried and R. Toledo (eds.) *The Biology of Echinostomes*,
DOI: 10.1007/978-0-387-09577-6_1,

1.1 Echinostomes: Systematics and Life Cycles

1.1.1 Introduction

Echinostomes are intestinal trematodes with an extensive literature dealing with them. In the last decade, three major reviews (Fried 2001; Fried and Graczyk 2004; Toledo et al. 2006), and the edited book by Fried and Graczyk (2000), provided ample evidence of the importance of these digeneans. Furthermore, the use of echinostomes as experimental models in host-parasite relationships between adult parasites and vertebrate hosts, and between larval trematodes and invertebrate and cold-blooded vertebrate hosts, has been highlighted by Toledo and Fried (2005) and Toledo et al. (2007), respectively.

In order to conduct studies in different areas of research, such as systematics, experimental, manipulative, ecological, physiological, biochemical, immunological, molecular, treatment, and control, two important features are needed: knowledge of which echinostome species are involved and the ability to maintain life cycles to do research on all echinostome stages, i.e., free-living, intramolluskan, and the adult stages.

Relative to these topics, this chapter provides an overview of the most relevant literature on the systematics and biology of echinostomes and echinostome-like digeneans. This review is mainly focused on the most significant literature of the last decade, although other previous papers are mentioned if needed; this review is mainly concerned with members of the genus *Echinostoma* because most research has focused on them; however, members of the other genera of the Echinostomatidae are also mentioned, including the echinostome-like digeneans.

1.1.2 Systematic Studies

Since the previous book edited by Fried and Graczyk (2000), some new papers and significant exhaustive revisions have appeared relative to the echinostomes. Of these, Cribb et al. (2001), and particularly Kostadinova and Jones (2005) and Kostadinova (2005a) in the recent volume 2 of the "Keys to the Trematoda," have defined the current classification of the Echinostomatoidea to the generic level. Among the eight families recognized by Kostadinova and Jones (2005), Echinostomatidae is characterized by the presence of a circumoral head-collar armed with one or two crowns of large spines interrupted ventrally; theses spines are larger than the tegumental spines (Fig. 1.1).

At the generic level, relatively few changes have been made, mainly in relation to the genus *Singhia*, reallocated to Echinostomatinae (see Table 1.1), whereas at the specific level, two new 37-collar spined species of *Echinostoma* belonging to the "*revolutum*" group have been described in the last decade. *E. friedi* was described from Spanish material on the basis of several morphological and biological

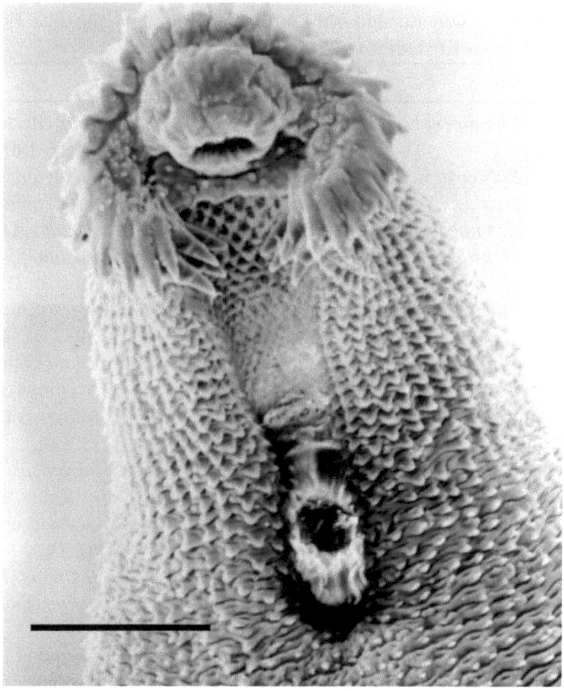

Fig. 1.1 Scanning electron micrograph of the anterior third region of an 11-day-old adult of *Euparyphium albuferensis* recovered from the intestine of an experimentally infected albino mouse. Scale bar: 100 μm

features of the life cycle stages and in its cercarial chaetotaxy (Toledo et al. 2000). *E. deserticum* was described from African materials (Kechemir et al. 2002). *E. luisreyi* has been described from Brazilian materials on the basis of the oral corner spines that increased in size from the latero-oral to the ventro-oral regions, as well as the excretory pore radially wrinkled and dorsally subterminal (Maldonado et al. 2003).

The remaining published studies refer to descriptions of new species belonging to other genera of Echinostomatidae, and to refined diagnoses and redescriptions of both type specimens and/or newly collected materials, and critical evaluation of the published data belonging to different species of genera, i.e., *Stephanoprora* (Ostrowski de Nuñez and Quintana 2007) and *Uroproctepisthmium* (Kostadinova and Gibson 2001a) (Echinochasminae), *Drepanocephalus* (Kostadinova et al. 2002), *Echinoparyphium* (Kostadinova et al. 2003), *Echinostoma* (Kostadinova et al. 2000a, b, 2003; Maldonado et al. 2001a), *Euparyphium* (Kostadinova and Gibson 2002; Kostadinova et al. 2003), *Hypoderaeum* (Kostadinova et al. 2003), *Isthmiophora* (Kostadinova and Gibson 2002; Kostadinova et al. 2003), *Paryphostomum* (Kostadinova et al. 2002), *Petasiger* (King and Van As 2000;

Table 1.1 Type species, genera, and subfamilies included in the last Echinostomatidae classification according to Kostadinova (2005a)

Subfamily	Genus	Type species
Chaunocephalinae	*Chaunocephalus* Dietz, 1909	*C. ferox* (Rudolphi, 1795) Dietz, 1909
	Balfouria Leiper, 1908	*B. monogama* Leiper, 1908
Echinochasminae	*Dissurus* Verma, 1936	*D. farukhabadi* Verma, 1936
	Echinochasmus Dietz, 1909	*E. coaxatus* Dietz, 1909
	Mehrastomum Saksena, 1959	*M. minutum* Saksena, 1959
	Microparyphium Dietz, 1909	*M. facetum* Dietz, 1909
	Pulchrosomoides Freitas et Lent, 1937	*P. elegans* Freitas et Lent, 1937
	Stephanoprora Odhner, 1902	*S. ornata* Odhner, 1902/ *S. spinulosa* Dietz, 1909
	Saakotrema Skrjabin et Bashkirova, 1956	*S. metatestis* (Saakova, 1952)
	Uroproctepisthmium Fischtal et Kuntz, 1976	*U. taiwanense* Fischtal et Kuntz, 1976
Echinostomatinae	*Bashkirovitrema* Skrjabin, 1944	*B. incrassatum* (Diesing, 1850)
	Drepanocephalus Dietz, 1909	*D. spathans* Dietz, 1909
	Echinodollfusia Skrjabin et Bashkirova, 1956	*E. stenon* (Dollfus, 1950)
	Echinoparyphium Dietz, 1909	*E. elegans* (Looss, 1899)
	Echinostoma Rudolphi, 1809	*E. revolutum* (Frölich, 1802)
	Euparyphium Dietz, 1909	*E. capitaneum* Dietz, 1909
	Hypoderaeum Dietz, 1909	*H. conoideum* (Bloch, 1782)
	Isthmiophora Lühe, 1909	*I. melis* (Schrank, 1788)
	Longicollia Bykhovskaya-Pavlovskaya, 1954	*L. echinata* Bykhovskaya-Pavlovskaya, 1954
	Lyperorchis Travassos, 1921	*L. lyperorchis* Travassos, 1921
	Moliniella Hübner, 1939	*M. anceps* (Molin, 1859)
	Neocanthoparyphium Yamaguti, 1958	*N. petrowi* (Nevostrueva, 1953)
	Pameileenia Wright et Smithers, 1956	*P. gambiensis* Wright et Smithers, 1956
	Parallelotestis Belopol'skaya, 1954	*P. horridus* Belopol'skaya, 1954
	Paryphostomum Dietz, 1909	*P. radiatum* (Dujardin, 1845)
	Petasiger Dietz, 1909	*P. exaeretus* Dietz, 1909
	Prionosoma Dietz, 1909	*P. serratum* (Diesing, 1850)
	Prionosomoides Freitas et Dobbin, 1967	*P. scalaris* Freitas et Dobbin, 1967
	Singhia Yamaguti, 1958	*S. thapari* (Singh, 1953)
Himasthlinae	*Acanthoparyphium* Dietz, 1909	*A. phoenicopteri* (Lühe, 1898)
	Aporchis Stossich, 1905	*A. croaticus* (Stossich, 1905)
	Artyfechinostomum Lane, 1915	*A. sufrartyfex* Lane, 1915
	Caballerotrema Prudhoe, 1960	*C. brasiliense* Prudhoe, 1960
	Cloeophora Dietz, 1909	*C. micata* Dietz, 1909
	Curtuteria Reimer, 1963	*C. numenii* Reimer, 1963
	Hismasthla Dietz, 1909	*H. rhigedana* Dietz, 1909
Ignaviinae	*Ignavia* Freitas, 1948	*I. venusta* Freitas, 1948
Neophrostominae	*Nephrostomum* Dietz, 1909	*N. ramosum* (Sonsino, 1895)

(continued)

Table 1.1 (continued)

Subfamily	Genus	Type species
	Patagifer Dietz, 1909	*P. bilobus* (Rudolphi, 1819)
Pegosominae	*Pegosomum* Ratz, 1903	*P. saginatum* (Ratz, 1897)
Pelmatostominae	*Pelmatostomum* Dietz, 1909	*P. episemum* Dietz, 1909
Ruffetrematinae	*Ruffetrema* Saxena et Singh, 1982	*R. indirae* Saxena et Singh, 1982
Sodalinae	*Sodalis* Kowalewski, 1902	*S. spathulatus* (Rudolphi, 1819)

Zamparo et al. 2005; Kostadinova and Skirnisson 2007), and *Singhia* (Kostadinova and Gibson 2001b) (Echinostomatinae), and *Acanthoparyphium, Caballerotrema, Curtuteria* and *Himasthla* (Himasthlinae) (Kostadinova and Gibson 2001b; Díaz and Cremonte 2004; Desclaux et al. 2006; Martorelli et al. 2006).

The echinostome-like trematode group has had a confusing and controversial taxonomic history. Numerous authors have commented on the close relationship between this group, basically the genus *Ribeiroia*, and echinostomatids based on morphological ("wish-bone" type of intestine with solid esophagus and cecae; excretory bladder shape, collecting tubule pattern and concretion character and amount; and flame cell pattern) and ecological (infecting amphibians, fishes, birds, and mammals) similarities. However, the genus *Ribeiroia* appears to be more species specific at the level of the first intermediate host and is not known to form metacercariae within snails (Johnson et al. 2004). In addition, *Ribeiroia* lacks a typical echinostome collar and circumoral spines, and possesses characteristic esophageal cecae or diverticula. All these features have allowed Kostadinova (2005b) to consider the genus *Ribeiroia*, and 12 other genera within the family Psilostomidae. Other genera also considered echinostome-like digeneans, such as *Parorchis* and *Philophthalmus*, are included within the family Philophthalmidae (Kanev et al. 2005).

1.1.2.1 Recent Information on Molecular Systematics and Morphometrics

The echinostomes constitute a group of digeneans characterized by great confusion regarding their systematic classification. This is attributable to a number of factors, including misidentified species or species that have been insufficiently described, as well as the existence of substantial interspecific homogeneity of the morphological characteristics of the adult stage. Such confusion is evidenced not only in the 37-collar-spined *Echinostoma* belonging to the "*revolutum*" group, but also in members of 43- and 45- collar-spined *Echinoparyphium* (*E. elegans* complex and *E. recurvatum* complex, respectively). However, the studies published in the last decade, and basically those of Kostadinova and coauthors, have contributed to further our understanding of the systematics of the echinostomes.

Kostadinova and Gibson (2000), in their exhaustive and rigorous review, contributed an update to our knowledge on the systematics of echinostomes, particularly in relation to the 37-collar-spined "*revolutum*" group of *Echinostoma*. These authors studied morphologically some of the materials used by different workers in their descriptions and analyzed the causes of the errors made in many of the diagnoses. In addition, they established the characteristic features of the species-rich genera (*Echinochasmus*, *Echinoparyphium*, *Echinostoma*, *Himasthla*, and *Stephanoprora*) and the features of the species recognized to date within the "*revolutum*" group: *E. revolutum* Frölich 1802, *E. echinatum* (Zeder 1803), *E. jurini* (Skvortsov 1924), *E. caproni* Richard 1964, *E. trivolvis* (Cort 1914), *E. paraensei* Lie et Basch 1967, *E. parvocirrus* Nassi et Dupouy 1988, *E. miyagawai* Ishii 1932, and *E. friedi* Toledo, Muñoz-Antolí et Esteban 2000.

Of relevance is the exhaustive analysis made to date by Kostadinova and Gibson (2000) on the main features used for differentiation at the specific level in the "*revolutum*" species group, that is: number and disribution of the paraesophageal gland-cell outlets in cercariae; patterns of the distribution of sensilla in cercariae; adult morphology; final host-parasite compatibility; behavior related to distribution of adults along the intestine of definitive host; mating behavior; geographical distribution; and finally genetic characteristics, related to karyotyping or chromosome counting and comparison, isoelectric focusing, and enzyme electrophoresis.

These criteria have been the subject of a recent critical comment by Fried and Toledo (2004), who point out that these characters considered individually are of limited value overall when trying to identify the 37-collar-spined *Echinostoma* group, and suggest that for the diagnosis and description of new species integrated and comparative studies involving adequate morphological and biological features are necessary.

However, morphological and biological studies are complicated *a priori*, since in many cases they imply the development and maintenance of the full life cycle, as well as the description of each of the different parasite stages in the cycle.

As an alternative to the classical morphological studies, molecular tools, especially primary sequence comparisons, are now being used for life cycle elucidation, the examination of potential cryptic species, the examination of species complexes and their phylogeographical genetic structure, and phylogenetic studies (Nolan and Cribb 2005; Vilas et al. 2005). Logically, the molecular data have the capacity to allow comparisons that remove confounding factors of age, host or geographically based variation; and the nature of molecular techniques means that it is possible to bypass studies of the biology of parasites.

However, relatively few molecular-based studies on echinostomes have been carried out to date. Earlier studies involved molecular techniques such as karyotyping and enzyme electrophoresis. More recent studies are based on the analysis of different DNA sequences and their utility for the species differentiation, to analyze the echinostome phylogenetics, and for the diagnosis of the echinostomes.

The random amplification of polymorphic DNA (RAPD) using polymerase chain reactions (PCR) is useful for identifying closely related echinostome species, i.e., *E. trivolvis* and *E. caproni* (Fujino et al. 1995), *E. caproni* and *E. paraensei*

(Petrie et al. 1996), *and Echinochasmus fujianensis* (conspecific to *E. liliputanus*) *and E. japonicus* (Cheng et al. 1999).

Sequence data from both mitochondrial DNA (mtDNA), nicotinamide adenine dinucleotide dehydrogenase subunit 1 (ND1) and cytochrome c oxidase subunit 1 (CO1), and nuclear ribosomal DNA (rDNA), internal transcribed spacers 1 and 2 (ITS1 and ITS2), have also been used as markers for identifying species of echinostomes and several strains of echinostomes isolated from different geographical locations, particularly for the four most studied species in the 37-collar-spined "*revolutum*" group, i.e., *E. caproni*, *E. paraensei*, *E. revolutum*, and *E. trivolvis* (Sorensen et al. 1998; Morgan and Blair 1995, 1998a,b, 2000; Maldonado et al. 2001a; Kostadinova et al. 2003).

Molecular studies on genera other than *Echinostoma*, or those on echinostome-like genera, are scant. Grabda-Kazubska et al. (1998) sequenced ITS1 from rediae and cercariae to distinguish *Echinoparyphium elegans, E. recurvatum, E. pseudorecurvatum, Pseudechinoparyphium echinatum, Neocanthoparyphium echinatoides*, and *Hypoderaeum conoideum.* These authors suggested a close relationship between *H. conoideum* and the three species of *Echinoparyphium.*

Special mention should be made of the results obtained by Kostadinova et al. (2003). These authors sequenced the ND1 and ITS genes of six species in the genera *Echinostoma* (including *E. revolutum*), *Echinoparyphium, Hypoderaeum and Isthmiophora,* and concluded that ND1 sequence provided useful information for the construction of a phylogenetic tree, which agrees with morphological identifications, and that the ITS gene provided insufficient resolution for distinguishing echinosome species. In addition, these authors reviewed the morphology of selected individuals and demonstrated congruence between morphological and molecular identification of the species examined.

In relation to the echinostome-like group, i.e., *Ribeiroia*, studies on ITS2 sequences have allowed for the recognition of three species within this genus: *R. ondatrae*, *R. congolensis*, and *R. marini* (Wilson et al. 2005). These results are in accord with the previous morphological studies and systematic classification of Yamaguti (1971, 1975). However, Wilson et al. (2005) noted that further molecular studies will be needed to elucidate the correct classification of *Ribeiroia* spp. into either Psilostomidae or Cathaemasiidae.

However, molecular techniques for the identification of echinostomes also have problems. Probably, the most relevant of which are questions related to materials used for these studies. The molecular studies require the availability of materials morphologically referred to a genus and species, in order to allow for adequate sequencing and an introduction of the data in the GenBank. It is therefore clear that the deposited type sequence of each echinostome species should be obtained from the "voucher" of type material, or alternatively from materials corresponding to the type locality ("terra typica"). This would allow comparative analysis of the results of the sequences of materials presumably based on morphological grounds to the same or other species, and from the same or different geographical origins. The rigorous retention of morphological vouchers for all sequenced samples, and ideally their placement in museums, would be the most correct approach.

However, some type specimens cannot be located, are inaccessible, or are in too poor a state of preservation to allow for detailed morphological examination. In this case, efforts must be made to obtain materials from the type locality, or alternatively, the material must be described morphologically before applying molecular studies since several problems may occur. It should possibly be considered that there exists more than one species in a sample and/or that two different samples represent the same species as a consequence of a misidentification error, mislabeling, or contamination. These aspects have been addressed by Kostadinova and coworkers, who observed that even the type material vouchers deposited in museums may in fact contain a pool of different species, rather than a single species only (Kostadinova and Gibson 2000; Kostadinova et al. 2000a,b, 2003).

Molecular studies are not usually accompanied by morphological studies to confirm the identity of the material. This represents an added problem, making it difficult to do comparative studies in situations where the application of molecular techniques may not be feasible.

Overall, molecular studies for species differentiation have not yielded results truly different from those already known through traditional morphological research – when the latter research has been exhaustive and rigorous. Moreover, in those cases where there is no agreement between schemes of classification based on morphology and those based on molecular techniques there are problems fundamentally related to voucher of type material and/or the incorrect identification of samples used.

Although the new technologies tend to displace the traditional systematic methods, the most reasonable approach in view of the aspects commented is that both systematic studies, i.e., based on morphological and molecular data, should be regarded as complementary and should coexist.

1.1.3 An Overview and Recent Advances in the Biology of Echinostomes and Echinostome-Like Digeneans

In the present section, information published from 2000 to date will be considered. Most of this information is concerned with four *Echinostoma* species of the "*revolutum*" group, i.e., *E. caproni*, *E. paraensei*, *E. trivolvis*, and *E. friedi*. For other *Echinostoma* species (*E. revolutum, E. miyagawai, and E. luisreyi*), other echinostomatid genera (*Echinoparyphium, Echinochasmus, Himasthla, Hypoderaeum, Petasiger, Euparyphium, Stephanoprora, Isthmiophora,* and *Acanthoparyphium*) or echinostome-related genera, such as *Parorchis, Philophthalmus*, and mainly *Ribeiroia*, the literature is not extensive.

The most relevant reviews on the biology of echinostomes and echinostome-like digeneans are as follow:

Huffman and Fried (1990) reviewed the significant earlier literature on *Echinostoma* and echinostomiasis to about 1988 and covered, among others, different aspects related with the biology of 37-collar-spined echinostomes, basically *on E. caproni*,

E. trivolvis, *E. paraensei*, *E. revolutum*, and *E. echinatum*, as well as the infectivity studies in first and second intermediate hosts, and the infectivity, growth, and development in vertebrate hosts. Later on, Fried and Huffman (1996) published an exhaustive monographic review of all the citations relating to *E. caproni* up to that date.

The review of Kanev et al. (2000), in the edited book by Fried and Graczyk (2000), compiled the echinostome citations from 1970 to 1998. Moreover, the most significant contributions to the biology of echinostomes from the eighteenth to the twentieth century are listed, and specific mention is made on the life cycle pattern of echinostomes and on the form and function for each of the different echinostome stages: egg, miracidium, sporocyst, redia, cercaria, metacercaria, and adult. The most recent reviews of Fried (2001) and Fried and Graczyk (2004) have served as a background for this overview of biology of echinostomes.

Among the echinostome-like digeneans, the genus *Ribeiroia* has the largest number of citations, because it can cause severe pathology and mortality in snails and amphibian intermediate hosts. The recent revision of Johnson et al. (2004) also serves as a compilation for this genus. For *Philophthalmus* and philophthalmiasis, the reviews of Nollen and Kanev (1995) and Nollen (1997) serve as an overview of this genus. Moreover, Graczyk and Fried (2007) discussed the transmission of human and animal philophthalmids by contaminated water.

1.1.3.1 Life Cycles

Members of echinostome and echinostome-related genera follow a three-host life cycle: a vertebrate definitive host, an invertebrate first intermediate host (usually an aquatic gastropod mollusk), and a second intermediate host carrying the encysted metacercarial stage. The life cycle consists of six phases (Fig. 1.2):

1. Passage of eggs from the definitive host to the outside environment and their subsequent development;
2. Hatching of miracidia, and search for and penetration of the first intermediate snail host;
3. Development and multiplication of the parasites inside the snail;
4. Emergence of the cercariae from the snails and search for the second intermediate host;
5. Penetration and encystment of the cercariae in the second intermediate host;
6. And, finally, ingestion of infective metacercariae by the final hosts and development to adult worms.

Some differences in relation to this general life cycle pattern may be observed. For example, fully developed eggs of *E. caproni* may hatch in the snail gut (Idris and Fried 1996), or even lymnaeid snails may become infected by ingestment of unhatched embryonated *E. revolutum* eggs (Davis 2005). The cercariae of *Himasthla quissetensis* infect directly domestic chicks through the cloaca, with worm development occurring in the ileum (Herman and Bacha 1978). Cercariae of *Echinochasmus* spp. infect humans and other mammals when they are ingested by

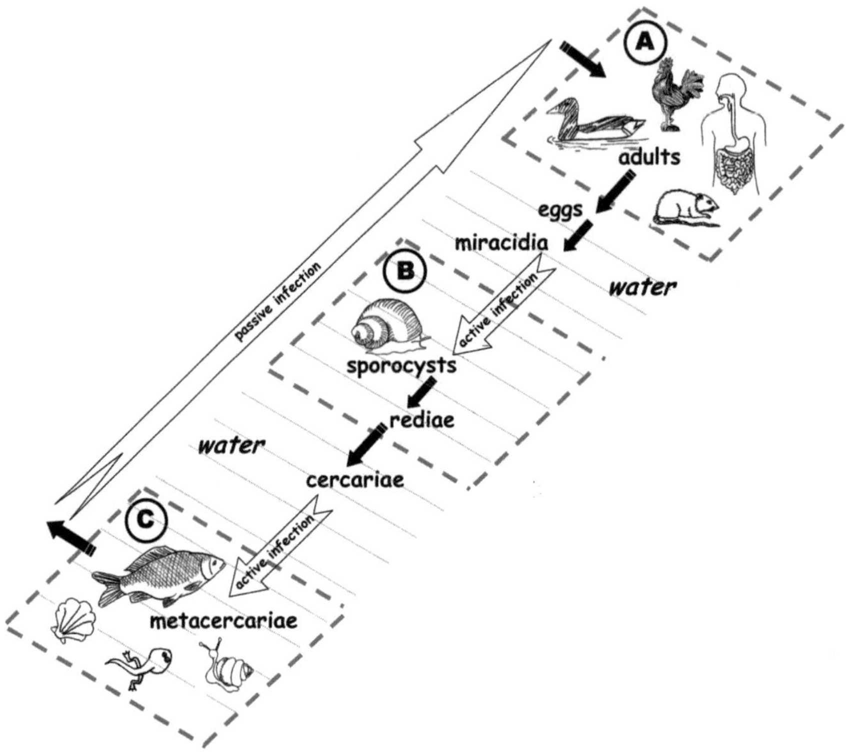

Fig. 1.2 Generalized life cycle of an echinostome. (**a**) definitive hosts; (**b**) first intermediate hosts; (**c**): second intermediate hosts

the host in contaminated water (Xiao et al. 1992, 1995). Cercariae of *Parorchis acanthus*, an echinostome-like species, emerge from infected dogwhelks (*Nucella lapillus*) and after a brief period of free swimming, they encyst on solid objects in the nearby locality (Rees 1966).

In the period reviewed, some studies have involved the description of life cycles: *E. friedi* in Spain, *E. miyagawai* in Bulgaria, *E. deserticum* in Niger, and *E. luisreyi* in Brazil (Toledo et al. 2000; Kostadinova et al. 2000b; Kechemir et al. 2002; Maldonado et al. 2003); *Echinoparyphium recurvatum* in Korea (Sohn 1998), and *E. megacirrus* and *Echinoparyphium* sp. in Argentina (Semenas et al. 1999; Prepelitchi and Ostrowski de Nuñez 2007); *Hypoderaeum conoideum* in Spain (Muñoz-Antolí et al. 2000); and *Petasiger variospinosus* in South Africa (King and Van As 2000), *P. combesi* in Costa Rica (Zamparo et al. 2005), and *P. islandicus* from Iceland (Kostadinova and Skirnisson 2007); *Stephanoprora uruguayense* and *S. aylacostoma* in Argentina (Ostrowski de Nuñez 2007; Ostrowski de Nuñez and Quintana 2007); *Acanthoparyphium tyosenense* in Korea (Chai et al. 2001; Kim et al. 2004) and *Himasthla escamosa* in Argentina (Diaz and Cremonte 2004); and *Philophthalmus lucipetus* and *P. distomatosa* in Israel (Radev et al. 1999, 2000).

Additional scant literature has covered reports of new first and second intermediate hosts of some echinostome species, i.e., *E. revolutum* in Finland (Vayrynen et al. 2000); *E. cinetorchis* in Korea (Chung et al. 2001a, b; Park et al. 2006), and *Curtuteria australis* in New Zealand (McFarland et al. 2003)

1.1.3.2 Free-Living Stages

Miracidia and cercariae are the two free-living stages in the life cycle of the echinostomes.

Miracidium

The echinostome egg is released by a definitive host in an aquatic environment. The egg is undeveloped when laid and consists of a fertilized ovum surrounded by yolk granules. It is typically oval in shape, variable in size and yellow, dark brown, or silver-white in color. It has an operculum at one end and a distinct knob at the abopercular end (Fig. 1.3a, b).

Fujino et al. (2000) provided a comparative study of ultrastructure of eggs in *E. paraensei*, *E. caproni*, and *E. trivolvis* using light microscopy, scanning electron microscopy, and transmission electron microscopy. Differences among these species are observed in relation to size and the aboperculum region, but no difference in eggshell structure was detected in any species.

The egg takes about 10–21 days at 22–28°C, depending on the echinostome species, to reach the fully developed miracidial stage. Several physicochemical factors influence embryonation, and the eggs can be maintained for at least 5 months at 4°C and still retain their ability to develop and hatch.

The miracidium is about 100 μm in length, broad anteriorly, and tapering posteriorly to a blunt end. The tegument is ciliated, and possesses a variable number of the ciliated epidermal plates and argentophilic papillae with a special arrangement. Retractile apical papillae followed by a pyriform gland have been described. Two dark-brown eye-spots, consisting of two pairs of pigmented bodies are located side by side, and the excretory system consists of two flame cells, two excretory ducts, and two excretory pores (Fig. 1.3c).

Pinheiro et al. (2004a) described the morphology and topography of *E. paraensei* miracidium by light and electron microscopy. The 19 papilla-like structures are arranged in three axes and four groups are observed at the terebratorium of this miracidium, which differs from the number of argentophilic papilla-like structures observed in *E. caproni*, *E. jurini* and *E. trivolvis*. The ultrastructural organization of the *E. paraensei* miracidium has been studied by Pinheiro et al. (2005) using transmission electron microscopy.

Among the large cells located in the posterior of *E. caproni* and *E. paraensei* miracidia are secretory cells, germinal cells, and undifferentiated cells. Secretory cells do not give rise to progeny, whereas germinal cells do. Undifferentiated cells

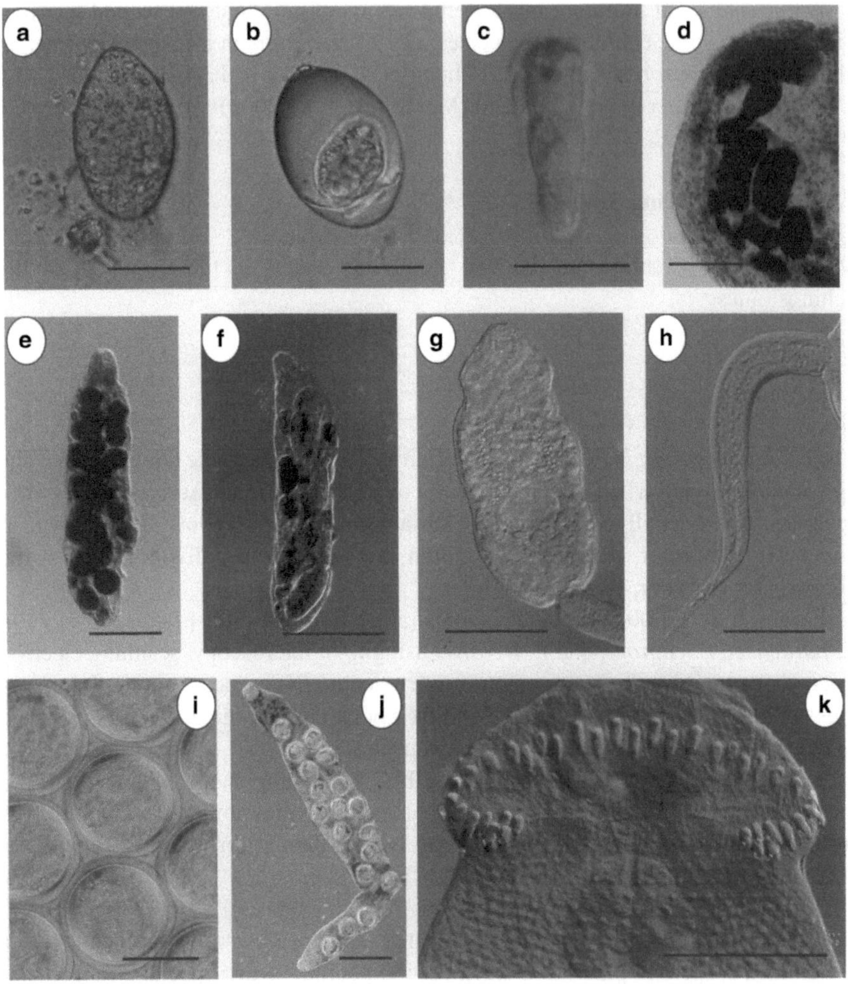

Fig. 1.3 Parasitic stages of the life cycle of *Echinostoma friedi*: (**a**) undeveloped egg; (**b**) embryonated egg; (**c**) miracidium recently hatched from an egg; (**d**) sporocyst in ventricle of snail; (**e**) first generation redia; (**f**) second generation redia; (**g**) cercarial body; (**h**) cercarial tail; (**i**) metacercariae; (**j**) redia containing encysted metacercariae; (**k**) anterior third of an adult. A, B, C, G, H, I, and J: fresh-living eggs, miracidia, cercaria, metacercariae, and metacercariae encysted in redia observed between slide and cover-slip; D, E, and F: fixed in Bouin's fluid under coverslip pressure, stained with Grenacher's borax carmine, mounted in Canada balsam, and observed with interference contrast microscopy; K: head collar region of a fresh-living adult observed between a slide and cover-slip with lactophenol fluid by interference contrast microscopy. Scale bars: A, B, and C: 50 μm; D, G, H, I, and J: 100 μm; E: 200 μm; F: 300 μm; K: 150 μm

develop into germinal cells that can also divide to produce embryos. Differences into the germinal cells of miracidia help to explain differences detected in the next of the intramolluskan echinostome stages (Ataev et al. 2001).

The behavioral patterns of echinostome miracidia have been reviewed by Haas (2000). The hatching of miracidia from eggs is stimulated by different factors; light is the major stimulus, although the pattern is different according to species. Whereas *E. paraensei* and *E. caproni* miracidia hatch in a strict diurnal pattern between 11.00 and 16.00 h, *E. trivolvis* does not show a daily hatching pattern.

Development, hatching, and infectivity of *E. caproni* eggs were described by Idris and Fried (1996). No significant differences in development were seen in the eggs maintained under conditions of light or darkness. The eggs maintained in the dark for 10 days and exposed to incandescent light produced a large synchronous hatch of miracidia within 3 h of exposure to light.

Maldonado et al. (2001b) found that miracidia of *E. paraensei* hatched after 10 days of incubation in the dark, and began to hatch 1 h after exposure to incandescent light, although most of them hatched between 3 and 4 h of exposure.

The effect of snail-conditioned water from *B. glabrata* snails on hatching rates of *E. caproni* miracidia was studied by Fried and Reddy (1999). Significantly greater hatching was obtained when snails were maintained in intact or perforated dialysis sacs in multiwell chambers as compared with sacs without snails.

The effects of salinity, pH, and temperature on the half-life and longevity of *E. caproni* miracidia have been analyzed by Ford et al. (1998). Miracidia tolerated pH ranges from 3 to 11 showing half-lives of 2.4 h, or greater under these conditions. At lower than ambient temperatures, *E. caproni* miracidia lived longer, the greatest being a half-life of 5.0 and a maximum life span of 15 h at 5°C.

After hatching, the miracidia swim and respond to environmental stimuli, such as light and gravity. The miracidia of *E. caproni* and *E. trivolvis* showed a negative geotaxis which was dominated by a positive phototaxis; meanwhile, *E. paraensei* miracidia showed no distinct geo-orientation and only weak photo-orientation.

The miracidia also respond to stimuli emanating from the host. The main host-finding signals for *E. caproni* miracidia detected have been complex macromolecules from snail mucus (Haberl et al. 2000).

The presence of nonhost snails, in a given snail community, may interfere with the ability of miracidia to orientate, reach, and successfully penetrate their normal host snail. These effects on echinostome miracidial host-finding behavior are poorly understood. A radioisotope assay system was used for testing the host finding of *E. revolutum* miracidium (Christensen 1980); two studies examined the response of *E. caproni* (Behrens and Nollen 1991) and *E. trivolvis* (Nollen 1994) miracidia to chemicals; a study was done on the different suitabilities of two lymnaeid snails for *H. conoideum* miracidia (Toledo et al. 1999a); a complete study was done on the host-finding process in *E. caproni* miracidia and cercariae (Haberl et al. 2000); and a recent study was done on the interactions related to host, nonhost snails, and water-conditioned snail on the host-finding processes of the miracidia of *Euparyphium albuferensis* and *E. friedi*; both species share the same natural habitat (Esteban et al. 1997; Muñoz-Antoli et al. 2003).

Muñoz-Antoli et al. (2002) studied the survival and infectivity of *E. friedi* miracidia, which was determined to be age dependent. This finding was in accord with the previous study on the survival of *H. conoideum* miracidia (Muñoz-Antolí et al. 2000). The maximum life span and the time to 50% mortality were determined to be 10.0 and 6.8 h, respectively. A gradual increase of infectivity in the few first hours after hatching was not observed, and the lack of a preinfective period in *E. friedi* miracidia was related to the low specificity of this echinostome species for its intermediate host (Muñoz-Antoli et al. 2002).

Toledo et al. (2004) found that the age of adult worms significantly influenced *E. friedi* miracidial infection in *Lymnaea peregra* snails. Infective miracidia only were obtained from adult worms in the age range from 4 to 9 weeks p.i., and the infectivity was maximal in those miracidia derived from adults collected 8 and 9 weeks p.i., which corresponds with maximum egg output. These authors suggested that adult worms producing viable eggs require additional maturation in order to yield eggs containing infective miracidia. However, these results differed from those obtained by Fried and Bandstra (2005) for *E. caproni* miracidia. These authors determined the percentage of fully developed miracidia in terms of eggs derived from adult worms obtained from laboratory mice at 2, 4, 6, and 8 wk postinfection, and obtained a percentage of fully developed miracidia >90% and 60–80% of the eggs hatched.

Although it was suggested that each species of echinostome infects only one or a few closely related snail species (Huffman and Fried 1990), more recent studies show marked differences in that the spectrum of first intermediate host species for each echinostome species may be broader than previously expected. Maldonado et al. (2001b) infected three sympatric snail species, belonging to the Planorbidae (*Biomphalaria glabrata*), Physidae (*Physa marmorata*), and Lymnaeidae (*Lymnaea columella*) families, with a Brazilian isolate of *E. paraensei*. Muñoz-Antolí et al. (2006) studied the infectivity of *E. friedi* miracidia in a range of sympatric and allopatric laboratory-reared snails, and the results showed that the miracidia were able to infect and develop in snail species belonging to three different families (Lymnaeidae, Planorbidae, and Bulinidae) and from different geographical origins, although the rates of infection were low for all these species.

In relation to echinostome-like digeneans, *Ribeiroia* is more species specific than echinostomes in its use of first intermediate hosts. The members of this genus use only 12 species of Planorbidae belonging to the genera *Biomphalaria, Helisoma*, and *Planorbella* (Johnson et al. 2004).

The effect of *E. friedi* miracidial infection on survival, growth, and fecundity of two susceptible first intermediate host snails, *Radix peregra* and *Biomphalaria glabrata*, has been studied by Muñoz-Antoli et al. (2007). Infected *B. glabrata* showed gigantism and had a shorter life span that *R. peregra*; infected *R. peregra* reduced normal development; *E. friedi* produced total parasitic castration of both infected snails species. These differences allowed the establishment of the level of host compatibility: *R. peregra* would be considered as the required snail host, and it is able to transmit the parasite and to maintain the parasite population without other snail hosts; *B. glabrata* would be considered only an adequate experimental snail host, being able to maintain the *E. friedi* life cycle in the absence of *R. peregra*.

Cercaria

The echinostome cercaria is typically distomate, gymnocephalus with an oral collar of spines, not always visible, and a simple tail (Fig. 1.3g, h). Differences in size among echinostome species are common. The main excretory ducts of the excretory system are simple and short, containing a large number of excretory granules. The number (from 10 to 100) and the arrangement (in pairs, by three or by four) of flame cells are of importance in the diagnosis echinostome cercariae. In addition to openings of penetration and cystogenous glands, some cercariae contain paraesophageal and esophageal glands. The number and distribution of tegumentary papillae, the so-called cercarial chaetotaxy patterns, has proved to be of taxonomic value for echinostomatid species (Toledo et al. 1998a, b, 2000; Kostadinova 1999; Nakano et al. 2003). The cercarial tail of most *Echinostoma* species has fin-fold structures, not described on the cercarial tail of other genera such as *Echinoparyphium, Euparyphium*, *Hypoderaeum*, and *Isthmiophora*. The general morphology of cercariae of different echinostome genera has been reviewed by Kanev et al. (2000).

The presence of two laterally projecting esophageal diverticula, located approximately halfway along the length of the esophagus, and the presence of a conspicuously rose-pink-colored structure located between the oral sucker and the pharynx, constitute two relevant morphological structures of the echinostome-like cercaria of *Ribeiroia* (Johnson et al. 2004).

Using indirect immunocytochemistry to demonstrate neuroactive substances and the phalloidin-fluorescence technique for staining myofibril F-actin, the muscle systems and aminergic and peptidergic innervation of *E. caproni* cercariae have been examined (Sebelová et al. 2004). Combined studies of high-performance thin-layer chromatography with lipid histochemistry of neutral lipids in cercariae of *E. caproni* showed the major lipid fraction to be free sterol, and this parasite stage had 6.5 times more free sterol than the encysted metacercaria (Marsit et al. 2000).

The free-swimming cercariae escape from their first intermediate host 4–6 weeks postinfection. The cercariae are short lived and rarely survive beyond 48 h at ambient temperatures.

Numerous papers have been published focusing on different aspects of echinostome cercariae. The movement patterns of cercariae (in *Echinoparyhphium, Himasthla, Pseudechinoparyphium, Echinostoma*, and *Hypoderaeum* species), the chemo-orientation of swimming cercariae (in *Echinostoma, Pseudechinoparyphium*, and *Hypoderaeum* species), and the cercarial attachment and host invasion (in *Echinostoma, Hypoderaeum, Pseudechinoparyphium*, and *Isthmiophora* species) have been extensively reviewed by Haas (2000).

McCarthy (1999a) studied phototactic responses of the cercariae of *E. recurvatum* from infected *L. peregra* snails and found that cercariae have an initially positively phototactic, low infectivity, dispersal phase followed by a negatively phototactic, maximally infective host location, and infection phase.

The effect of light and gravity on orientation of cercariae was studied in *Echinostoma, Hypoderaeum, Pseudechinoparyphium*, and *Isthmiophora* species (Loy et al. 2001). These authors demonstrated that each of the 4 species studied has an individual behavior pattern of horizontal photo-orientation and geo-orientation, with distinct changes during the time after emergence. This diversity of behavioral responses corresponds with that of their chemo-orientation toward the intermediate hosts. *Hypoderaeum* approaches its host snails by direct chemotactic orientation along concentration gradients of snail-emitted peptides, whereas *Pseudechinoparyphium* and *Echinostoma* swim back when the concentration of snail-emitted amino acids, urea, and ammonia decreases, and *Isthmiophora* seems to show no chemo-orientation at all toward its amphibian and fish hosts. The geo-orientation was controlled differently in each species by the intensity and the direction of light radiation, and the different orientation patterns suggest functions such as leaving the habitats of the host snails emitting the cercariae, dispersal, and frequenting the microhabitats of potential hosts (Loy et al. 2001).

Temperature is a determinant factor in survival and longevity, and survival is shorter at higher temperatures (Evans 1985). Schmidt and Fried (1996) studied the emergence of *E. trivolvis* cercariae from naturally infected *H. trivolvis* snails. Of the numerous laboratory conditions tested, the only significant factors that had an impact on cercarial emergence were temperature related. Recently, Fried and Ponder (2003) observed that the effects of temperature on cercarial survival in artificial spring water and on infectivity of *E. caproni* are temperature dependent, and the infective life span of the cercariae is shorter than the total life span. Moreover, the temperature has also a major effect on in vitro encystment of the cercariae of *E. caproni*.

Schmidt et al. (1996) studied the effect of storage on the survival of the intramolluskan stages of *E. trivolvis* in *H. trivolvis* (Pennsylvania strain) snails. Twenty snails were stored at 4°C in artificial spring water for 10 months; three snails were alive at 10 months and released cercariae. The number of cercariae released and their infectivity to a Colorado strain of *H. trivolvis* snails was significantly lower than that of cercariae from freshly collected snails.

The survival and infectivity characteristics of echinostome cercariae have been the subject of several studies. McCarthy (1999b) studied the influence of temperature on the survival and infectivity of *E. recurvatum* cercariae; cercarial survival was temperature dependent with the maximum survival time being reduced from 68 h at 10°C to 12 h at 30°C. However, the survival and infectivity of *E. friedi* cercariae was found to be markedly age dependent (Muñoz-Antoli et al. 2002). The maximum life span at 20°C and time to 50% mortality was determined at 28.00 and 23.5 h, respectively. These results are consistent with those observed for other sympatric echinostome species, such as *H. conoideum* and *E. albuferensis* (Toledo et al. 1999b), but the times observed in the latest two species were significantly shorter than those reported for other echinostome cercariae such as *E. caproni*, *E. trivolvis*, and *E. recurvatum* for which the time to 50% mortality ranged from 29 to 31 h at similar temperatures. Moreover, the cercarial infectivity of *E. friedi* gradually increased during the first few hours, and reached a peak after a prior period of aging. The existence of this preinfective period may represent a dispersal phase that aids cercarial dissemination, thus reducing superinfection and parasite associated mortality of the first intermediate host (Muñoz-Antoli et al. 2002).

The emergence pattern under different conditions has been analyzed in the *E. trivolvis-H. trivolvis*, *E. recurvatum-L. peregra*, and *E. albuferensis-G.chinensis* systems. Schmidt and Fried (1996) described the number of *E. trivolvis* cercariae released during 1 h under different conditions, whereas McCarthy (1999c) studied the photoperiodic cercarial emergence of *E. recurvatum* from *L. peregra*. Cercariae emerged during the light phase of the experiment from 08.00 to 20.00 h. Toledo et al. (1999d) studied the production and chronobiology of the emergence of cercariae of *E. albuferensis* from *G. chinensis* experimentally infected with a single miracidium. These authors noted that although the daily cercarial shedding rates were very variable, a progressive increase in cercarial production was observed in the first several weeks of cercarial shedding. Under a 12-h light-dark cycle, the cercariae emerged in the light and the rhythm was circadian. A sudden change in the light-dark cycle resulted in corresponding changes in the emergence patterns, showing that cercarial emergence in this species is correlated to light-dark changes.

Nevertheless, in the echinostome-like *Ribeiroia*, the emergence is rhythmic under continuous darkness. The cercariae emerge at night, usually between 7 PM and 3 AM, and may swim actively for at least 12 h or more. Infected snails typically release around 300–400 cercariae per night for 3–7 months (Théron and Moné 1986), though some individuals produced more than 1,000 cercariae per night. However, significant diurnal emergence of cercariae was also observed (Fried pers. com. to Johnson).

Maldonado et al. (2001b) studied the kinetics of cercarial emergence for the sympatric snails infected with cercariae of a Brazilian isolate of *Echinostoma paraensei*. Differences in relation to the miracidial doses (one vs five miracidia) were detected, and the length of the prepatent period differed significantly between snail species. Fried, LaTerra and Kim (2002) examined various physicochemical factors related to optimal release of *E. caproni* from experimentally *B. glabrata* snails. Among the numerous conditions tested, e.g., the addition of lettuce, the use of snail-conditioned water from *B. glabrata*, and a high temperature (35°C) significantly increased the shedding of *E. caproni* cercariae.

Snail size is another relevant factor in determining the mortality associated with echinostome cercarial penetration and encystment in the second intermediate host (Ponder and Fried 2004b; Schneck and Fried 2004). In general, neonatal snails are more susceptible to infection but show a significant decrease in survival.

Studies on the effects of pollutants on echinostome cercariae are limited. Evans (1982) and more recently Morley et al. (2002) studied the toxic effects of copper and zinc, and cadmium and zinc, respectively, on the transmission of *E. recurvatum* cercariae toward the second intermediate host snails. These authors detected a differential response in infectivity of cercariae dependent on the snail species to be infected; the exposure of different snails to toxic pollutants caused a different susceptibility to *E. recurvatum* cercariae depending on the snail species exposed. Similar results were obtained in the study on effects of tributyltin and copper on cercariae of *Parorchis acanthus* (Bennett et al. 2003). Reddy et al. (2004) analyzed the effects of copper toxicity on cercariae of *E. caproni* and *E. trivolvis*, as well as on the survival of *B. glabrata* snails, and suggested that copper sulfate, used in concentrations sufficient to kill juvenile snails, was also sufficient to eliminate the cercariae of both echinostomes. More recently, Koprivnikar et al. (2006) analyzed the effects of the herbicide Atrazine, commonly used in North America, on longevity, activity, and infectivity of *E. trivolvis* cercariae, and observed that the viability of these cercariae is compromised by exposure to this herbicide.

Free-swimming cercariae normally come in contact with a compatible second intermediate host. The specificity of echinostomes toward the second intermediate host is low, and usually numerous species of snails, clams, tadpoles, frogs, and even fishes, other invertebrates, and natural products such as snail mucus may serve as second intermediate hosts of echinostomes (Huffman and Fried 1990). Recently, an ectosymbiotic flatworm, *Temnocephala chilensis*, was found naturally parasitized with metacercariae of *Echinoparyphium megacirrus* (Viozzi et al. 2005).

Cercarial preference toward second intermediate hosts has been the subject of some studies (Anderson and Fried 1987; McCarthy 1990d; Fried 2001). Studies on *E. albuferensis* and *E. friedi* cercarial infectivity toward different snail communities composed of combinations of 4 sympatric snail species were performed to evaluate the level of parasite-snail host compatibility. The results obtained from single- to four-host exposures showed that high densities of low compatible hosts may reduce the level of parasite transmission; however, the presence of gastropods of high compatibility in the communities contributes to an increase in the susceptibility of those snail species showing low compatibility (Muñoz-Antolí et al. 2008).

Effects of snail diet on transformation from cercaria to metacercaria in the second intermediate host have been poorly studied. Recent studies have demonstrated that snail diet may affect larval development in the first intermediate host (Sandland and Minchella 2003). Glucose added to artificial spring water extended the survival time of *E. trivolvis* and *E. caproni* cercariae (Fried et al. 1998; Ponder and Fried 2004a), whereas in *E. caproni*, the glucose decreased the cercarial ability to infect snails or move in a linear direction. Fried et al. (1997a) showed that the lipophilic fraction of a *Helisoma trivolvis* dialysate significantly enhanced cercarial chemoattraction and penetration in this planorbid by *E. caproni* and *E. trivolvis*. However, Ponder and Fried (2004b) did not detect an apparent difference in *E. caproni* encystment in juvenile *H. trivolvis* fed either with hen's egg yolk or with Romaine leaf lettuce diet; in consequence, lipophilicity was not a factor in cercarial encystment. Moreover, the diet did not enhance larval development in the snail.

In the absence of a second intermediate host, in vitro encystment of echinostome cercariae can occur. *E. revolutum* and *E. liei* have been reported to encyst on snail mucus (Fried and Bennett 1979; Christensen et al. 1980), *E. revolutum* in Locke's solution with or without glucose (Fried and Bennett 1979; Fried et al. 1997c), *E. paraensei* and *E. caproni* in cultures with *B. glabrata* embryonic cells (Stein and Basch 1977; Loker et al. 1999), *E. caproni* in Locke's-ASW (1:1) medium with or without copper sulfate (Fried and LaTerra 2002; Fried and Schneck 2004), *Echinostoma cinetorchis* in RPMI 1640 plus 10% fetal bovine serum (Park et al. 2006), *Himasthla quissetensis* in casein hydrolysate supplemented with glucose (Laurie 1974), and *Echinochasmus liliputanus* in different solutions (Xiao et al. 2005). The results showed differences in relation to echinostome species and conditions of culture, but the cysts formed in vitro became infective to the definitive host. However, the infectivity was lower than those obtained from experimentally infected snails (Fried and LaTerra 2002; Park et al. 2006).

Recently, Schotthoefer et al. (2007) tested the cercariophagic activity of several freshwater invertebrates on the echinostome-like species, *Ribeiroia ondatrae*. From the species tested, *Hydra* sp., damselfly (Odonata, Coenagrionidae) larvae, dragonfly (Odonata, Libellulidae) larvae, and copepods (Cyclopoida) consumed cercariae. In some cases, 80–90% of the cercariae exposed to damselfly and dragonfly larvae were consumed within 10 min.

1.1.3.3 Parasitic Stages in the Invertebrate Host

Sporocysts and rediae in the first intermediate host and metacercariae in the second intermediate host are the parasitic stages in the invertebrate host.

Sporocysts and Rediae

These intramolluskan stages are formed after the invasion of an aquatic snail by miracidia and their later transformation. Kanev et al. (2000) compiled the morphological characteristics of these parasitic stages in different echinostome genera, whereas Ataev and coauthors contributed greatly to the knowledge of echinostome stage development in the first intermediate host (Ataev et al. 1997, 1998, 2001, 2005, 2006). Miracidia usually enter the head foot region of the snail, shed their ciliated plates, and transform into sporocysts at the site of penetration, typically the mantle collar, the foot and head covering (including velum and tentacles), the mantle cavity, and the oral cavity. These sporocysts are sac-like structures, about 100 mm long, that after 5–8 days, develop and produce the next parasite stage, the rediae. The ventricle and the common aorta are the final sites of infection for the mother sporocysts after migration (Fig. 1.3d). The behavior patterns and host cues responsible for this site-finding remain unknown, though it was suggested that the ventricle and aorta may be recognized by sporocysts (Haas 2000).

In vitro studies by Ataev et al. (1998) showed that the presence of *B. glabrata* embryonic cells in the cultures is essential for the development of *E. caproni* mother sporocyst in the snail host. During *E. caproni* sporocyst development, every primary germinal cell gave rise to a redial embryo, whereas undifferentiated cells gave rise to both somatic and secondary generative cells. Each mother sporocyst produced about 15 rediae. Intramolluskan development of the *E. caproni* mother sporocysts consists of 5 development steps (resting, migration, growth, reproduction, and degeneration) that are not temperature dependent (Ataev et al. 1997). The mother sporocyst produces the first generation, which consists of maternal rediae forming only redoid embryos (Fig. 1.3e). These initially reside within the ventricle and aorta of the snail, although when these sites became filled, these rediae begin to colonize the ovotestis of the snail. The next generations are represented by daughter rediae (Fig. 1.3f). In the beginning of their life they produce redoid embryos, forming later cercariae. The germinal masses, located caudally in the mother sporocyst, are responsible for multiplication and development of generative elements in all generations (Ataev et al. 2005, 2006). In the case of *E. paraensei*, development of the so-called precocious mother redia has been observed under laboratory conditions. This precocious mother redia attaches to the ventricle and remains with the sporocyst for 31 days or more. This developmental stage has not been observed in other echinostome species; the presence of this precocious mother redia reduces the success rate of subsequent *E. paraensei* sporocysts to establish in the ventricle of the snail host (Sapp et al. 1998).

Echinostome rediae are elongate structures with an anterior mouth, pharynx, sac-like gut, a collar-like tegumentary ring at the anterior end, and two posteriorly located ambulatory buds. Rediae are more active than sporocysts. Sebelová et al. (2004) investigated the musculature and associated innervation of *E. caproni* rediae by confocal microscopy and detected a somatic and pharyngeal muscle and a well-developed muscular body wall comprising a mesh of numerous, well-organized outer circular fibers and fewer, but thicker, inner longitudinal fibers. The neuroactive elements were mainly restricted to the central nervous system, comprising essentially a bilobed cerebral ganglion or brain, and paired longitudinal nerve cords.

Rediae migrate within the snail body and prefer particular habitats, mainly the digestive gland and gonads. Rediae use their mouth and digestive system to ingest host tissues and hemolymph, but they may also attack and feed upon other rediae and sporocysts of echinostome and nonechinostome trematodes. *E. trivolvis* and *E. caproni* rediae moved toward other individual rediae, including those conspecifics (Reddy and Fried 1996). The attractants are used to recognize that the intramolluskan stages were contained in the free sterol fractions and free fatty acid fractions of the redial excretory-secretory products (Reddy et al. 1997). Biochemical studies on the rediae of *E. caproni* showed free sterol as the major lipid fraction, with lesser amounts of triacylglycerols and steryl esters (Marsit et al. 2000). Recently, *E. revolutum* has been used to control fasciolosis in West Java through antagonism between larvae of *Fasciola gigantica* and *E. revolutum* in *Lymnaea rubiginosa* (Suhardono et al. 2006).

Fried and Awatramani (1992) in their light and scanning electron microscopical studies on the daughter rediae of *E. trivolvis* reported that this stage is hematophagous and also noted the presence of black particulate material in the redial gut, presumably derived from the presence of hemoglobin degradation. Pinheiro et al. (2004b) studied the morphology of *E. paraensei* rediae using light, scanning, and transmission electron microscopy, and noted considerable morphological variation in the papilliform process in relation to the echinostome species. Folds were the only structures observed in the tegument of *E. paraensei* rediae, and the presence of mitochondria and secretory bodies in the outer syncytium indicated that this tegument plays a role in nutrient absorption.

Pisciotta et al. (2005) detected that *E. trivolvis* daughter rediae produced ovoid hemozoin crystals within the first intermediate gastropod host, *H. trivolvis*, with insolubility characteristics identical to *Plasmodium falciparum* hemozoin. However, the rediae of *E. caproni* did not produce measurable pigment even though the snail host, *B. glabrata*, utilizes hemoglobin; these findings indicated that hemozoin production varies according to the echinostome species.

Little is known about the intramolluskan parasitic stages of *Ribeiroia*, but the pattern of sporocyst and rediae development is similar to that of the echinostomes. Mature sporocysts are primarily found in the renal and pulmonary veins near the kidney, and the germinal cells/germ balls of the posterior portion of the miracidium are retained in the sporocyst and produce mother rediae. These rediae escape from the sporocyst and migrate via the heart posteriorly, and the germinal cells within the mother rediae produce daughter rediae, which become concentrated in the area of the ovotestis, and also in the gizzard, albumen gland, and between lobules of the digestive gland. Beaver (1939) noted that mother rediae can produce both daughter rediae and cercariae concurrently.

Detailed aspects relating to the dynamics of intramolluskan development and the characterization of the echinostome capacity to interfere with the function of first intermediate host defense cells (the so-called hemocytes) may be found in Chap. 2 of this book.

Metacercariae

Emerging echinostome cercariae infect the second intermediate host. This role may be accomplished by several species of gastropods, bivalves, tadpoles, and even fishes, annelids, or temnocephalads. These cercariae encyst in different sites as metacercariae (Fig. 1.3i). The cyst wall is composed of several layers of different thickness. By transmission electron microscopy the cysts may show two or three layers, with the inner cyst layer being lamellated. The metacercarial body is enclosed inside the cyst, and numerous studies have focused on the structure, ultrastructure, and histochemistry of both cyst walls and the larva within the cyst (Kanev et al. 2000). Musculature and associated innervation of metacercariae of *E. caproni* have been exhaustively studied by Sebelová et al. (2004).

The larva within the cyst is transparent in viable organisms, and excretory concretions and cephalic spines are visible; nonviable cysts are opaque and granular. These cysts are different in shape and size in relation to the echinostome species; penetration sites, migration routes, and sites of localization of echinostome cercariae and metacercariae have been the subject of different papers. Significant differences in the cyst diameter, thickness of outer and inner cysts, and diameter of the excretory concretions between *E. caproni* and *E. trivolvis* metacercariae were observed by Frazer and Fried (1998). Furthermore, in concurrent infections of *E. trivolvis* and *E. caproni*, *E. trivolvis* were localized exclusively in the saccular kidney, while *E. caproni* occurred in both the saccular kidney and the pericardium.

The morphological features that are diagnostic for echinostome-like metacercariae of *Ribeiroia* include: lateral esophageal diverticula, sinuous excretory ducts containing numerous, large concretions, spinose tegument, oral sucker with a longitudinal slit opening to the outside, and a ventral sucker located in the posterior half/third of the body.

Cysts may remain viable within the second intermediate hosts for months. Survival of *E. caproni* and *E. trivolvis* reached 4 and 6 months within planorbid intermediate host snails, respectively. Furthermore, cysts of *Echinostoma* spp. can be removed from snails and stored for more than 1 year in 1:1 Locke's solution at 4°C, retaining their infectivity toward the definitive host. McCarthy (1999d) examined the potential influence of the second intermediate host species on the infectivity of metacercarial cysts of *Echinoparyphium recurvatum* and concluded that the species of second intermediate hosts did not influence the infectivity of the metacercarial cyst of *E. recurvatum*, nor the subsequent reproductive status, growth, or development of this echinostome in an experimentally infected definitive host.

The localization of echinostome metacercariae is variable. In gastropods, the cercariae have a predilection for the kidney-pericardial region; in bivalves the gills and gonads serve as a preferred site; in amphibians the kidney and cloaca are preferred sites; and the skin, subtegument, and musculature are preferred in fishes (Huffman and Fried 1990). Toledo et al. (1999c) studied the encystment sites of metacercariae of *H. conoideum* in different freshwater gastropods, and metacercariae were found primarily in the lumen of the pericardial cavity and also in the kidney. No other organs or tissues were found infected, although the heart has been usually reported a site of metacercarial encystment.

In the case of *Ribeiroia*, numerous species of amphibians and freshwater fishes may become infected, but metacercariae do not form within snails (Johnson et al. 2004). In amphibians, metacercariae are found primarily around the base of the hind limbs, cloaca, and to a lesser extent within the posterior celom near the urinary bladder, gill resorption sites, nares, and along the ventral margin of the mandible; whereas in fishes, the metacercariae locations include the lateral line canals of the head, the lateral line scales along the body, the gills, lips, and internal nares.

Metacercariae of some echinostomes have also been found in freshwater snails harboring rediae and cercariae and serving as the first intermediate host; some metacercariae encyst within the rediae of the same echinostome species (Fig. 1.3j). Metacercariae of some echinostome species become infective to definitive hosts within 4 h of encystment in the second intermediate host; and other species require a longer period (from 7 to 20 days) of cyst maturation before the metacercaria is infective to the definitive host.

Although echinostome cercariae typically encyst in the heart, pericardium, and kidney, some ectopic encystment does occur, mainly on snail mucus or in vitro in a tissue culture medium. These ectopic cysts become infective to definitive hosts, but infections with cultured metacercariae of *E. caproni* and *E. cinetorchis* showed a lower recovery rate of adult worms than those from experimental snails (Fried and LaTerra, 2002; Park et al. 2006). These results suggest that faster maturation of metacercariae occurs in the snail's hosts. Xiao et al. (2005) studied in vivo and in vitro encystment of *Echinochasmus liliputanus* cercariae released from *Bellamya aeruginosa*. In vivo encystment of cercariae occurred in the gills of goldfish, and Locke's or NaCl solutions were shown to be appropriate for in vitro encystment. The cysts formed in vivo and in vitro were equally infective to experimental rabbit definitive hosts.

The neutral lipids present in encysted metacercariae are free sterol, but they are present in considerably lesser amounts than in the cercariae (Marsit et al. 2000).

Morley et al. (2004b) observed in the *E. recurvatum* – *L. peregra* system an increase in the mean number of cysts per host in relation to host size and time of exposure. Aggregation resulting from host and habitat-related factors increased with host size and time of exposure; the encystment within the host was restricted to the peripheral organs in smaller juvenile snails, but as snails increased in size, metacercariae were distributed throughout other tissues.

The pathology induced by echinostomes in their second intermediate hosts has been the subject of few studies, mainly in their amphibian hosts, despite the fact that this infection causes edema, mortality, and growth inhibition in tadpoles (Fried et al.1997b). Recently, Schotthoefer et al. (2003) studied this topic using *E. trivolvis* and tadpoles of *Rana pipiens* as model. They noted that success of infection was significantly influenced by tadpole stage, exposure dose, and interactions of both factors, which agrees with the previous report by Fried et al. (1997b). Moreover, the tadpoles died only when metacercariae were concentrated in the functional portion of the kidney with the most limited renal capacity. Thiemann and Wassersug (2000) reported a biased distribution of *E. trivolvis* metacercariae in the nephric system of *Rana* tadpoles. The distribution of encysted metacercariae was biased in favor of the right kidney. The authors suggested that the high frequency of unilateral encystment in anurans may be the result of a coevolved relationship that benefits both the host and the parasite by ensuring host survival.

Echinostome metacercarial cysts often need to be stored for several months to a year in laboratory. Keeler et al. (2007) examined the bacteria associated with encysted metacercariae of *E. caproni* stored at 4°C. Different Gram-negative and Gram-positive organisms were isolated. *Aeromonas hydrophila*, *Pseudomonas* sp., *Enterobacter cloaceae*, and *Alcaligenes* sp. were the Gram-negative bacteria isolated from the stored cysts, and *A. hydrophila*, *Citrobacter freundii*, and *E. cloaceae* were isolated from the infected and noninfected snails. The number of species of bacteria associated with *E. caproni* metacercariae increased during the first 4 months of storage, and the bacterial contamination may contribute to the decreased excystation of the *E. caproni* cysts over time. For this reason, the authors suggested storage of the metacercariae in media containing antibiotics and refreshing the storage solutions monthly.

The effects of pollutants on echinostome and echinostome-like genera cysts have recently been studied. Bennett et al. (2003) analyzed the effects of tributyltin and copper on metacercariae of *Parorchis acanthus* and concluded that tributyltin and copper had a detrimental effect on *P. acanthus* metacercariae at high concentrations, but the cyst wall afforded an element of protection if formed in unpolluted seawater before the larval stages were subjected to the pollutants. A similar effect on *P. acanthus* metacercariae was previously attributed to Cd, Zn, and Cd/Zn mixtures by Morley et al. (2002).

Recently, Morley et al. (2004a) investigated the viability of metacercarial cysts of *E. recurvatum* parasitizing two common freshwater snails, *Lymnaea peregra* and *Physa fontinalis*, and demonstrated that exposure to antifouling biocides, especially

TBTO and copper, reduced the viability of these metacercarial cysts after prolonged exposure periods, principally through death of the snail host; this may, in turn, have detrimental effects on successful parasite transmission. These authors concluded that the survival of metacercariae and host is interlinked and that host mortality has a subsequent influence on the long-term viability of the metacercarial population. Furthermore, the copper molluskicide, in the form of copper sulfate, was capable of killing excysted metacercariae of *E. caproni* and *E. trivolvis*, but not the encysted metacercariae (Reddy et al. 2004).

Further details on echinostomes in the second intermediate host are covered in Chap. 3 of this book.

1.1.3.4 Parasitic Stages in the Vertebrate Hosts

The definitive host is a vertebrate, which finally becomes infected when metacercariae are ingested with the second intermediate host. Within the definitive host, metacercarial cysts excyst usually in the ileum of the vertebrate host, and newly emerged juveniles establish in the intestine within 4 h postfeeding on the cysts; the excysted metacercariae gradually mature into adults.

Much is known on the activation and excystment of metacercariae (Fried 1994, 2000). During excystation, the organism breaches the lamellated layer prior to final disruption of the outer cyst. Conditions in the intestine, i.e., alkaline pH, trypsin-bile salts, and a temperature of 37–43°C, influence the excystation process. Excystation *in vitro* in an alkaline bile salts-trypsin medium at 37–41°C in the absence of acid-pepsin pretreatment has been reported for several species of echinostomes. Different host cues stimulating the metacercarial excystment behavior have been described, such as pH conditions, bile components, carbon dioxide, and various enzymes (Fried 1994). It is probable that the various host cues activate the coordinated emergence behavior via sensory receptors.

Rossi et al. (2001) provided information on excystation and infectivity of *E. caproni* metacercarial cysts after prolonged storage. The percentage of excystation in trials with fresh cysts was 98.3% compared to 51.5% for cysts stored in Locke's 1:1 solution for 5 months. Cysts stored for 8 or 12 months did not excyst in an artificial trypsin-bile salts excystation medium.

The effects of tonicity, digestive enzymes and bile salts, and various nutrients added to Locke's solution have been studied on the chemically excysted metacercariae of *E. caproni* (Fried et al. 2004). The findings suggest that the acidic pepsin environment in the stomach of a definitive host would be detrimental to the survival of excysted metacercariae, but prolonged survival in alkaline trypsin-bile salts would facilitate establishment of this larval stage in the mucosa of the host small intestine. Furthermore, the addition of exogenous glucose to the incubation medium of Locke's solution significantly increased the survival of the excysted metacercariae, because this glucose provided an energy source for the enhanced survival of this larval stage. Similar results were obtained with Locke's solution supplemented with hen's egg yolk. Fried and Peoples (2007) demonstrated that *E. caproni* metacercariae excystation was strongly affected by vinegar treatment, boiling for 1 min, and increased potassium permanganate concentrations.

Adult echinostomatids are predominantly found in birds, and also parasitize mammals, including man, and occasionally reptiles and fishes (Huffman and Fried 1990). The main characteristic feature of the echinostome adult is the cincumoral collar armed with one or two ventrally interrupted crowns of large spines (Fig. 1.3k). Other morphological features of echinostome genera containing numerous species of the Echinostomatidae have been summarized by Fried (2001). This author provided the most relevant morphological criteria of *Echinostoma, Echinoparyphium, Echinochasmus, Stephanoprora*, and *Himasthla* at level of body shape and size, forebody, collar and collar spines, and male and female reproductive systems.

The echinostome-like genus *Ribeiroia* utilizes a broad range of definitive hosts, including numerous avian and several mammal host species. Among the birds, *Ribeiroia* has been recorded from 40 avian species, with the highest prevalence among ciconiforms, falconiforms, and pelicaniforms. The raccoon (*Procyon lotor*), the muskrat (*Ondatra zibethica*), the black and Norwegian rats (*Rattus rattus* and *R. norvegicus*), and the badger (*Taxidea taxus*) are the recorded mammalian definitive host species. Experimentally, adult *Ribeiroia* have been recovered from hamsters, muskrats, mice and rats. An exhaustive list of natural and experimental definitive hosts in the life cycle of *Ribeiroia* has been compiled by Johnson et al. (2004).

The adult stage is the most studied in the life cycle of echinostomes and has been the subject of extensive reviews (Huffman and Fried 1990; Fried and Huffman 1996; Fried 2001; Fried and Graczyk 2000, 2004; Toledo et al. 2006).

The establishment of adults in the definitive host varies greatly for each host-echinostome species system. Factors such as host species, immunological status, and age of the host, the infective metacercarial doses, and coinfection with other parasites or with other echinostome species influence the course of infection. Although the general pattern for the distribution of several species of echinostomes has been well established, there are several factors that may alter the distribution of echinostome species. The species of echinostome, egg production, population density, age of infection, and host species are factors that affect the fecundity of echinostomes. More recently, *in vitro* and *in vivo* investigations of trematocidal activity of different drugs (praziquantel, artemisinin derivatives, and tribendimidine) on echinostomes have been the subject of some papers.

These aforementioned topics are not further reviewed here since they are analyzed in more detail in Chap. 4 of this book.

1.1.4 Concluding Remarks

In this chapter, salient information on the systematic and biological studies done on echinostomes and echinostome-like digeneans in the past 10 years has been provided. Although the available significant literature pertains mainly to *Echinostoma* species, in particular to *E. caproni, E. trivolvis, E. paraensei*, and *E. friedi*, some relevant information on genera of echinostomatids other than *Echinostoma* is also covered. Information on some echinostome-like digenean genera was also included because the organisms selected are important in human and veterinary medicine, in wildlife disease, or serve as models for biological research. Specifically, the genus *Ribeiroia* and some literature on *Parorchis* and *Philophthalmus* have been considered in the review.

Although molecular studies are proving decisive in certain aspects related to echinostome's phylogenetics and diagnostics, systematic studies based on morphological data are both necessary and essential, and must be updated in a rigorous and exhaustive manner.

From the biological perspective, and although a great amount of information is available in the literature, a number of areas of biological research still remain to be explored. The echinostomes and echinostome-like genera possess complicated life cycles in which three hosts are normally involved: two intermediate hosts and one definitive host. However, the easy maintenance of echinostome life cycles in the laboratory makes these digeneans good experimental models for studies on host-parasite relationships.

References

Anderson, J.W. and Fried, B. 1987. Experimental infection of *Physa heterostropha, Helisoma trivolvis* and *Biomphalaria glabrata* (Gastropoda) with *Echinostoma revolutum* (Trematoda) cercariae. Journal of Parasitology 73: 49–54.

Ataev, G.L., Dobrovol'skii, A.A., Fournier, A. and Jourdane, J. 1997. Migration and development of mother sporocysts of *Echinostoma caproni* (Digenea: Echinostomatidae). Journal of Parasitology 83: 444–453.

Ataev, G.L., Fournier, A. and Coustau, C. 1998. Comparison of *Echinostoma caproni* mother sporocyst development in vivo and in vitro using *Biomphalaria glabrata* snails and a *B. glabrata* embryonic cell line. Journal of Parasitology 84: 227–235.

Ataev, G.L., Dobrovol'skii, A.A., Avanessian, A.V. and Loker, E.S. 2001. Germinal elements and their development in *Echinostoma caproni* and *Echinostoma paraensei* (Trematoda) miracidia. Journal of Parasitology 87: 1160–1164.

Ataev, G.L., Dobrovol'skii, A.A. and Isakova, N.P. 2005. [Development of the partenitae infrapopulation of *Echinostoma caproni* (Trematoda: Echinostomatidae)]. Parazitologiia 39: 124–136.

Ataev, G.L., Isakova, N.P. and Dobrovol'skii, A.A. 2006. [Development of mother sporocysts of *Echinostoma caproni* (Trematoda: Echinostomatidae)]. Parazitologiia 40: 47–56.

Beaver, P.C. 1939. The morphology and life history of *Psilostomum ondatrae* Price 1931 (Trematoda: Psilostomidae). Journal of Parasitology 25: 383–393.

Behrens, A.C. and Nollen, P.M. 1991. Responses of *Echinostoma caproni* miracidia to gravity, light, and chemicals. International Journal for Parasitology 22: 673–675.

Bennett, S.C., Irwin, S.W.B. and Fitzpatrick, S.M. 2003. Tributylin and copper effects on encystment and in vitro excystment of *Parorchis acanthus* larvae. Journal of Helminthology 77: 291–296.

Chai, J.Y., Han, E.T., Park, Y.K., Guk, S.M. and Lee, S.H. 2001. *Acanthoparyphium tyosenense*: the discovery of human infection and identification of its source. Journal of Parasitology 87: 794–800.

Cheng, Y.Z., Zhang, Y.J., Lin, C.X., Zhan, Z.H., Yu, X.Z., Lin, J.X., Cai, M.J. and Che, W.H. 1999. Taxonomic studies on *Echinochasmus fujianensis* and the related species by random amplified polymorphic DNA analysis and experimental infection. Chinese Journal of Parasitology and Parasitic Diseases 17: 135–139.

Christensen, N.Ø. 1980. A review of the influence of host- and parasite-related factors and environmental conditions on the host-finding capacity of the trematode miracidium. Acta Tropica 37: 303–318.

Christensen, N.Ø., Frandsen, F. and Roushdy, M.Z. 1980. The influence of environmental conditions and parasite-intermediate host related factors on the transmission of *Echinostoma liei*. Zeitschrift für Parasitenkunde 63: 47–63.

Chung, P.R., Jung, Y. and Park, Y.K. 2001a. *Segmentina hemisphaerula*: a new molluscan intermediate host for *Echinostoma cinetorchis* in Korea. Journal of Parasitology 87: 1169–1171.

Chung, P.R., Jung, Y., Park, Y.K., Hwang, M.G. and Soh, C.T. 2001b. *Corbicula fluminea* (Bivalvia: Corbiculidae): a possible second molluscan intermediate host of *Echinostoma cinetorchis* (Trematoda: Echinostomatidae) in Korea. The Korean Journal of Parasitology 39: 329–332.

Cribb, T.H., Bray, R.A., Littlewood, D.T.J., Pichelin, S. and Herniou, E.A. 2001. The Digenea. In Interrelationships of the platyhelminthes. D.T.J. Littlewood and R.A. Bray (eds.), pp. 168–185. London: Taylor and Francis.

Davis, N.E. 2005. Storage and incubation of *Echinostoma revolutum* eggs recovered from wild *Branta canadensis*, and their infectivity to *Lymnaea tomentosa* snails. Journal of Helminthology 79: 321–326.

Desclaux, C., Russell-Pinto, F., de Montaudouin, X. and Bachelet, G. 2006. First record and description of metacercariae of *Curtuteria arguinae* n. sp. (Digenea: Echinostomatidae), parasite of cockles *Cerastoderma edule* (Mollusca: Bivalvia) in Arcachon Bay, France. Journal of Parasitology 92: 578–587.

Díaz, J.I. and Cremonte, F. 2004. *Himasthla escamosa* n.sp. (Digenea: Echinostomatidae) from the kelp gull, *Larus dominicanus* (Caradriformes: Laridae), on the Patagonian coast, Argentina. Journal of Parasitology 90: 308–314.

Esteban, J.G., Toledo, R., Sénchez, L. and Muñoz-Antolí, C. 1997. Life-cycle of *Euparyphium albuferensis* n. sp. (Trematoda: Echinostomatidae) from rats in Spain. Systematic Parasitology 38: 211–219.

Evans, N.A. 1982. Effects of copper and zinc upon the survival and infectivity of *Echinoparyphium recurvatum* cercariae. Parasitology 85: 295–303.

Evans, N.A. 1985. The influence of environmental temperature upon transmission of the cercariae of *Echinostoma liei* (Digenea: Echinostomatidae). Parasitology 90: 269–275.

Ford, D.M., Nollen, P.M. and Romano, M.A. 1998. The effects of salinity, pH and temperature on the half-life and longevity of *Echinostoma caproni* miracidia. Journal of Helminthology 72: 325–330.

Frazer, B.A. and Fried, B. 1998. Single-species infections of *Echinostoma caproni* cercariae in pulmonate snails and concurrent infections of *E. caproni* and *Echinostoma trivolvis* cercariae in *Biomphalaria glabrata*. International Journal for Parasitology 28: 595–597.

Fried, B. 1994. Metacercarial excystment of trematodes. Advances in Parasitology 33: 91–144.

Fried, B. 2000. Maintenance, cultivation, and excystation of echinostomes. In Echinostomes as experimental models for biological research. B. Fried and T.K. Gracyzk (eds.), pp. 99–118. Dordrecht: Kluwer.

Fried, B. 2001. Biology of echinostomes except *Echinostoma*. Advances in Parasitology 49: 163–210.

Fried, B. and Awatramani, R. 1992. Light and scanning electron microscopic observations of the daughter rediae of *Echinostoma trivolvis* (Trematoda). Parasitology Research 78: 257–259.

Fried, B. and Bandstra, S.R. 2005. Age of adult worms of *Echinostoma caproni* does not affect development of miracidia. Journal of Parasitology 91: 1220–1221.

Fried, B. and Bennett, M.C. 1979. Studies on encystment of *Echinostoma revolutum* cercariae. Journal of Parasitology 56: 38–40.

Fried, B. and Graczyk, T.K. (eds) 2000. Echinostomes as experimental models for biological research. Dordrecht: Kluwer, p. 273.
Fried, B. and Graczyk, T.K. 2004. Recent advances in the biology of *Echinostoma* species in the "*revolutum*" group. Advances in Parasitology 58: 139–195.
Fried, B. and Huffman, J.E. 1996. The biology of the intestinal trematode *Echinostoma caproni*. Advances in Parasitology 38: 311–368.
Fried, B. and LaTerra, R. 2002. In vitro and in vivo encystment of the cercariae of *Echinostoma caproni*. Journal of Parasitology 88: 1124–1129.
Fried, B. and Peoples, R.C. 2007. Effects of various physical and chemical factors on excystation of the encysted metacercariae of *Echinostoma caproni*. Journal of Parasitology 93: 1388–1391.
Fried, B. and Ponder, E.L. 2003. Effects of temperature on survival, infectivity and in vitro encystment of the cercariae of *Echinostoma caproni*. Journal of Helminthology 77: 235–238.
Fried, B. and Reddy, A. 1999. Effects of snail-conditioned water from *Biomphalaria glabrata* on hatching of *Echinostoma caproni* miracidia. Parasitology Research 85: 155–157.
Fried, B. and Schneck, J.L. 2004. Effects of copper sulphate on *in vitro* encystment of the cercariae of *Echinostoma caproni*. Journal of Helminthology 78: 371–372.
Fried, B. and Toledo, R. 2004. Criteria for species determination in the "*revolutum*" group of *Echinostoma*. Journal of Parasitology 90: 917.
Fried, B., Frazer, B.A. and Reddy, A. 1997a. Chemoattraction and penetration of *Echinostoma trivolvis* and *E. caproni* cercariae in the presence of *Biomphalaria glabrata, Helisoma trivolvis*, and *Lymnaea elodes* dialysate. Parasitology Research 83: 193–197.
Fried, B., Pane, P.L. and Reddy, A. 1997b. Experimental infection of *Rana pipiens* tadpoles with *Echinostoma trivolvis* cercariae. Parasitology Research 83: 666–669.
Fried, B., Schmidt, K.A. and Sorensen, R. 1997c. In vivo and ectopic encystment of *Echinostoma revolutum* and excystation of the metacercariae. Journal of Parasitology 83: 251–254.
Fried, B., Eyster, L.S. and Pechenik, J.A. 1998. Histochemical glycogen and neutral lipid in *Echinostoma trivolvis* cercariae and effects of exogenous glucose on cercarial longevity. Journal of Helminthology 72: 83–85.
Fried, B., LaTerra, R. and Kim, Y. 2002. Emergence of cercariae of *Echinostoma caproni* and *Schistosoma mansoni* from *Biomphalaria glabrata* under different laboratory conditions. Journal of Helminthology 76: 369–371.
Fried, B., Schneck, J.L. and Ponder, E.L. 2004. Effects of tonicity, digestive enzymes and biles salts, and nutrient media on the survival of excysted metacercariae of *Echinostoma caproni*. Parasitology Research 93: 1–4.
Fujino, T., Takahashi, Y. and Fried, B. 1995. A comparison of *Echinostoma trivolvis* and *E. caproni* using random amplified polymorphic DNA analysis. Journal of Helminthology 69: 263–264.
Fujino, T., Nakano, T., Washioka, H., Tonosaki, A., Ichikawa, H. and Fried, B. 2000. Comparative ultrastructure of eggs in *Echinostoma paraensei, E. caproni* and E. *trivolvis* (Trematoda: Echinostomatidae). Parasitology Research 86: 427–430.
Grabda-Kazubska, B., Borsuk, P., Laskowski, Z. and Moné, H. 1998. A phylogenetic analysis of trematodes of the genus *Echinoparyphium* and related genera based on sequencing of internal transcribed spacer region of rDNA. Acta Parasitologica 43: 116–121.
Graczyk, T.K. and Fried, B. 2007. Human waterborne trematode and protozoan infections. Advances in Parasitology 64: 111–160.
Haas, W. 2000. The behavioral biology of echinostomes. In Echinostomes as experimental models for biological research. B. Fried and T.K. Gracyzk (eds.), pp. 175–197. Dordrecht: Kluwer.
Haberl, B., Körner, M., Spengler, Y., Hertel, J., Kalbe, M. and Haas, W. 2000. Host-finding in *Echinostoma caproni*: miracidia and cercariae use different signals to identify the same snail species. Parasitology 120: 470–496.
Herman, S.M. and Bacha, W.J. Jr. 1978. Growth, development, and site location of the trematode *Himasthla quissetensis* in the chick following cloacal drop infections with cercariae. Journal of Parasitology 64: 827–830.

Huffman, J.E. and Fried, B. 1990. *Echinostoma* and echinostomiasis. Advances in Parasitology 29: 215–269.
Idris, N. and Fried, B. 1996. Development, hatching and infectivity of *Echinostoma caproni* (Trematoda) eggs, and histologic and histochemical observations of the miracidia. Parasitology Research 82: 136–142.
Johnson, P.T.J., Sutherland, D.R., Kinsella, J.M. and Lunde, K.B. 2004. Review of the trematode genus *Ribeiroia* (Psilostomidae): ecology, life history and pathogenesis with special emphasis on the amphibian malformation problem. Advances in Parasitology 57: 191–253.
Kanev, I., Sterner, M., Radev, V. and Fried, B. 2000. An overview of the biology of echinostomes. In Echinostomes as experimental models for biological research. B. Fried and T.K. Gracyzk (eds.), pp. 1–29. Dordrecht: Kluwer.
Kanev, I., Radev, V. and Fried, B. 2005. Family Philophthalmidae Looss, 1899. In Keys to the Trematoda, Vol. 2. A. Jones, R.A. Bray and D.I. Gibson (eds.), pp. 87–97, London: CABI Publishing and The Natural History Museum.
Kechemir, N., Jourdane, J. and Mas-Coma, S. 2002. Life cycle of a new African echinostome species reproducing by parthenogenesis. Journal of Natural History 36: 1777–1784.
Keeler, S.P., Fried, B. and Huffman, J.E. 2007. Association of bacteria with metacercarial cysts of *Echinostoma caproni*. Parasitology Research 101: 1131–1133.
Kim, Y.G., Yu, J.E., Chung, E.Y. and Chung, P.R. 2004. *Acanthoparyphium tyosenense* (Digenea: Echinostomatidae): experimental confirmation of the cercaria and its complete life history in Korea. Journal of Parasitology 90: 97–102.
King, P.H. and Van As, J.G. 2000. Morphology and life history of *Petasiger variospinosus* (Trematoda: Echinostomatidae) in the Free State, South Africa. Journal of Parasitology 86: 312–318.
Koprivnikar, J., Forbes, M.R. and Baker, R.L. 2006. Effects of Atrazine on cercarial longevity, activity, and infectivity. Journal of Parasitology 92: 306–311.
Kostadinova, A. 1999. Cercarial chaetotaxy of *Echinostoma miyagawai* Ishii, 1932 (Digenea: Echinostomatidae), with a review of the sensory patterns in the "*revolutum*" group. Systematic Parasitology 44: 201–209.
Kostadinova, A. 2005a. Family Echinostomatidae Looss, 1899. In Keys to the Trematoda, Vol. 2. A. Jones, R.A. Bray and D.I. Gibson (eds.), pp. 9–64, London: CABI Publishing and The Natural History Museum.
Kostadinova, A. 2005b. Family Psilostomatidae Looss, 1899. In Keys to the Trematoda, Vol. 2. A. Jones, R.A. Bray and D.I. Gibson (eds.), pp. 99–118, London: CABI Publishing and The Natural History Museum.
Kostadinova, A. and Gibson, D.I. 2000. The systematics of the echinostomes. In Echinostomes as experimental models for biological research. B. Fried and T.K. Gracyzk (eds.), pp. 31–57. Dordrecht: Kluwer.
Kostadinova, A. and Gibson, D.I. 2001a. A redescription of *Uroproctepisthmium bursicola* (Creplin, 1837) n. comb. (Digenea: Echinostomatidae), and re-evaluations of the genera *Episthmium* Lühe, 1909 and *Uroproctepisthmium* Fischthal & Kuntz, 1976. Systematic Parasitology 50: 63–67.
Kostadinova, A. and Gibson, D.I. 2001b. Redescription of two echinotomes from freshwater fishes, with comments on *Singhia* Yamaguti, 1958 and *Caballerotrema* Prudhoe, 1960 (Digenea: Echinostomatidae). Systematic Parasitology 49: 195–204.
Kostadinova, A. and Gibson, D.I. 2002. *Isthmiophora* Lühe, 1909 and *Euparyphium* Dietz, 1909 (Digenea: Echinostomatidae) re-defined, with comments on their nominal species. Systematic Parasitology 52: 205–217.
Kostadinova, A. and Jones, A. 2005. Superfamily Echinostomatoidea Looss, 1899. In Keys to the Trematoda, Vol. 2. A. Jones, R.A. Bray and D.I. Gibson (eds.), pp. 5–8, London: CABI Publishing and The Natural History Museum.
Kostadinova, A. and Skirnisson, K. 2007. *Petasiger islandicus* n.sp. (Digenea: Echinostomatidae) in the horned grebe *Podices auritus* (L.) (Aves: Podicipedidae) from Iceland. Systematic Parasitology 68: 217–223.

Kostadinova, A., Gibson, D.I., Biserkov, V. and Ivanova, R. 2000a. A quantitative approach to the evaluation of the morphological variability of two echinostomes, *Echinostoma miyagawai* Ishii, 1932 and *E. revolutum* (Frölich, 1802) from Europe. Systematic Parasitology 45: 1–15.

Kostadinova, A., Gibson, D.I., Biserkov, V. and Chipev, N. 2000b. Re-validation of *Echinostoma miyagawai* Ishii, 1932 (Digenea: Echinostomatidae) on the basis of the experimental completion of its cycle. Systematic Parasitology 45: 81–108.

Kostadinova, A., Vaucher, C. and Gibson, D.I. 2002. Redescription of two echinostomes from birds in Paraguay, with comments on *Drepanocephalus* Dietz, 1909 and *Paryphostomum* Dietz, 1909 (Digenea: Echinostomatidae). Systematic Parasitology 53: 147–158.

Kostadinova, A., Herniou, E.A., Barret, J. and Littlewood, D.T.J. 2003. Phylogenetic relationships of *Echinostoma* Rudolphi, 1809 (Digenea: Echinostomatidae) and related genera re-assessed via DNA and morphological analyses. Systematic Parasitology 54: 159–176.

Laurie, J.S. 1974. *Himasthla quissetensis*: induced in vitro encystment of cercariae and ultrastructure of the cyst. Experimental Parasitology 35: 350–362.

Loker, E.S., Coustau, C., Ataev, G.L. and Jourdane, J. 1999. In vitro culture of rediae of *Echinostoma caproni*. Parasite 6: 169–174.

Loy, Ch., Motzel, W. and Haas, W. 2001. Photo- and geo-orientation by echinostome cercariae results in habitat selection. Journal of Parasitology 87: 505–509.

Maldonado, A. Jr., Loker, E.S., Morgan, J.A., Rey, L. and Lanfredi, R.M. 2001a. Description of the adult worms of a new Brazilian isolate of *Echinostoma paraensei* (Platyhelminthes: Digenea) from its natural vertebrate host *Nectomys squamipes* by light and scanning electron microscopy and molecular analysis. Parasitology Research 87: 840–848.

Maldonado, A. Jr., Vieria, G.O., Garcia, J.S., Rey, L. and Lanfredi, R.M. 2001b. Biological aspects of a new isolate of *Echinostoma paraensei* (Trematoda: Echinostomatidae): susceptibility of sympatric snails and the natural vertebrate host. Parasitology Research 87: 853–859.

Maldonado, A. Jr., Vieira, G.O. and Lanfredi, R.M. 2003. *Echinostoma luisreyi* n.sp. (Platyhelminthes: Digenea) by light and scanning electron microscopy. Journal of Parasitology 89: 800–808.

Marsit, C.J., Fried, B. and Sherma, J. 2000. Neutral lipids in cercariae, encysted metacercariae, and rediae of *Echinostoma caproni*. Journal of Helminthology 74: 365–367.

Martorelli, S.R., Poulin, R. and Mouritsen, K.N. 2006. A new cercaria and metacercaria of *Acanthoparyphium* (Echinostomatidae) found in an intertidal snail *Zeacumantus subcarinatus* (Batillaridae) from New Zealand. Parasitology International 55: 163–167.

McCarthy, A.M. 1999a. Phototactic responses of the cercaria of *Echinoparyphium recurvatun* during phases of sub-maximal and maximal infectivity. Journal of Helminthology 73: 63–65.

McCarthy, A.M. 1999b. The influence of temperature on the survival and infectivity of the cercariae of *Echinoparyphium recurvatun* (Digenea: Echinostomatidae). Parasitology 118: 383–388.

McCarthy, A.M. 1999c. Photoperiodic cercarial emergence patterns of the digeneans *Echinoparyphium recurvatun* and *Plagiorchis* sp. from a mixed infection in *Lymnaea peregra*. Journal of Helminthology 73: 63–65.

McCarthy, A.M. 1999d. The influence of second intermediate host species on the infectivity of metacercarial cysts of *Echinoparyphium recurvatun*. Journal of Helminthology 73: 143–145.

McFarland, L.H., Mouritsen, K.N. and Poulin, R. 2003. From first to second and back to first intermediate host: the unusual transmission route of *Curtuteria australis* (Digenea: Echinostomatidae). Journal of Parasitology 89: 625–628.

Morgan, J.A.T. and Blair, D. 1995. Nuclear rDNA ITS sequence variation in the trematode genus *Echinostoma*: an aid to establishing relationships within the 37-collar-spine group. Parasitology 111: 609–615.

Morgan, J.A.T. and Blair, D. 1998a. Mitochondrial ND1 gene sequence used to identify echinostome isolates from Australia and New Zealand. International Journal for Parasitology 28: 493–502.

Morgan, J.A.T. and Blair, D. 1998b. Relative merits of nuclear ribosomal internal transcribed spacers and mitochondrial CO1 and ND1 genes for distinguishing among *Echinostoma* species (Trematoda). Parasitology 116: 289–297.

Morgan, J.A.T. and Blair, D. 2000. Molecular biology of echinostomes. In Echinostomes as experimental models for biological research. B. Fried and T.K. Gracyzk (eds.), pp. 245–266. Dordrecht: Kluwer.

Morley, N.J., Crane, M. and Lewis, J.W. 2002. Toxic effects of cadmium and zinc on the transmission of *Echinoparyphium recurvatum* cercariae. Journal of Helminthology 76: 157–163.

Morley, N.J., Leung, K.M.Y., Morritt, D. and Crane, M. 2004a. Toxicity of anti-fouling biocides to encysted metacercariae of *Echinoparyphium recurvatum* (Digenea: Echinostomatidae) and their snail hosts. Chemosphere 56: 353–358.

Morley, N.J., Lewis, J.W. and Adam, M.E. 2004b. Metacercarial utilization of a naturally infected single species (*Lymnaea peregra*) snail community by *Echinoparyphium recurvatum*. Journal of Helminthology 78: 51–56.

Muñoz-Antolí, C., Toledo, R. and Esteban, J. 2000. The life cycle and transmission dynamics of the larval stages of *Hypoderaeum conoideum*. Journal of Helminthology 74: 165–172.

Muñoz-Antolí, C., Trelis, M., Espert, A., Toledo, R. and Esteban, J.G. 2002. Survival and infectivity of *Echinostoma friedi* (Trematoda: Echinostomatidae) miracidia and cercariae under experimental conditions. Helminthologia 39: 149–154.

Muñoz-Antolí, C., Trelis, M., Gozalbo, M., Toledo, R., Haberl, B. and Esteban, J.G. 2003. Interactions related to non-host snails in the host-finding process of *Euparyphium albuferensis* and *Echinostoma friedi* (Trematoda: Echinostomatidae) miracidia. Parasitology Research 91: 353–356.

Muñoz-Antolí, C., Trelis, M., Toledo, R. and Esteban, J.G. 2006. Infectivity of *Echinostoma friedi* miracidia to different snail species under experimental conditions. Journal of Helminthology 80: 323–325.

Muñoz-Antolí, C., Marin, A., Toledo, R. and Esteban, J.G. 2007. Effect of *Echinostoma friedi* (Trematoda: Echinostomatidae) experimental infection on longevity, growth and fecundity of juvenile *Radix peregra* (Gastropoda: Lymnaeidae) and *Biomphalaria glabrata* (Gastropoda: Planorbidae) snails. Parasitology Research 101: 1663–1670.

Muñoz-Antolí, C., Marin, A., Vidal, A., Toledo, R. and Esteban, J.G. 2008. *Euparyphium albuferensis* and *Echinostoma friedi* (Trematoda: Echinostomatidae): experimental cercarial transmission success in sympatric snail communities. Folia Parasitologica, 55: 122–126.

Nakano, T., Fujino, T., Washioka, H., Tonosaki, A., Goto, K. and Fried, B. 2003. Tegumentary papillae of *Echinostoma caproni* cercariae (Trematoda: Echinostomatidae). Parasitology Research 89: 446–450.

Nolan, M.J. and Cribb, T.H. 2005. The use and implications of ribosomal DNA sequencing for the discrimination of digenean species. Advances in Parasitology 60: 101–163.

Nollen, P.M. 1994. The hatching behavior of *Echinostoma trivolvis* miracidia and their responses to gravity, light, and chemicals. International Journal for Parasitology 24: 637–642.

Nollen, P.M. 1997. Reproductive physiology and behavior of digenetic trematodes. In Advances in trematode biology. B. Fried and T.K. Gracyzk (eds.), pp. 117–147. Boca Raton, Florida: CRC Press.

Nollen, P.M. and Kanev, I. 1995. The taxonomy and biology of philophthalmid eyeflukes. Advances in Parasitology 36: 205–269.

Ostrowski de Núñez, M. 2007. Life cycle of *Stephanoprora uruguayense* (Digenea: Echinostomatidae) in Argentina. Journal of Parasitology 93: 1090–1096.

Ostrowski de Núñez, M. and Quintana, M.G. 2007. The life cycle of *Stephanoprora aylacostoma* n.sp. (Digenea: Echinostomatidae), parasite of the threatened snail *Aylacostoma chloroticum* (Prosobranchia, Thiaridae), in Argentina. Parasitology Research 102: 647–655.

Park, Y.K., Hwang, M.K. and Chung, P.R. 2006. Encystment and metacercariae development of *Echinostoma cinetorchis* cercariae in an in vitro culture system. Journal of Parasitology 92: 1010–1013.

Petrie, J.L., Burg, E.F. and Cain, G.D. 1996. Molecular characterization of *Echinostoma caproni* and *E. paraensei* by random amplification of polymorphic DNA (RAPD) analysis. Journal of Parasitology 82: 360–362.

Pinheiro, J., Maldonado, A. and Lanfredi, R.M. 2004a. Light and scanning electron microscopy of the miracidium of *Echinostoma paraensei* (Trematoda, Echinostomatidae). Veterinary Parasitology 121: 265–275.

Pinheiro, J., Maldonado, A. Jr., Atias, M. and Lanfredi, R.M. 2004b. Morphology of the rediae of *Echinostoma paraensei* (Trematoda, Echinostomatidae) from its intermediate host *Lymnaea columella* (Mollusca, Gastropoda). Parasitology Research 93: 171–177.

Pinheiro, J., Maldonado, A., Atias, M. and Lanfredi, R.M. 2005. Ultrastructure of the miracidium of *Echinostoma paraensei* Lie and Basch, 1967 (Trematoda, Echinostomatidae). Parasitology Research 97: 367–372.

Pisciotta, J.M., Ponder, E.L., Fried, B. and Sullivan, D. 2005. Hemozoin formation in *Echinostoma trivolvis* rediae. International Journal for Parasitology 35: 1037–1042.

Ponder, E.L. and Fried, B. 2004a. Effects of glucose on survival, infectivity and linear movement of the cercariae of *Echinostoma caproni*. Journal of Helminthology 78: 185–187.

Ponder, E.L. and Fried, B. 2004b. Effects of snail size and diet on encystment of *Echinostoma caproni* cercariae in juvenile *Helisoma trivolvis* (Colorado strain) and observations on survival of infected snails. Journal of Parasitology 90: 422–424.

Prepelitchi, L. and Ostrowski de Nuñez, M. 2007. Echinostomatid larval stages in *Lymnaea viatrix* (Gastropoda: Pulmonata) from Southwest Patagonia, Argentina. Journal of Parasitology 93: 323–327.

Radev, V., Kanev, I., Nollen, P.M. and Gold, D. 1999. Life history and identification of *Philophthalmus lucipetus* from Israel. Journal of Parasitology 85: 291–294.

Radev, V., Kanev, I. and Gold, D. 2000. Life cycle and identification of an eyefluke from Israel transmitted by *Melanoides tuberculata* (Müller, 1774). Journal of Parasitology 86: 773–776.

Reddy, A. and Fried, B. 1996. In vitro studies on intraspecific and interspecific chemical attraction in daughter rediae on *Echinostoma trivolvis* and *E. caproni*. International Journal for Parasitology 26: 1081–1085.

Reddy, A., Frazer, B.A., Fried, B. and Sherma, J. 1997. Chemoattraction of *Echinostoma trivolvis* (Trematoda) rediae to lipophilic excretory-secretory products and thin layer-chromatographic analysis of redial lipids. Parasite 4: 37–40.

Reddy, A., Ponder, E.L. and Fried, B. 2004. Effects of copper sulfate toxicity on cercariae and metacercariae of *Echinostoma caproni* and *Echinostoma trivolvis* and on the survival of *Biomphalaria glabrata* snails. Journal of Parasitology 90: 1332–1337.

Rees, G. 1966. Light and electron microscope studies of the redia of *Parorchis acanthus* Nicoll. Parasitology 56: 589–602.

Rossi, M.L., Balfour, C.D. and Fried, B. 2001. Excystation and infectivity of *Echinostoma caproni* metacercariae after prolonged storage. Comparative Parasitology 68: 126–129.

Sandland, G.J. and Minchella, D.J. 2003. Effects of diet and *Echinostoma revolutum* infection on energy allocation patterns in juvenile *Lymnaea elodes* snails. Oecologia 134: 479–486.

Sapp, K., Meyer, K.A. and Loker, E.S. 1998. Intramolluscan development of the digenean *Echinostoma paraensei*: rapid production of a unique mother redia that adversely affects development of conspecific parasites. Invertebrate Biology 117: 20–28.

Schmidt, K.A. and Fried, B. 1996. Emergence of cercariae of *Echinostoma trivolvis* from *Helisoma trivolvis* under different conditions. Journal of Parasitology 82: 674–676.

Schmidt, K.A., Fried, B. and Reddy, A. 1996. Maintenance of *Helisoma trivolvis* naturally infected with *Echinostoma trivolvis* in spring water at 4 C for 300 days. Journal of Helminthology 72: 91–92.

Schneck, J.L. and Fried, B. 2004. Effects of snail size on encystment of *Echinostoma caproni* in juvenile *Biomphalaria glabrata* (NMRI strain) and observations on survival of infected snails. Journal of Helminthology 78: 277–279.

Schotthoefer, A.M., Cole, R.A. and Beasley, V.R. 2003. Relationship of tadpole stage to location of echinostome cercariae encystment and the consequences for tadpole survival. Journal of Parasitology 89: 475–482.

Schotthoefer, A.M. Labak, K.M and Beasley, V.R. 2007. *Ribeiroia ondatrae* cercariae are consumed by aquatic invertebrate predators. Journal of Parasitology 93: 1240–1243.

Sebelová, S., Stewart, M.T., Mousley, A., Fried, B., Marks, N.J. and Halton, D.W. 2004. The musculature and associated innervation of adult and intramolluscan stages of *Echinostoma caproni* (Trematoda) visualised by confocal microscopy. Parasitology Research 93: 196–206.

Semenas, L., Brugni, N. and de Nuñez, M.O. 1999. Metacercariae of echinostomatidae in *Diplodon chilensis* (Unionacea, Pelecypoda) and description of *Echinoparyphium megacirrus* sp.n. in Patagonia (Argentina). Acta Parasitologica 44: 63–67.

Sohn, W.M. 1998. Life history of *Echinoparyphium recurvatum* (trematoda: Echinostomatidae) in Korea. The Korean Journal of Parasitology 36: 91–98.

Sorensen, R.E., Curtis, J. and Minchella, D.J. 1998. Intraspecific variation in the rDNA ITS loci of 37-collar-spined echinostomes from North America: implications for sequence-based diagnoses and phylogenetics. Journal of Parasitology 84: 992–997.

Stein, P.C. and Basch, P.F. 1977. Metacercarial cyst formation in vitro of *Echinostoma paraensei*. Journal of Parasitology 63: 1031–1040.

Suhardono, X., Roberts, J.A. and Copeman, D.B. 2006. Biological control of *Fasciola gigantica* with *Echinostoma revolutum*. Veterinary Parasitology 140: 166–170.

Théron, A. and Moné, H. 1986. Shedding patterns of *Schistosoma mansoni* and *Ribeiroia marini* cercariae from a mixed infection of *Biomphalaria glabrata*. Journal of Helminthology 60: 255–259.

Thiemann, G.W. and Wassersug, R.J. 2000. Biased distribution of trematode metacercariae in the nephric system of *Rana* tadpoles. Journal of Zoology 252: 534–538.

Toledo, R. and Fried, B. 2005. Echinostomes as experimental models for interactions between adult parasites and vertebrate hosts. Trends in Parasitology 21: 251–254.

Toledo, R., Muñoz-Antolí, C. and Esteban, J.G. 1998a. Cercarial chaetotaxy differences among *Hypoderaeum conoideum* (Bloch, 1782) (Trematoda: Echinostomatidae) from various European localities. Research and Reviews in Parasitology 58: 83–89.

Toledo, R., Muñoz-Antolí, C., Sénchez, L., Bayssade-Dufour, Ch. and Esteban, J.G. 1998b. Cercarial chaetotaxy of *Euparyphium albuferensis* Esteban et al., 1997 (Trematoda: Echinostomatidae), with a review of some genera of the Echinostomatinae. Systematic Parasitology 39: 35–44.

Toledo, R., Muñoz-Antolí, C., Pérez, M. and Esteban, J.G. 1999a. Miracidial infectivity of *Hypoderaeum conoideum* (Trematoda: Echinostomatidae): differential susceptibility of two lymnaeid species. Parasitology Research 85: 212–215.

Toledo, R., Muñoz-Antolí, C., Pérez, M. and Esteban, J.G. 1999b. Survival and infectivity of *Hypoderaeum conoideum* and *Euparyphium albuferensis* cercariae under laboratory conditions. Journal of Helminthology 73: 177–182.

Toledo, R., Muñoz-Antolí, C., Pérez, M. and Esteban, J.G. 1999c. Encystment sites of metacercariae of *Hypoderaeum conoideum* (Trematoda: Echinostomatidae) in freshwater gastropods. Folia Parasitologica 46: 157–158.

Toledo, R., Muñoz-Antolí, C. and Esteban, J.G. 1999d. Production and chronobiology of emergence of the cercariae of *Euparyphium albuferensis* (Trematoda: Echinostomatidae). Journal of Parasitology 85: 263–267.

Toledo, R., Muñoz-Antolí, C. and Esteban, J.G. 2000. The life-cycle of *Echinostoma friedi* n.sp. (Trematoda: Echinostomatidae) in Spain and a discussion on the relationships within the "*revolutum*" group based on cercarial chaetotaxy. Systematic Parasitology 45: 199–217.

Toledo, R., Espert, A., Carpena, I., Trelis, M., Muñoz-Antolí, C. and Esteban, J.G. 2004. *Echinostoma friedi*: the effect of age of adult worms on the infectivity of miracidia. Journal of Helminthology 78: 91–93.

Toledo, R., Esteban, J.G. and Fried, B. 2006. Immunology and pathology of intestinal trematodes in their definitive hosts. Advances in Parasitology 63: 285–365.

Toledo, R., Muñoz-Antolí, C. and Fried, B. 2007. The use of echinostomes to study host-parasite relationships between larval trematodes and invertebrate and cold-blooded vertebrate hosts. Parasitology Research 100: 1177–1185.

Vayrynen, T., Siddall, R., Valtonen, E.T. and Taskinen, J. 2000. Patterns of trematode parasitism in lymnaeids snails from northern and central Finland. Annales Zoologici Fennici 37: 189–199.

Vilas, R., Criscione, C.D. and Blouin, M.S. 2005. A comparison between mitochondrial DNA and the ribosomal internal transcribed regions in prospecting for cryptic species of plathyhelminth parasites. Parasitology 131: 839–846.

Viozzi, G., Flores, V. and Rauque, C. 2005. An ectosymbiotic flatworm, *Temnocephala chilensis*, as second intermediate host for *Echinoparyphium megacirrus* (Digenea: Echinostomatidae) in Patagonia (Argentina). Journal of Parasitology 91: 229–231.

Wilson, W.D., Johnson, P.T.J., Sutherland, D.R., Moné, H. and Loker, E.S. 2005. A molecular phylogenetic study of the genus *Ribeiroia* (Digenea): trematodes known to cause limb malformations in amphibians. Journal of Parasitology 91: 1040–1045.

Xiao, X., Wang, T.P. and Lu, D.B. 1992. The first record of human natural infection of *Echinochasmus liliputanus*. Chinese Journal of Parasitology and Parasitic Diseases 10: 132–136.

Xiao, X., Lu, D., Wang, T.P., Gao, J., Zhu, C., Zhang, B., An, J., Pong, H., Xu, M. and Wu, W. 1995. Studies on the mode of human infection with *Echinochasmus liliputanus*. Chinese Journal of Parasitology and Parasitic Diseases 13: 197–199.

Xiao, X., Wang, T., Zheng, X., Shen, G. and Tian, Z. 2005. In vivo and in vitro encystment of *Echinochasmus liliputanus* cercariae and biological activity of the metacercariae. Journal of Parasitology 91: 492–498.

Yamaguti, S. 1971. Synopsis of digenetic trematodes of vertebrates, Vol. I. Tokyo: Keigaku Publishing Company, 1074 p.

Yamaguti, S. 1975. A synoptical review of life histories of digenetic trematodes of vertebrates with special reference to the morphology of their larval forms. Tokyo: Keigaku Publishing Company, 590 p.+219 pls.

Zamparo, D., Overstreet, R.M. and Brooks, D.R. 2005. A new species of *Petasiger* (Digenea: Echinostomiformes: Echinostomatidae) in the brown pelican, *Pelecanus occidentalis,* (Aves: Pelecaniformes: Pelecanidae), from the area of conservation Guanacaste, Costa Rica. Journal of Parasitology 91: 1465–1467.

Chapter 2
Echinostomes and Snails: Exploring Complex Interactions

Christine Coustau, Benjamin Gourbal, Guillaume Mitta, and Coen Adema

Contents

2.1 Echinostomes and Snails: Exploring Complex Interactions 35
 2.1.1 Introduction 35
 2.1.2 The Mystery of Specificity for Snail Intermediate Hosts 37
 2.1.3 Immune Evasion Strategies of Larval Echinostomes 40
 2.1.4 Echinostomes as Tools for Exploring Snail Innate Immunity 44
2.2 Concluding Remarks 51
References 53

Abstract Echinostomes rely on a molluskan host to achieve the asexual multiplication phase of larval stages in their complex life cycle. Here we present prominent characteristics of the *Biomphalaria-Echinostoma* systems and detail recent advances in our understanding of their molecular interactions. Remarkably, studies using these echinostomes, pathogens with relatively modest medical impact have shed light on important aspects of parasite strategies for immune evasion, invertebrate internal defenses, and general immunology.

2.1 Echinostomes and Snails: Exploring Complex Interactions

2.1.1 Introduction

Echinostomes, together with most of the 6,000 known digenean species, rely on a molluskan host to achieve the asexual multiplication phase of larval stages in their complex life cycle. This obligatory association is often highly specific. An echinostome that invades a nonhost species usually fails to complete its life cycle and faces rapid immune elimination. Depending on the echinostome species, one or a small number

---

C. Coustau (✉) B. Gourbal, G. Mitta, and C. Adema
U547 Inserm, Institut Pasteur de Lille, Lille, France

B. Fried and R. Toledo (eds.) *The Biology of Echinostomes*,
DOI: 10.1007/978-0-387-09577-6_2, © Springer Science+Business Media, LLC 2009

of freshwater or marine snail species may serve as an effective first intermediate host (Kanev et al. 2000). Echinostome miracidia penetrate the host gastropod mollusk and rapidly transform into mother sporocyst. According to species, sporocysts may settle near the site of miracidia entry, or migrate through the host tissues and settle in the heart area within 1–2 days (Lo 1995; Adema and Loker 1997; Ataev et al. 1998; Sapp et al. 1998; Kanev et al. 2000). There, they undergo an asexual multiplication process resulting in the production of generations of mother rediae (see Fig. 2.1), daughter rediae, and finally cercariae (see Chap. 1) that will emerge from this host and infect the secondary intermediate host.

This fascinating development, often occurring in a host-specific association, raises a number of questions. Why are gastropods uniquely suited to serve as intermediate hosts for echinostomes? How does the parasite regulate the complex differentiation steps between its larval stages? What are the physiological interactions between host and parasite? What determines the specificity of this parasite host interaction? How do nonhost gastropods manage to rapidly eliminate the parasite? How can the parasite avoid destruction by the immune system of a suitable snail host? These issues are not only relevant for a general understanding of all parasites that have a complex life cycle, but they also relate to all host-pathogen interactions. Even considering results from other, more frequently studied vertebrate-pathogen model systems, it is obvious that these general parasitological questions remain largely unanswered to this date. Work continues to elaborate at least partial answers for echinostomes and snails.

The main model for study of snail-digenean interactions involves the snail *Biomphalaria glabrata* and the human blood fluke *Schistosoma mansoni* (*Strigeidida*,

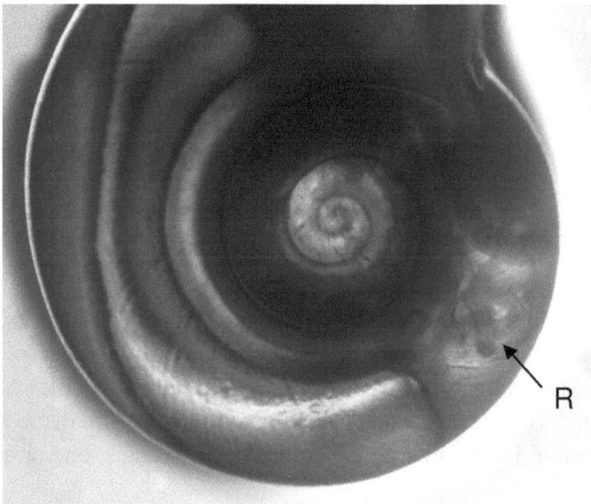

Fig. 2.1 Rediae (R) of *Echinostoma paraensei* in the pericardical cavity of a *Biomphalaria glabrata* snail

Digenea). This is logically motivated by the continued negative impact that schistosomiasis exerts on global health (Fenwick et al. 2006; King et al. 2006).

By comparison, the number of studies of associations of *B. glabrata* with echinostome flukes such as *Echinostoma paraensei* and *Echinostoma caproni* (*Echinostomida*, Digenea) is modest. However, several biological aspects of echinostome infection in *B. glabrata* are more obvious for observation relative to schistosome infection. As described later these studies have provided invaluable insights in immunoparasitology, several of which transcend beyond the field of snail-digenean immunoparasitology. Comparative study of echinostome infections in *B. glabrata* will likely facilitate functional interpretation of factors that characterize digenean-snail interactions in general. This is highly relevant, especially now that factors that take part in the parasite-host interactions are being cataloged in unprecedented numbers as sequence data by gene discovery efforts with modern molecular (transcriptomic and proteomic) techniques and as part of genome projects for digeneans such as *S. mansoni* and *Schistosoma haematobium*) and for *B. glabrata* as snail intermediate host.

Here we will present prominent characteristics of the *Biomphalaria-Echinostoma* systems and detail recent advances in our understanding of their molecular interactions. Remarkably, studies using these echinostomes, pathogens with relatively modest medical impact, have shed light on important aspects of parasite strategies for immune evasion, invertebrate internal defenses, and general immunology.

2.1.2 The Mystery of Specificity for Snail Intermediate Hosts

When considering the complex ontological development that takes place in the snail host, including complex migration paths through the tissues, the intensive multiplication that gives rise to hundreds of larvae, and the tremendous metabolic requirements to fuel these parasite activities, it seems easy to intuitively assume that echinostomes must be highly adapted to exploit a particular species of host snail and that they cannot develop in other species. Surprisingly, it seems that digeneans are highly permissive of varying physiological conditions, and that such factors contribute only modestly to compatibility. A number of studies have investigated the various steps possibly involved in determining this host specificity, and following is an overview of these contributions.

2.1.2.1 Host-Finding Behavior of Miracidia

The first requirement to be met in any digenean life cycle is that the free-swimming larval miracidium (see Fig. 2.2) encounters and penetrates an appropriate molluskan species. Therefore, potential variations in the attractiveness of various potential snail hosts to miracidia, and in the ability to perforate the tegument of a host via the

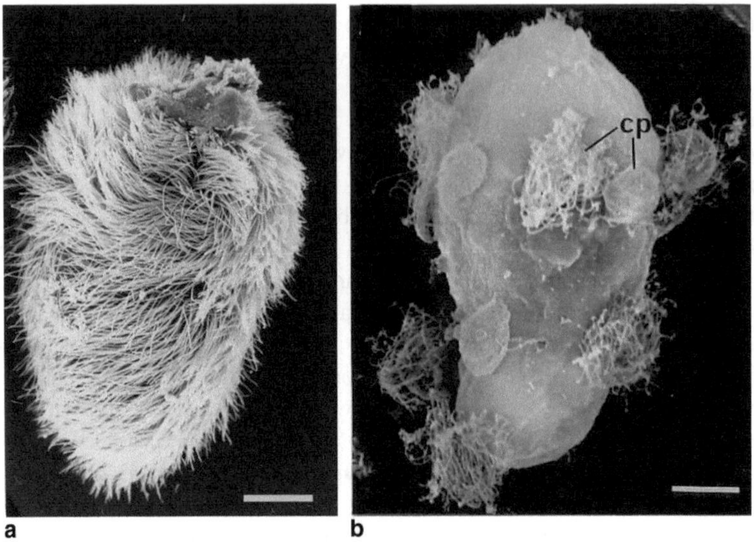

Fig. 2.2 Scanning electron micrograph of *Echinostoma caproni* newly hatched miracidium (**a**) and transforming miracidium after one day in culture (**b**) Scale bar is 10 mm. The loss of ciliate plates (cp) typical of the transformation process normally occurring shortly after snail infection is induced under in vitro conditions. Such in vitro transformed sporocysts were used in the cocultivation experiments, and the culture supernatants were used as source of excretory-secretory products

secretion of cytolytic enzymes from the apical glands of the miracidium may govern specificity. Early studies on snail-trematode specificity (reviewed by Cheng 1968) showed that chemotaxy does exist but that it operates only within very restricted distances and in a nonspecific manner. Miracidia can detect and utilize gradients of macromolecular components that diffuse from surface mucus-covering snails to locate potential host snail in the aquatic environment (Haas et al. 1995; Haberl et al. 2000). Sapp and Loker (2000a, b) compared the miracidial attachment and penetration behavior of four digenean species including two echinostome species (*Echinostoma trivolvis, E. paraensei)* and two schistosome species by arranging encounters with host and nonhost snail species (*Biomphalaria glabrata*, *Helisoma trivolvis*, *Lymnaea stagnalis*, *Stagnicola elodes*, and *Helix aspersa*) representing 3 gastropod families. Miracidia from all digenean species attached to the tentacles and head-foot region of snails and attempted penetration of both compatible (specific host) and noncompatible (nonhost) snail species. This demonstrated that miracidia of these parasites, including echinostomes, do not discriminate between compatible and incompatible snails (Sapp and Loker 2000a, b). When assessing the number of miracidia achieving snail penetration (determined by the number of free miracidia remaining after the exposure interval), the results showed significant differences between host and nonhost snails with a higher penetration rate of host species.

However, half of the miracidia from each digenean species did penetrate nonhost snail species (Sapp and Loker 2000a, b). In support of these observations, Haberl et al. (2000) concluded that differently from schistosome miracidia (from an Egyptian isolate of *S. mansoni*), the miracidia of *E. caproni* were not able to differentiate between different snail species. Altogether these results show that preferential attachment or penetration of miracidia does not explain the high degree of specificity of echinostomes for their snail hosts.

2.1.2.2 Host Physiological Suitability vs. Host-Parasite Immunobiological Interactions

Theoretically, once in the host tissues, parasites face two major types of challenge: (1) exploiting host internal milieu for their growth and development; and (2) evading host immune response. Successful host exploitation depends on complex physiological conditions including the presence and availability of appropriate nutriments as well as physicochemical parameters such as oxygen tension or osmolarity. It is an *a priori* expected fact that variations in these parameters from species to species may play a major role in digenean-snail specificity (Cheng 1968).

Only a limited number of past biochemical studies have investigated physiological interactions between mollusks and digeneans. While the free-living larval miracidia usually employ aerobic respiration, intramolluskan stages of digeneans are able to switch after penetration of the snail host and rely mostly on anaerobic metabolism (Tielens et al. 1991, 1992; Boyunaga et al. 2001), likely to accommodate the occasional hypoxic environment that may occur within snail hosts.

Digeneans utilize carbohydrates as energy source and also take up free amino acids from host hemolymph and lipids, in the form of short-chain fatty acids (Cheng 1968). Studies on *S. mansoni/B. glabrata* interactions indicate that host glutamine may play a central role in this interaction (Coustau et al. 2003; Kayath et al. 2006), as it promotes sporocyst growth and differentiation through its use for neosynthesis of glucose, in a minor part, and of glycerol, in a more important part (Kayath et al. 2006).

Although differences in physicochemical properties of the internal milieu of mollusks probably exist and the physiological requirements of different digeneans may vary, there is no current evidence for the notion that such parameters figure in determining parasite-host specificity. Conversely, there is intriguing indirect evidence to suggest that the importance of physiological parameters for intramolluskan development of digenean parasites is actually limited. Several so-called cocultivation studies have investigated the ability of digeneans to develop under in vitro conditions (Fig. 2.1), in the presence of *Biomphalaria glabrata* embryonic (Bge) cells. The Bge cell line (Hansen 1976; ATCC CRL 1494; Rockville, MD) is unique because it represents the only cell line that is available for mollusks to date. These studies have involved *E. caproni*, an echinostome species that normally develops in *B. glabrata*, but not other echinostome species that employ other snail species as intermediate host. However, comparisons of in vitro development of sporocysts from *S. mansoni, E. caproni, Schistosoma japonicum, Schistosoma mattheei, Schistosoma intercalatum*, and *Fascioloides magna*, revealed that advanced development of intramolluskan larvae of all these digeneans was achieved, regardless of the phylogenetical distance separating *B. glabrata* from the natural snail host (Yoshino and Laursen 1995; Coustau et al. 1997; Ataev et al. 1998; Coustau and Yoshino 2000). For example, both *S. japonicum* and *F. magna*, normally developing in *Oncomelania* (prosobranchia) and *Lymnaea* (gastropoda) snails, produced daughter sporocysts/rediae in the presence of Bge cells (Coustau and Yoshino 2000). Yet another study showed that *S. mansoni* sporocysts develop in vitro in culture medium independent of host factors, under low oxygen pressure. Parasite yield was sharply reduced under ambient oxygen levels, except when sporocyst-conditioned culture medium was applied. Likely, antioxidant properties of the secreted-excreted products of sporocysts are critical for parasite development and survival. Axenic culture did not yield cercariae, however, leading to the hypothesis that cercariogenesis may need a trigger from host components (Bixler et al. 2001).

These observations are consistent with the idea that although the physiological environment and nutrients is critical to support parasite development, it probably plays a limited role in determining snail/digenean (including echinostome) specificity. Therefore, the success or failure of an infection may largely depend on immunobiological interactions, that is, interactions between parasite immune evasion strategies and host immune responses.

2.1.3 Immune Evasion Strategies of Larval Echinostomes

All parasites are known to evade host immune responses through mechanisms such as molecular mimicry, antigenic variation, immune modulation, or immune suppression (see Würzner 1999; Zambrano-Villa et al. 2002; Maizels et al. 2004). Studies focused mainly on *E. paraensei* and *E. caproni* suggest that echinostomes' immune evasion strategy largely relies on suppression of host cellular immune functions.

2.1.3.1 In Vivo Studies

In the 1970s Lie and coworkers performed a series of landmark studies that provided evidence for immune suppression by larval echinostomes of host defense responses in experimentally infected snails. They showed, for example, that *B. glabrata* snails that are normally resistant to *S. mansoni*, no longer encapsulate *S. mansoni* when they harbored a previous *E. paraensei* infection (Lie et al. 1976). Similar observations were made with several other echinostome species (Lie and Heyneman 1977a), suggesting echinostomes actively interfere with their snail host's internal defense system (Lie 1982).

2.1.3.2 Ex Vivo Studies

The notion that echinostomes aggressively interfere with their host immune response was then supported by *ex vivo* studies. Hemocytes collected from *E. paraensei*-infected *B. glabrata* exhibited significantly lower adhesion, spreading and phagocytic abilities than hemocytes from control snails (Noda and Loker 1989a, b). Hemocytes from uninfected *B. glabrata* snails do not encapsulate live sporocysts of *E. paraensei*, but vigorously encapsulate dead sporocysts (Loker et al.1989). This parasite must be metabolically active to maintain suppression of hemocyte function. Time-lapse videomicroscopy showed that *B. glabrata* hemocytes exposed to live *E. paraensei* sporocysts or rediae move away from the parasites (Adema et al., 1994a). Hemocytes closest to sporocysts were affected before more distant cells, suggesting that the parasite continuously releases active factor(s) to generate gradient of interference around the sporocyst. Indeed, the potency of excretory-secretory (ES) products collected from in vitro transformed *E. paraensei* sporocysts to cause rounding and loss of adherent properties of *B. glabrata* hemocytes in vitro, was correlated with the infectivity rate for *B. glabrata* snails of miracidia derived from the same batch of *E. paraensei* eggs (DeGaffe and Loker 1998). Whereas the mechanism by which ES products inhibit hemocyte function remains to be resolved, interference may be effected through interaction with signal transduction pathways of the snail (Walker 2006).

The in vitro exposure of *B. glabrata* hemocytes to *E. paraensei* sporocysts or their ES products yields a calcium wave in the cytoplasm of adherent hemocytes (indicative of activation of signaling pathways that employ Ca ions as second messenger) prior to the rounding up of these cells. Both processes are not necessarily related because they could be evoked independently by several different bioactive molecules. Thus, *E. paraensei* may affect hemocyte function in different ways (Hertel et al. 2000). Similarly, excretory-secretory (ES) products collected from in vitro transformed *E. caproni* sporocysts (Fig. 2.2) induced a total loss of *B. glabrata* hemocyte defense functions encompassing adhesion, spreading, and phagocytosis (Fig. 2.3) (Humbert and Coustau 2001).

Interestingly, this immuno-suppressive effect of *E. paraensei* and *E. caproni* ES products may significantly contribute to the specificity of host-parasite compatibility, as the potency of ES products to affect hemocytes correlated with echinostome infectivity (DeGaffe and Loker 1998), the degree to which hemocyte function was interfered was less in snails exposed to but not infected with echinostome (Noda and Loker 1989b), and ES products failed to affect hemocytes from nonhost species (Adema et al. 1994a, b) or from genetically selected resistant strains of host species (Humbert and Coustau 2001).

2.1.3.3 Insights from Proteomic Studies

In order to start the characterization of the molecular processes underlying echinostome immune evasion, Guillou et al. (2007b) performed a comparative proteomic

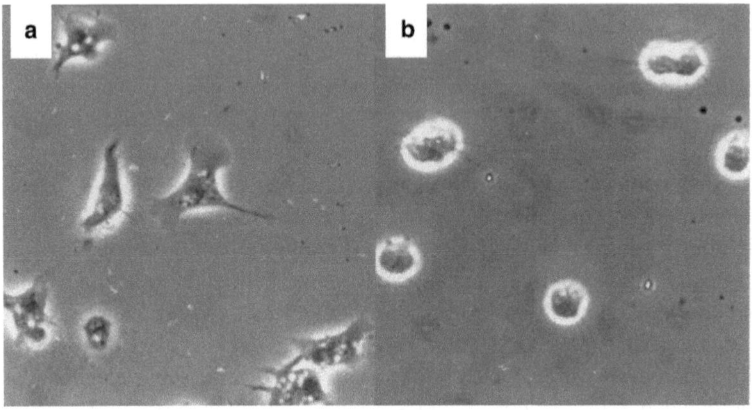

Fig. 2.3 Representative effect of echinostome excretory/secretory products on *Biomphalaria glabrata* hemocytes. (**a**) Control hemocytes after 3 h in saline solution. They are well spread and present pseudopodia. (**b**) Hemocytes after 3 h in saline solution containing excretory/secretory products from *Echinostoma caproni*

analysis of mother sporocyst excretory-secretory (ES) products from *E. caproni* and from *S. mansoni*. Because ES products from *S. mansoni* have not been shown to suppress hemocytes' immune functions in ways comparable to the dramatic effects from echinostome ES products (reviewed by Loker 1994; Loker and Adema 1995), this comparison was expected to help toward identification of particular ES factors that contribute to immune suppression from among the complex mixture of ES proteins, components of which are likely involved also in other aspects of the host-parasite interactions.

Interestingly, after 24 h of in vitro culture, most proteins recovered from the ES products of both *S. mansoni* and *E. caproni* sporocysts belonged to the same functional categories. Abundantly present in both ES proteomes were (a) proteins that function in the detoxification of oxidative stress; and (b) glycolytic enzymes involved in the Embden-Meyerhof-Parnas pathway. Based on previous observations and experimental data, it is likely that these two groups of proteins function as antioxidants and protect sporocysts from oxidative damage (Guillou et al. 2007a, b). Protozoan parasites for instance employ glycolysis to counter oxidative stress (Roth 1990; Cordeiro et al. 2004), and glycolysis may be indirectly involved in the glutathione red-ox cycle through the production of NADH (Kum-Tatt et al. 1975, Guillou et al. 2007a, b).

In addition, other glycolytic enzymes (enolase and GAPDH) identified from the ES of both *E. caproni* and *S. mansoni* could bind host plasminogen and thus facilitate invasion and migration of the parasite in host tissues (Jolodar et al. 2003; Bergmann et al. 2004; Kolberg et al. 2006). Finally, a Mical-like protein was identified in the ES proteome of *E. caproni*. This protein, known to play a key role in repulsive neuronal guidance, could be involved in immune evasion strategies, particularly in the observed inhibition of adhesion and phagocytosis capabilities of hemocytes (Adema et al. 1994a, b; DeGaffe and Loker 1998; Humbert and Coustau 2001; Guillou et al. 2007a, b).

In light of the dramatically different strategies for immune evasion of schistosomes (immune evasion through molecular mimicry) and echinostomes (interference) (Loker 1994; Loker and Adema 1995), it is especially intriguing that the ES products of both *E. caproni* and *S. mansoni* include similar sets of proteins that provide protection against oxidative stress by scavenging (detoxification) reactive oxygen species (ROS). Several studies have shown that the cytotoxic armament of hemocytes includes the respiratory burst; immune activation causes hemocytes to generate and release toxic ROS that play a crucial role in the killing of parasites (Dikkeboom et al. 1988; Adema et al. 1994a, b; Bayne et al. 2001; Hahn et al. 2001; Bender et al. 2005). The release of antioxidant factors may protect the postmiracidium larval stage of *S. mansoni* and *E. caproni* from initial host responses during the first hours of infection. The parasite is likely to be highly vulnerable during this stage as it is remodeling anatomy and metabolism to transform into the sporocyst stage. An immediate, early immunoprotective strategy against hemocyte-mediated cytotoxicity could help digenean parasites to mount a secondary immune evasion strategy relying either on molecular mimicry or immunosuppressive strategy, thus avoiding the recognition and elimination of the parasite for the rest of its intramolluskan cycle (Loker et al. 1986; Damian 1997; Abu-Shakra et al. 1999; Guillou et al. 2007a, b). Most likely, these disparate strategies are affected by entirely different molecular mechanisms. This difference is underscored by the realization that none of the gene transcripts specific to the sporocyst stage of *E. paraensei* sporocysts, for which in vivo expression was confirmed by hybridization involving cDNA extracted from parasites that had developed in *B. glabrata*, had significant similarity to any of over 200.000 sequences derived from *S. mansoni.* Moreover, only 21 of 69 sequences from *E. paraensei* displayed similarity to any of the known sequences present in GenBank (Nowak and Loker 2005).

While this clearly indicates that much remains to be learned about the biology of echinostome sporocysts, it is also evident that the biology of the echinostome and schistosome digeneans differs considerably. Failure to establish an effective secondary means for immune evasion leads to elimination of the (noncompatible) parasite. Typically, noncompatible digenean parasites are met almost immediately (within hours after penetration) with vigorous encapsulation responses, but actual elimination by hemocyte-mediated cytotoxicity may take up to 72 h.

Trematodes are exposed to ROS throughout their life cycle and are known to suppress or modulate the immune response of their vertebrate hosts (Van der Knaap and Loker 1990; LoVerde 1998; Maizel et al. 2004). ES products from adult schistosomes were described to counteract host adaptive response (LoVerde 1998; Perez-Sanchez et al. 2006). Antioxidant protective systems and proteins that scavenge ROS were described in schistosomula and adult worms (Mkoji et al. 1988a, b; Mei and LoVerde 1997), and were found in the ES products of cercariae and adults of echinostomes and schistosomes (Knudsen et al. 2005; Bernal et al. 2006; Perez-Sanchez et al. 2006). In addition, antioxidant proteins are involved in immune evasion in other host-parasite systems involving both invertebrate and vertebrate hosts (Mei and LoVerde 1997; Bernal et al. 2004, 2006; Knudsen et al. 2005; Kolberg et al. 2006; Perez-Sanchez et al. 2006; Zhang et al. 2006).

The aforementioned observations illustrate that ROS-scavenging molecules figure importantly in immune evasion strategies that are applied by parasites across the phylogenetic spectrum. The recent insights from larvae of *E. caproni* and *S. mansoni* (Guillou et al. 2007a, b) are consistent with the idea that also these parasites rely on a set of factors with scavenging activity that counter oxidative stress resulting from ROS to gain a measure of protection against initial defense responses from a snail host. Building on this idea, the successful establishment of these digeneans in their intermediate host may rely on a two-stage deployment of survival strategies, a first general approach to temporarily counter ROS toxicity followed by parasite-specific strategies that emphasize either molecular mimicry or interference.

While molecular approaches are beginning to facilitate the functional interpretation of digenean survival mechanisms, the identification of the factors involved so far is restricted to abundant proteins with known function. A multitude of novel sequences (proteins, but possibly also other compounds) with unknown function indicates the need for substantial additional efforts toward characterize the active factors that allow digeneans to exert their survival strategies, for a more complete understanding of specific parasite-host compatibility.

2.1.4 Echinostomes as Tools for Exploring Snail Innate Immunity

A future understanding of the interactions between trematodes and their first intermediate host also requires a comprehensive knowledge of snail immune system. Considerable efforts have been made in the last decade to start characterizing molecular processes underlying *B. glabrata* immune response, and some major findings have been achieved while exploring *B. glabrata*-echinostome interactions. Here, we present a brief chronological overview of studies on *B. glabrata* immunity.

2.1.4.1 General Features of *Biomphalaria Glabrata* Immune System: Insights from "PreMolecular" Era

The immune system of *B. glabrata* relies on both cellular and humoral factors that cooperate in the recognition and elimination of nonself-invaders (Loker and Bayne 1986; Yoshino and Vasta 1996; Lardans and Dissous 1998).

Circulating hemocytes are the major effector cells responsible for parasite encapsulation and killing. The major hemocyte type, referred to as granulocytes, type A cells, or adherent hemocytes (Yoshino and Granath 1985; Loker and Bayne 1986; Coustau C, & Yoshino T.P. 1994 Surface), are actively phagocytic; they rapidly spread on glass and are involved in the formation of the multilayer hemocytic capsules around multicellular invaders (Loker and Bayne 1986). Also present are two other cell types: hyalinocytes (function unknown) and so-called round cells, nonadherent spherical cells (Loker and Bayne 1986), and it is not established whether they represent an undifferentiated stage of young granulocytes or an ontologically different cell type. In addition to the key role of cellular effectors, many studies have demonstrated that humoral factors contribute to promoting or modulating cellular responses or function in inducing parasite degeneration and death (Granath and Yoshino 1984; Loker and Bayne 1986; Fryer and Bayne 1989, 1996; Ataev and Coustau 1999). In particular, a multitude of soluble plasma opsonins and agglutinins or lectins was evident, but experimental characterization was not usually developed beyond molecular weight of protein fractions, binding activity of particular carbohydrates, or the ability to agglutinate experimental target cells. As a consequence the identity of such factors has largely remained unknown (e.g., Jeong et al. 1981; Loker and Hertel 1987).

2.1.4.2 Historically Perceived Simplicity of Invertebrate Immunity

For a long time, the innate defenses of invertebrates have been considered to be rather unsophisticated, especially compared to the immune capabilities of jawed vertebrates. In addition to innate type immunity for initial protection to pathogens, this latter group also disposes overacquired immunity; somatic mutations are employed for rearranging genes to generate increasingly specific antigen receptors (antibodies and T-cell receptors) that can efficiently direct immune responses to pathogens. Clonal expansion of B cells that produce effective antibodies amplifies the immune response, and creation of memory cells leads to an immunological memory (acquired immunity) such that defense responses are immediate if the same pathogen is encountered again. The absence of obvious homologs of lymphocytes, lymphocytic tissues, and antibodies from invertebrates, along with arguments that most invertebrates had only limited life spans and bodies too small to accommodate clonal expansion of antigen receptor-producing cells, led to the consideration that invertebrates were not able to generate diverse and specific antigen receptors and that they not possess acquired immune capabilities (e.g., Klein 1989).

As alternative to antibodies, nonself-recognition of pathogens by invertebrates was mediated by lectins. These are nonenzymatic proteins that specifically bind particular carbohydrates (e.g., Vasta et al. 2004). The pattern recognition model (Medzhitov and Janeway 2000) proposed that even a limited set of nonchanging lectins (encoded by a modest number of genes in the genome of invertebrates) could afford an effective system for nonself-recognition in the context of (invertebrate) innate immunity. Groups of pathogens are characterized by so-called pathogen-associated molecular patterns (PAMPs), repetitive carbohydrate structures such as LPS for Gram – bacteria and peptidoglycan for Gram + bacteria. Thus, lectins with nonchanging specificity for a particular PAMP can serve as pattern recognition receptors (PRRs) not only to detect invaders, but also identify the type of pathogen and activate an innate immune response. Recently, it is becoming clear that innate immunity systems function far less rigidly and display a remarkable level of sophistication. Important contributions to these new insights regarding innate immunity in general stem from studies of the interactions between echinostomes and their invertebrate host snail *B. glabrata*.

2.1.4.3 Immunological Memory in Invertebrates?

The exploration of the echinostome-mediated interference with snail host defenses by Lie, Heyneman and coworkers led to experiments that involved irradiated echinostome miracidia to produce nonviable infection in the snail host. The irradiated parasites penetrated *B. glabrata* normally, but subsequently failed to develop and died. This experimental model yielded the observation that previous exposure with nonviable echinostomes rendered normally susceptible *B. glabrata* refractory to infection when exposed to untreated, viable miracidia of the same parasite (Lie and Heyneman 1976, 1979; Lie et al. 1975; Jeong et al. 1980; Lie et al. 1982). Thus, a previous exposure caused *B. glabrata* to gain an "acquired resistance" that also provided some crossprotection against closely related echinostome species. Intriguingly, irradiated echinostome larvae did not induce resistance to *S. mansoni* infection (Lie and Heyneman 1977a, b), and an effort to induce acquired resistance in a lymnaied snail proved unsuccessful (Dondero et al. 1977). The methodologies available in that era did not yet permit molecular characterization of the parameters underlying the acquired immunity, but snails with acquired resistance responded to parasite infection with a more rapid increase in the number of circulating hemocytes as compared to naïve *B. glabrata* (Lie et al. 1982).

Possibly, previous exposure to nonviable (noncompatible) parasites caused a state of "immune priming" where innate immune defenses remain at a heightened activation level for some interval of time after the pathogen has been eliminated. Newly entering parasites are then met by a more rapid defense response. This hypothesis does not readily account for the specificity of "acquired resistance" in *B. glabrata*, however. Alternatively, the previous immune interaction did give rise to some as yet unresolved form of immunological memory to counter specific parasites. Whatever the underlying mechanism, the phenomenon of acquired resistance in *B. glabrata* deserves continued investigative attention, especially given recent reports of specific immune memory also from other invertebrate organisms such as the copepod *Macrocyclops albidus* and *Drosophila* (Kurtz 2005; Little et al. 2005; Pham et al. 2007).

With the realization that noncompatible echinostome parasites (and other digeneans?) may continuously penetrate and challenge *B. glabrata* in its natural habitat, the likelihood of induction of potentially crossprotective "acquired resistance" must be considered as a significant modifier of host-parasite compatibility patterns in the field.

2.1.4.4 AntiDigenean Responses in *Biomphalaria Glabrata* are Complex and Dynamic

Likely due to their highly effective interference strategy aimed at inhibition of the function of host defense cells, echinostomes seem to not try to avoid recognition as pathogen by the snail host. In fact ES products of *E. paraensei* actually stimulate the amoebocyte-producing organ (Noda 1992) where the hemocytes (defense cells) originate, and echinostome infection leads to higher numbers of circulating hemocytes as well as increased concentration of plasma proteins. By comparison, the responses of *B. glabrata* to schistosome infection are very modest (see Loker and Adema 1995).

Echinostome parasites have proven to be a good experimental tool for immune challenge of *B. glabrata* and have provided valuable insights into the plasma polypeptides that are part of the humoral responses to infection.

Initial experiments revealed that the response of *B. glabrata* to *E. paraensei* infection comprised a complex mixture of plasma proteins with a wide range of molecular weights (Loker and Hertel 1987).

A variety of carbohydrate-binding and agglutinating activities identified several of these plasma factors as lectins (Monroy et al. 1992; Monroy and Loker 1993; Hertel et al. (1994). potential nonself-recognition factors of *B. glabrata.*

A potential nonself-recognition function was further indicated by showing that several plasma polypeptides bound parasite antigens; a precipitate formed when ES products from in vitro cultured *E. paraensei* sporocysts were mixed with the cell-free plasma from *B. glabrata* that had been exposed to *E. paraensei*. SDS-PAGE analysis revealed that the snail-derived components of this precipitate fell into 3 different groups of proteins, that grouped as fuzzy bands around molecular weights of ~200kD, 80–120 kD, and ~65kD, respectively (Adema et al. 1997).

This precipitation reaction depended on lectin-like properties of the snail polypeptides; the formation of the precipitate was reversible by addition of excess antigens and proved inhibitable by the monosaccharide fucose. The ~65-kD snail plasma lectins, produced in response to echinostome infection and reactive with parasite antigens, were selected for further study with molecular techniques. Molecular sequence analysis revealed that these snail plasma lectins were FREPs or fibrinogen-related proteins, consisting of N-terminal immunoglobulin superfamily (IgSF)-domains and a fibrinogen (FBG)-like domain at the C-terminus (Adema et al. 1997).

The exact function of FREPs is subject of ongoing research, but considering the pattern-recognition hypothesis (Medzhitov and Janeway 2000), as parasite-reactive lectins, FREPs may well serve to mediate nonself-recognition and immune activation in *B. glabrata*. At this time however other functions have not been excluded. Once identified, FREP-encoding sequences were also detected from other (families of) snail species. A comparative study evidenced that both *B. glabrata* and *H. trivolvis*, a related planorbid, each react to both compatible and incompatible echinostome parasites with similar complex response patterns of multiple plasma polypeptides, including FREPs. Because such responses do not automatically correspond with elimination of the infecting echinostome parasites, the FREP response is considered to represent a best-effort response to an invading pathogen. The ultimate outcome of the parasite-host interaction likely depends on the success of the interference capabilities of the parasite (Adema et al. 1999).

FREPs have never been obvious from snail plasma protein profiles derived from *B. glabrata/S. mansoni* interactions. With the availability of nucleic acid sequence it became possible to track FREP activity with alternative methods. Reverse transcription PCR did not detect FREP expression in *B. glabrata* following infection with a compatible *S. mansoni*, but resistant *B. glabrata* do deploy FREPs in response to incompatible *S. mansoni* (Hertel et al. 2005).

It remains to be resolved whether detection on protein gels of FREPs produced in response to schistosomes is hampered by low concentrations of these lectins or whether FREPs do not circulate freely in the plasma but perhaps more closely associate with parasites and/or snail cells and tissues.

As foretold by the fuzzy banding of FREPs as bands on SDS-PAGE gels, FREPs form a highly diverse gene family. To date at least 13 subgene families of highly similar FREP-encoding sequences have been documented, the members of which are distinguished by sequence differences, intron-exon structure, and the presence of either one single or two tandemly arranged IgSF domains upstream of the single FBG domain. Based on predicted molecular weight it is likely that FREP proteins with two IgSF domains fall in the 80–120kD molecular weight range, similar to that observed from another component of the parasite-reactive plasma polypeptides produced by *B. glabrata* (Léonard et al. 2001; Zhang et al. 2001).

Full-length characterization of FREP cDNA transcripts and FREP gene sequences (genomic DNA) identified additional levels of diversity due to alternative splicing (Zhang and Loker 2003) as well as the frequent occurrence of single nucleotide polymorphisms (SNPs) in otherwise identical FREP-encoding sequences (Léonard et al. 2001; Zhang et al. 2001).

This level of diversity presented an apparent inconsistency with the notion that pattern recognition in the context of (invertebrate) innate immunity relied only a limited, unchanging set of broad specificity lectins for nonself-recognition (Klein 1989; Medzhitov and Janeway 2000). To investigate this issue, genomic DNA from two individual *B. glabrata* was used to PCR amplify a sequence region from exon 2 of FREP3, the resulting amplicons were shotgun cloned and multiple inserts were sequenced for each snail. The sequences obtained all were clearly derived from FREP3 (≥84% sequence identify), but 45 and 37 different sequences were recovered from the respective individual snails, and only one sequence was held in common by both snails. This inter- and intraindividual diversity was all the more surprising given that experimental estimation predicted no more than five FREP loci in the genome of *B. glabrata.* Similar levels of diversity of FREP3 sequences were evident at both DNA and (mRNA) cDNA levels. Computer-aided biomathematical analysis revealed that the complete range of sequence diversity of FREP3 sequences could be explained by somatic mutation (involving gene conversion or point mutations or both) of a limited number of so-called source sequences, similar to the estimated number of FREP3 loci. The resulting diversity of FREP lectins is assumed to afford *B. glabrata* with an increase in the range of antigens that can be recognized. Also, diversification over time would prevent pathogens from tracking a susceptible genotype (Zhang et al. 2004).

In practice, a continually diversifying repertoire of FREP sequences might change the immune-recognition capabilities and possibly with that the compatibility of individual *B. glabrata* for particular digeneans as a function of time. This report was among the first to invoke somatic diversification of genes with relevance for innate immunity in invertebrate metazoans. Currently, no mechanisms have been described from *B. glabrata* to explain somatic mutation. However, diversification of various immune genes has since been reported from a variety of both protostome and deuterostome animals that do not have lymphocyte-based capabilities of acquired immunity such as shrimp, mosquito, *Drosophila*, sea urchins, amphioxus, and from lamprey, a jawless vertebrate. These findings have blurred the distinction between innate and adaptive immunity. The "pattern recognition model" (Medzhitov and Janeway 2000) with innate immunity type recognition relying on limited sets of broad spectrum lectins may now be updated with the notion that diversification of recognition proteins is a common feature of animal immunity in general (see reviews by Loker et al. 2004; Litman et al. 2005b; Litman et al. 2005a).

It is remarkable to realize that through revealing diversity of FREP sequences, the study of interactions between *E. paraensei* and the snail *B. glabrata* has indirectly contributed to a change in the paradigm of how innate immunity functions across animal phylogeny.

2.1.4.5 Insights from Transcriptomic and Proteomic Studies

In the last decade, tremendous efforts have been mounted to identify immune-relevant genes or gene products from *B. glabrata*. Complementary strategies have been developed by various laboratories, such as random sequencing of hemocyte expressed genes (Mitta et al. 2005), comparative analysis of ESTs pre- and postimmune challenge (Miller et al. 2001; Raghavan et al. 2003; Lockyer et al. 2004; Nowak et al. 2004; Guillou et al. 2007a, b; Lockyer et al. 2007), or comparative analysis of ESTs in susceptible or resistant snails (Schneider and Zelck 2001; Bouchut et al., 2006b; Bouchut et al. 2007; Lockyer et al., 2007b). In addition, proteomic techniques have been focused on characterizing *B. glabrata* proteins that are differentially represented in the plasma or hemocytes from *E. caproni*-susceptible and -resistant *B. glabrata* (Vergote et al. 2005; Bouchut et al. 2006a, b).

Altogether, these studies yielded a multitude of immune-relevant genes that fall in several major functional categories (see Table 2.1). Although more work is needed to gain a comprehensive understanding, the efforts to date provide considerable insights into the immune capabilities of *B. glabrata*. For the first time sequence data

Table 2.1 Predicted identity of immune-relevant genes (Other than freps) identified in published gene discovery studies

Nonself-recognition	Antimicrobial proteins
Peptidoglycan Recognition Proteins (Mitta et al. 2005; Zhang et al. 2007)	LBP/BPIs (Mitta et al. 2005; Guillou et al. 2007)
Cell-cell or cell-matrix adhesion	Theromacins (Mitta et al. 2005)
Cadherin like (Bouchut et al. 2006a, b)	Achacin (Guillou et al. 2007)
Dermatopontins (Mitta et al. 2005; Bouchut et al. 2006a, b)	Aplysianin (Jung et al. 2005)
Focal adhesion kinase (Lockyer et al. 2007a,b)	Immune regulators
F-spondin precursor (Mitta et al. 2005)	Macrophage Migration Inhibitory Factor (MIF) (Mitta et al. 2005)
Galectins (Mitta et al. 2005)	Allograft inflammatory Factor (AIF) (Mitta et al. 2005)
Integrin b-related (2) (Davids et al. 1999; Mitta et al. 2005)	Oxidative stress
Integrin a-2 (Bouchut et al. 2006a, b)	Dual oxidases (Mitta et al. 2005; Lockyer et al. 2007a)
Integrin a-3 precursor (Mitta et al. 2005; Lockyer et al. 2007a)	Glutathione S Transferase (Guillou et al. 2007)
Integrin a-5 precursor (Mitta et al. 2005)	Peroxidase (Lockyer et al. 2007a)
Matrilins (Mitta et al. 2005; Bouchut et al. 2006a)	Peroxidasin (Lockyer et al. 2007a)
Selectin (Duclermortier et al. 1999)	Peroxinectin (Lockyer et al. 2007a)
Thrombospondin (Bouchut et al. 2006a)	Cu/Zn Superoxide dismutases (Jung et al. 2005; Goodall et al. 2006; Guillou et al. 2007)
von Willebrand domain-containing proteins (8) (Bouchut et al. 2006a, b)	Mn SOD (Jung et al. 2005)
Galectin IIa (Mitta et al. 2005; Yoshino et al. 2008)	Thioredoxin (Guillou et al. 2007)
Galectin IIIa (Mitta et al. 2005)	Thioredoxin peroxidase (Mitta et al. 2005)
Matrilin 1 (Mitta et al. 2005)	Others
F-spondin precursor (Mitta et al. 2005)	C-type Lectins (Guillou et al. 2007)
Proteases and protease inhibitors	Ca-Binding proteins (Vergote et al. 2005; Guillou et al. 2007)
Cystatins (Mitta et al. 2005, Guillou et al. 2007)	Ca-Binding Protein (1 EF-HAND) (Guillou et al. 2007)
Inter a-trypsin inhibitor (Mitta et al. 2005)	Cyclophilin domain-containing Protein (Guillou et al. 2007)
Kunitz protease inhibitor (Coustau et al. 2003)	Endo-1,4-b-mannanases (2) (Vergote et al. 2005
α 2-macroglobulin (Lockyer et al. 2007a)	Ferritin (Lockyer et al. 2006a, b)
Metalloproteinases (Mitta et al. 2005)	Heat Sock Proteins 2 (Laursen et al. 1997; Lockyer et al. 2004; Guillou et al. 2007)
Serine proteases (Lockyer et al. 2007b)	Serine/threonine protein Kinase (Guillou et al. 2007)
Serine proteinase inhibitor (Mitta et al. 2005)	

have identified various types of putative nonself-recognition proteins, immune regulators, adhesion proteins, and antimicrobial proteins (Table 2.1). Importantly, genes involved in toll and MAPK-associated signaling pathways now open possibilities to investigate the activation and regulation of defense responses (Humphries et al. 2001; Lockyer et al. 2007a, b), also in the context of parasite-snail interactions.

2.1.4.6 *Echinostoma vs. Schistosoma:* Tools for Functional Validation of Candidate Immune Genes

The *B. glabrata/S. mansoni* interaction is of particular interest due to the medical and veterinary importance of the parasite. To date, however, study of this host-parasite combination at the transcriptomic level has not provided evidence for strong activation of a snail immune response. It is possible that *S. mansoni*, when it is recognized as nonself, is rapidly eliminated by constitutively expressed immune molecules and circulating defense cells, either before or without eliciting a substantial induction of transcription of immune-relevant genes.

In contrast to the modest responses to *S. mansoni*, infections with *E. caproni* (or *E. paraensei*) significantly modify expression of a variety of (candidate) immune genes of *B. glabrata* genes (Mitta et al. 2005; see Fig. 2.4). In all cases, encapsulation of *E. caproni* correlates with significant increases in transcripts levels, whereas encapsulation of *S. mansoni* does not. Such results further underscore differences in the immunobiological interactions between *B. glabrata* and these two trematodes, and make it abundantly clear that echinonostomes represent valuable biological tools for investigating the immune response of *Biomphalaria*. It is anticipated that their use in future larger-scale studies will help deciphering molecular processes underlying antiparasitic response.

2.2 Concluding Remarks

In this day and age, only few organisms are generally considered to have the status of model organism. The (imminent) availability of a completed genome sequence seems a prerequisite for this status. Due to the considerable costs and effort required, it is expected that only organisms that are of great economic value or that have medical impact will be selected for full genome characterization within the near future. It is due to this reasoning that the genomes of *S. mansoni* and *S. japonicum* are currently being characterized, while it is far less likely that comprehensive genomic sequence data for any echinostome species will become available in the near-to-distant future. However, as described, echinostomes are in fact valuable model organisms for the study of general parasitological and immunological questions. Insights derived from study of echinostome biology and snail host interactions may guide and direct studies aimed at further general understanding of transmission of digean parasites by snails. As one example, the discovery of FREPs as antidigenean

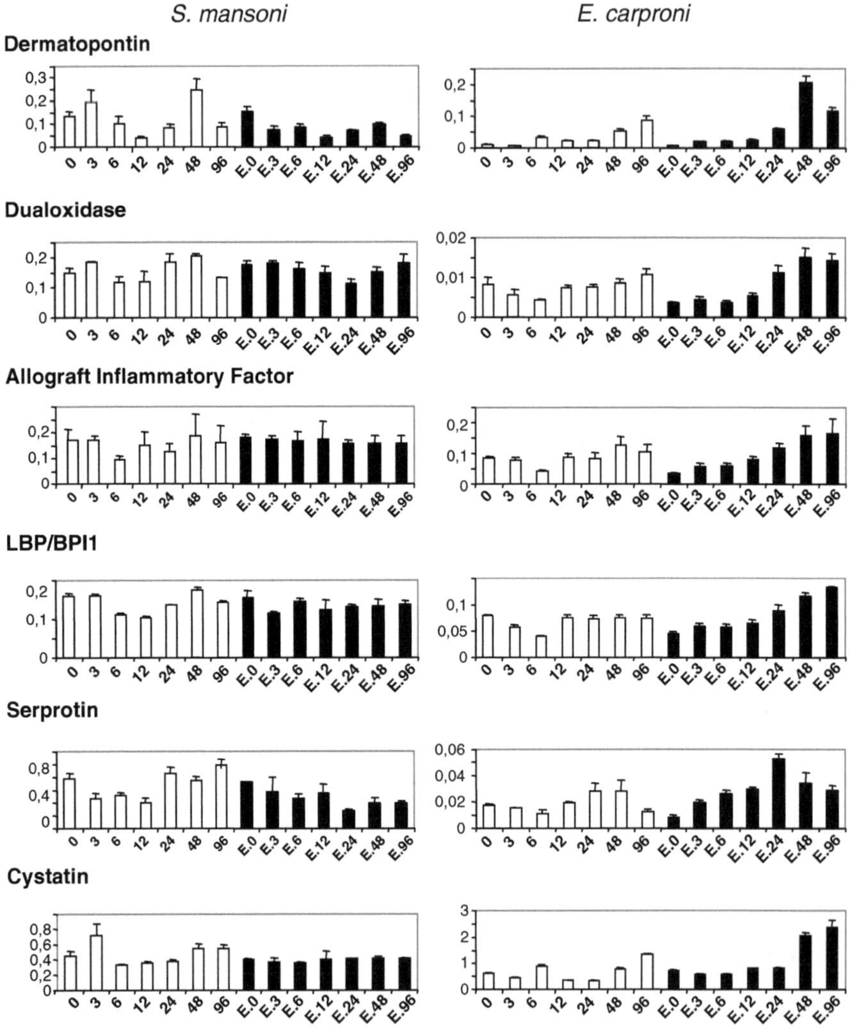

Fig. 2.4 Expression profiles of 6 immune-relevant genes after exposure to *Schistosoma mansoni* and *Echinostoma caproni* at 3, 6, 12, 24, 48, and 96 h after miracidial exposure. In both cases white boxes indicate expression in compatible combinations (successful development of the sporocysts), whereas black boxes indicate expression during an encapsulation response. Expression ratios were determined using real-time quantitative PCR and are expressed relative to S19 expression levels

response factors is a direct result of experimental use of echinostomes. The investigation of the mechanisms that govern expression and diversification of FREPs will inform considerably on basic immunology. This issue has figured importantly alongside several other arguments used to motivate, organize (http://biology.unm.edu/biomphalaria-genome/index.html), and successfully arrange for funding the characterization of the full genome sequence of *B. glabrata*. (see http://genome.

wustl.edu/ancillary/data/whitepapers/Biomphalaria_glabrata_WP.pdf, and http://genome.wustl.edu/genome.cgi?GENOME = Biomphalaria%20glabrata). Whereas the main rationale for obtaining the genome sequence of this snail species is to allow continued modern investigation of the *B. glabrata-S.mansoni* association, preliminary experiments have shown that the BB02 strain of *B. glabrata* (Adema et al. 2006) is also compatible with *E. paraensei* (unpublished results). Thus the genomic data forthcoming from the *B. glabrata* genome project will be valid also in the context of studying echinostome-snail host interactions. With or without these genomics-era developments, echinostomes will continue to remain important model organisms in the field of general immunoparasitology.

Acknowledgments CMA was supported by NIH grants R01 AI024340, AI052363, and RR-1P20RR18754 from the Institute Development Award (IDeA) Program from the National Center for Research Resources.

References

Abu-Shakra, M., Buskila, D., and Shoenfeld, Y. 1999. Molecular mimicry between host and pathogen: examples from parasites and implication. Immunology Letters 67: 147–152.

Adema, C.M., and Loker, E.S. 1997. Specificity and immunobiology of larval digenean-snail associations. In Trematode biology, B. Fried and T. K. Graczyk (eds.), pp 229–263. Boca Raton: CRC Press.

Adema, C.M., Arguello, D.F., Striker, S.A., and Loker, E.S. 1994a. A time-lapse study of interactions between *Echinostoma paraensei* intramolluscan larval stages and adherent hemocytes from *Biomphalaria glabrata* and *Helix aspersa*. Journal of Parasitology 80: 719–727.

Adema, C.M., Van Deutekom-Mulder, E.C., Van der Knaap, P.W., and Sminia, T. 1994b. Schistosomicidal activities of *Lymnaea stagnalis* hemocytes: the role of oxygen radicals. Parasitology 109: 479–485.

Adema, C.M., Hertel, L.A., and Loker, E.S. 1997. Infection with *Echinostoma paraensei* (Digenea) induces parasite-reactive polypeptides in the hemolymph of the gastropod host *Biomphalaria glabrata*. In parasite effects on host physiology and behavior, N. Beckage (ed.), pp. 77–99. New York: Chapman Press.

Adema, C.M., Hertel, L.A., Miller, R.D., and Loker, E.S. 1997. A family of fibrinogen-related proteins that precipitates parasite-derived molecules is produced by an invertebrate after infection. Proceedings of the National Academy of Sciences of the United States of America 94: 8691–8696.

Adema, C.M., Hertel, L.A., and Loker, E.S. 1999. Evidence from two planorbid snails of a complex and dedicated response to digenean (echinostome) infection. Parasitology 119: 395–404.

Adema, C.M., Mei-Zhong, L., Hanelt, B., Hertel, L.A., Marshall, J.J., Zhang, S.-M., DeJong, R.J., Kim, H.R., Kudrna, D., Wing, R.A., Soderlund, C., Knight, M., Lewis, F.A., Caldeira, R.L., Jannotti-Passos, L.K., dos Santos Carvalho, O., and Loker, E.S. 2006. A bacterial artificial chromosome library for *Biomphalaria glabrata*, intermediate snail host of *Schistosoma mansoni*. Memórias do Instituto Oswaldo Cruz 101: S167–S177.

Ataev, G., and Coustau, C. 1999. Cellular response to *Echinostoma caproni* infection in *Biomphalaria glabrata* strains selected for susceptibility/resistance. Developmental and Comparative Immunology 23: 187–198.

Ataev, G., Fournier, A., and Coustau, C. 1998. Comparison of *Echinostoma caproni* mother sporocyst development in vivo and in vitro using *Biomphalaria glabrata* and *B. glabrata* embryonic cell line. Journal of Parasitology 84: 227–235.

Bayne, C.J., Hahn, U.K., and Bender, R.C. 2001. Mechanisms of molluscan host resistance and of parasite strategies for survival. Parasitology 123: S159–S167.

Bender, R.C., Broderick, E.J., Goodall, C.P., and Bayne, C.J. 2005. Respiratory burst of *Biomphalaria glabrata* hemocytes: *Schistosoma mansoni*-resistant snails produce more extracellular H2O2 than susceptible snails. Journal of Parasitology 91: 275–279.

Bergmann, S., Rohde, M., and Hammerschmidt, S. 2004. Glyceraldehyde-3-phosphate dehydrogenase of *Streptococcus pneumoniae* is a surface-displayed plasminogen-binding protein. Infection and Immunity 72: 2416–2419.

Bernal, D., de la Rubia, J.E., Carrasco-Abad, A.M., Toledo, R., Mas-Coma, S., and Marcilla, A. 2004. Identification of enolase as a plasminogen-binding protein in excretory-secretory products of *Fasciola hepatica*. FEBS Letters 563: 203–206.

Bernal, D., Carpena, I., Espert, A.M., De la Rubia, J.E., Esteban, J.G., Toledo, R., and Marcilla, A. 2006. Identification of proteins in excretory/secretory extracts of *Echinostoma* friedi (Trematoda) from chronic and acute infections. Proteomics 6: 2835–2843.

Bixler, L.M., Lerner, J.P., Ivanchenko, M., McCormick, R.S., Barnes, D.W., and Bayne, C.J. 2001. Axenic culture of *Schistosoma mansoni* sporocysts in low 02 environments. Journal of Parasitology 87: 1167–1168.

Bouchut, A., Roger, E., Coustau, C., Gourbal, B., and Mitta, G. 2006a. Compatibility in the *Biomphalaria glabrata/Echinostoma caproni* model: potential involvement of adhesion genes. International Journal for Parasitology 36: 175–184.

Bouchut, A., Sautiere, P.E., Coustau, C., and Mitta, G. 2006b. Compatibility in the *Biomphalaria glabrata/Echinostoma caproni* model: potential involvement of proteins from hemocytes revealed by a proteomic approach. Acta Tropica 98: 234–246.

Bouchut, A., Coustau, C., Gourbal, B., and Mitta, G. 2007. Compatibility in the *Biomphalaria glabrata/Echinostoma caproni* model: new candidate genes evidenced by a suppressive subtractive hybridization approach. Parasitology. 134: 575–588

Boyunaga, H., Schmitz, M.G., Brouwers, J.F., Van Hellemond, J.J., and Tielens, A.G. 2001. *Fasciola hepatica* miracidia are dependent on respiration and endogenous glycogen degradation for their energy generation. Parasitology 122: 169–173.

Cheng, T.C. 1968. The compatibility and incompatibility concept as related to trematodes and molluscs. Pacific Science 12: 141–160.

Cordeiro, A.T., Michels, P.A., Delboni, L.F., and Thiemann, O.H. 2004. The crystal structure of glucose-6-phosphate isomerase from *Leishmania mexicana* reveals novel active site features. European Journal of Biochemistry 271: 2765–2772.

Coustau, C., and Yoshino, T.P. 1994. Surface membrane –associated polypeptides from adherent and nonadherent hemocyte subpopulations in Schistosoma mansoni-susceptible and resistant strains of *Biomphalaria glabrata* (Gastropoda). Journal of Invertebrate Pathology 63:82–89.

Coustau, C., and Yoshino, T.P. 2000. Flukes without snails: advances in the in vitro cultivation of intramolluscan stages of trematodes. Experimental Parasitology 94: 62–66.

Coustau, C., Ataev, G., Jourdane, J., and Yoshino, T.P. 1997. *Schistosoma japonicum*: in vitro cultivation of miracidium to daughter sporocysts using a *Biomphalaria* embryonic cell line. Experimental Parasitology 8: 77–87.

Coustau, C., Mitta, G., Dissous, C., Guillou, F., Galinier, R., Alliene, J.F., and Modat, S. 2003. *Schistosoma mansoni* and *Echinostoma caproni* differentially affect gene expression in *Biomphalaria glabrata* embryonic cells. Parasitolog 127: 533–542.

Damian, R.T. 1997. Parasite immune evasion and exploitation: reflections and projections. Parasitology 115: S169–S175.

Davids, B.J., Wu, X.J., and Yoshino, T.P. 1999. Cloning of a beta integrin subunit cDNA from an embryonic cell line derived from the freshwater mollusc, *Biomphalaria glabrata*. Gene 228: 213–223.

DeGaffe, G., and Loker, E.S. 1998. Susceptibility of *Biomphalaria glabrata* to infection with *Echinostoma paraensei*: correlation with the effect of parasite secretory-excretory products on host hemocyte spreading. Journal of Invertebrate Pathology 71: 64–72.

Dikkeboom, R., Bayne, C.J., Van der Knaap, W.P., and Tijnagel, J.M. 1988. Possible role of reactive forms of oxygen in in vitro killing of *Schistosoma mansoni* sporocysts by hemocytes of *Lymnaea stagnalis*. Parasitology Research 75: 148–154.

Dondero, T.J. Jr., Ow-Yang, C.K., and Lie, K.J. 1977. Failure of irradiated *Echinostoma audyi* and *Hypoderaeum dingeri* to sensitize *Lymnaea rubiginosa* snails. Southeast Asian Journal of Tropical Medicine and Public Health 8: 359–363.

Duclermortier, P., Lardans, V., Serra, E., Trottein, F., and Dissous, C. 1999. *Biomphalaria glabrata* embryonic cells express a protein with a domain homologous to the lectin domain of mammalian selectins. Parasitology Research 85: 481–486.

Fenwick, A., Rollinson, D., and Southgate, V. 2006. Implementation of human schistosomiasis control: challenges and prospects. Advances in Parasitology 61: 567–622.

Fryer, S.E., and Bayne, C.J. 1989. Opsonisation of yeast by the plasma of *Biomphalaria glabrata* (gastropoda): a strain-specific, time-dependent process. Parasite Immunology 11: 269–278.

Fryer, S.E., and Bayne, C.J. 1996. Phagocytosis of latex beads by *Biomphalaria glabrata* hemocytes is modulated in a strain-specific manner by adsorbed plasma components. Developmental and Comparative Immunology 20: 23–37.

Goodall, C.P., Bender, R.C., Brooks, J.K., and Bayne, C.J. 2006. *Biomphalaria glabrata* cytosolic copper/zinc superoxide dismutase (SOD1) gene: association of SOD1 alleles with resistance/susceptibility to *Schistosoma mansoni*. Molecular and Biochemical Parasitology 147: 207–210.

Granath, W.O., and Yoshino, T.P. 1984. *Schistosoma mansoni*: passive transfer of resistance by serum in the vector snail, *Biomphalaria glabrata*. Experimental Parasitology 58: 188–193.

Guillou, F., Mitta, G., Galinier, R., and Coustau, C. 2007a. Identification and expression of transcripts generated during an anti-parasitic response in *Biomphalaria glabrata*. Developmental and Comparative Immunology 31: 657–671.

Guillou, F., Roger, E., Moné, Y., Rognon, A., Grunau, C., Thèron, A., Mitta, G., Coustau, C., and Gourbal, B. 2007b. Excretory-secretory proteome of larval *Schistosoma mansoni* and *Echinostoma caproni*, two parasites of *Biomphalaria glabrata*. Molecular and Biochemical Parasitology 155: 45–56.

Haas, W., Körner, M., Hutterer, E., Wegner, M., and Haberl, B. 1995. Finding and recognition of the snail intermediate hosts by 3 species of echinostome cercariae. Parasitology 110: 133–142.

Haberl, B., Körner, M., Spengler, Y., Hertel, J., Kalbe, M., and Haas, W. 2000. Host-finding in *Echinostoma caproni*: miracidia and cercariae use different signals to identify the same snail species. Parasitology 120: 479–486.

Hahn, U.K., Bender, R.C., and Bayne, C.J. 2001. Involvement of nitric oxide in killing of *Schistosoma mansoni* sporocysts by hemocytes from resistant *Biomphalaria glabrata*. Journal of Parasitology 87: 778–785.

Hansen, E.L. 1976. A cell line from embryos of *Biomphalaria glabrata* (Pulmonata): establishment and characteristics. In Invertebrate tissue culture, E. Kurstak and K. Maramorosch (eds.), pp. 75–99. New York: Academic.

Hertel, L.A., Stricker, S.A., Monroy, F.P., Wilson, W.D., and Loker, E.S. 1994. *Biomphalaria glabrata* hemolymph lectins: binding to bacteria, mammalian erythrocytes, and to sporocysts and rediae of *Echinostoma paraensei*. Journal of Invertebrate Pathology 64: 52–61.

Hertel, L.A., Stricker, S.A., and Loker, E.S. 2000. Calcium dynamics of hemocytes of the gastropod *Biomphalaria glabrata*: effects of digenetic trematodes and selected bioactive compounds. Invertebrate Biology 119: 27–37.

Hertel, L.A., Adema, C.M., and Loker, E.S. 2005. Differential expression of FREP genes in two strains of *Biomphalaria glabrata* following exposure to the digenetic trematodes *Schistosoma mansoni* and *Echinostoma paraensei*. Developmental and Comparative Immunology 29: 295–303.

Humbert, E., and Coustau, C. 2001. Refractoriness of host *hemocytes* to parasite immunosuppressive factors as a putative resistance mechanism in the *Biomphalaria glabrata* - *Echinostoma caproni* system. Parasitology 122: 651–660.

Humphries, J.E., Elizondo, L., and Yoshino, T.P. 2001. Protein kinase C regulation of cell spreading in the molluscan *Biomphalaria glabrata* embryonic (Bge) cell line. Biochemica et Biophysica Acta 1540: 243–252.

Jeong, K.H., Lie, K.J., and Heyneman, D. 1980. Leucocytosis in *Biomphalaria glabrata* sensitized and resensitized to *Echinostoma lindoense*. Journal of Invertebrate Pathology 35: 9–13.

Jeong, K.H., Sussman, S., Rosen, S.D., Lie, K.J., and Heyneman, D. 1981. Distribution and variation of hemagglutinating activity in the hemolymph of *Biomphalaria glabrata*. Journal of Invertebrate Pathology 38: 256–263.

Jolodar, A., Fischer, P., Bergmann, S., Büttner, D.W., Hammerschmidt, S., and Brattig, N.W. 2003. Molecular cloning of an alpha-enolase from the human filarial parasite *Onchocerca volvulus* that binds human plasminogen. Biochemica et Biophysica Acta 1627: 111–120.

Jung, Y., Nowak, T.S., Zhang, S.-M., Hertel, L.A., Loker, E.S., and Adema, C.M. 2005. Manganese superoxide dismutase from *Biomphalaria glabrata*. Journal of Invertebrate Pathology 90: 59–63.

Kanev, V., Radev, V., Sterner, M., and Fried, B. 2000. An overview of the biology of echinostomes. In Echinostomes as experimental models for biological research, B. Fried and T. K. Graczyk (eds.), pp. 1–29. Dordrecht: Kluwer.

Kayath, N., Mithieux, G., Zitoun, C., Coustau, C., Vicogne, J., Tielens, A.G., and Dissous, C. 2006. Glyceroneogenesis: an unexpected metabolic pathway for glutamine in *Schistosoma mansoni* sporocysts. Molecular and Biochemical Parasitology 147: 145–153.

King, C.H., Sturrock, R.F., Kariuki, H.C., and Hamburger, J. 2006. Transmission control for schistosomiasis - why it matters now. Trends in Parasitology 22: 575–582.

Klein, J. 1989. Are invertebrates capable of anticipatory immune responses? Scandinavian Journal of Immunology 29: 499–505.

Knudsen, G.M., Medzihradszky, K.F., Lim, K.C., Hansell, E., and McKerrow, J.H. 2005. Proteomic analysis of *Schistosoma mansoni* cercarial secretions. Molecular and Cell Proteomics 4: 1862–1875.

Kolberg, J., Aase, A., Bergmann, S., Herstad, T.K., Rødal, G., Frank, R., Rohde, M., and Hammerschmidt, S. 2006. *Streptococcus pneumoniae* enolase is important for plasminogen binding despite low abundance of enolase protein on the bacterial cell surface. Microbiology 152: 1307–1317.

Kum-Tatt, L., Tan, I.K., and Seet, A.M. 1975. A new colorimetric method for the determination of NADH/NADPH dependent glutathione reductase in erythrocytes and in plasma. Clinica Chimica Acta 58: 101–108.

Kurtz, J. 2005. Specific memory within innate immune systems. Trends in Immunology 26: 186–192.

Lardans, V., and Dissous, C. 1998. Snail control strategies for reduction of schistosomiasis transmission. Parasitology Today 14: 413–417.

Laursen, J.R., di Liu, H., Wu, X.J., and Yoshino, T.P. 1997. Heat-shock response in a molluscan cell line: characterization of the response and cloning of an inducible HSP70 cDNA. Journal of Invertebrate Pathology 70: 226–233.

Léonard, P.M., Adema, C.M., Zhang, S.-M., and Loker, E.S. 2001. Structure of two FREP genes that encode IgSF and fibrinogen domains, with comments on diversity of the FREP gene family in the snail *Biomphalaria glabrata*. Gene 269: 155–165.

Lie, K.J. 1982. Survival of *Schistosoma mansoni* and other trematode larvae in the snail *Biomphalaria glabrata*. A discussion of the interference hypothesis. Tropical and Geographical Medicine 34: 111–122.

Lie, K.J., and Heyneman, D. 1976. Studies on resistance in snails. 3. Tissue reactions to *Echinostoma lindoense* sporocysts in sensitized and resensitized *Biomphalaria glabrata*. Journal of Parasitology 62: 51–58.

Lie, K.J., and Heyneman, D. 1977a. Studies on resistance in snails: interference by nonirradiated echinostome larvae with natural resistance to *Schistosoma mansoni* in *Biomphalaria glabrata*. Journal of Invertebrate Pathology 29: 118–125.

Lie, K.J., and Heyneman, D. 1977b. *Schistosoma mansoni, Echinostoma lindoense*, and *Paryphostomum segregatum*: interference by trematode larvae with acquired resistance in snails, *Biomphalaria glabrata*. Experimental Parasitology 42: 343–347.

Lie, K.J., and Heyneman, D. 1979. Acquired resistance to echinostomes in four *Biomphalaria glabrata* strains. International Journal for Parasitology 9: 533–537.

Lie, K.J., Heyneman, D., and Lim, H.K. 1975. Studies on resistance in snails: specific resistance induced by irradiated miracidia of *Echinostoma lindoense* in *Biomphalaria glabrata* snails. International Journal for Parasitology 5: 627–631

Lie, K.J., Heyneman, D., and Jeong, K.H. 1976. Studies on resistance in snails. 7. Evidence of interference with the defense reaction in *Biomphalaria glabrata* by trematode larvae. Journal of Parasitology 62: 608–615.

Lie, K.J., Jeong, K.H., and Heyneman, D. 1982. Further characterization of acquired resistance in *Biomphalaria glabrata*. Journal of Parasitology 68: 529–531.

Litman, G.W., Cannon, J.P., and Rast, J.P. 2005a. New insights into alternative mechanisms of immune receptor diversification. Advances in Immunology 87: 209–236.

Litman, G.W., Cannon, J.P., and Dishaw, L.J. 2005b. Reconstructing immune phylogeny: new perspectives. Nature Reviews Immunology 5: 866–879.

Little, T.J., Hultmark, D., and Read, A.F. 2005. Invertebrate immunity and the limits of mechanistic immunology. Nature Immunology 6: 651–654.

Lockyer, A.E., Noble, L.R., and Rollinson, D.S. 2004. *Schistosoma mansoni*: resistant specific infection-induced gene expression in *Biomphalaria glabrata* identified by fluorescent-based differential display. Experimental Parasitology 107: 97–104.

Lockyer, A.E., Spinks, J.N., Walker, A.J., Kane, R.A., Noble, L.R., Rollinson, D., Dias-Neto, E., and Jones, C.S. 2007a. *Biomphalaria glabrata* transcriptome: identification of cell-signalling, transcriptional control and immune-related genes from open reading frame expressed sequence tags (ORESTES). Developmental and Comparative Immunology 31: 763–782.

Lockyer, A.E., Spinks, J., Noble, L.R., Rollinson, D., and Jones, C.S. 2007b. Identification of genes involved in interactions between *Biomphalaria glabrata* and *Schistosoma mansoni* by suppression subtractive hybridization. Molecular and Biochemical Parasitology 151: 18–27.

Lo, C.T. 1995. *Echinostoma macrorchis*: life history, population dynamics of intramolluscan stages, and the first and second intermediate hosts. Journal of Parasitology 8: 569–576.

Loker, E.S. 1994. On being a parasite in an invertebrate host: a short survival course. The Journal of Parasitology 80: 728–747.

Loker, E.S., and Adema, C.M. 1995. Schistosomes, echinostomes and snails: comparative immunobiology. Parasitology Today 11: 120–124.

Loker, E.S., and Bayne, C.J. 1986. Immunity to trematode larvae in the snail *Biomphalaria*. Zoological Symposium 56: 199–220.

Loker, E.S., and Hertel, L.A. 1987. Alterations in *Biomphalaria glabrata* plasma induced by infection with the digenetic trematode *Echinostoma paraensei*. Journal of Parasitology 73: 503–513.

Loker, E.S., Bayne, C.J., and Hui, M.A. 1986. *Echinostoma paraensei*: hemocytes of *Biomphalaria glabrata* as targets of echinostome mediated interference with host snail resistance to *Schistosoma mansoni*. Experimental Parasitology 62: 149–154.

Loker, E.S., Boston, M.E., and Bayne, C.J. 1989. Differential adherence of M-line *Biomphalaria glabrata* hemocytes to *Schistosoma mansoni* and *Echinostoma paraensei* larvae, and experimental manipulation of hemocyte binding. Journal of Invertebrate Pathology 54: 260–268.

Loker, E.S., Adema, C.M., Zhang, S.-M., and Kepler, T.B. 2004. Invertebrate immune systems: not homogeneous, not simple, not well understood. Immunological Reviews 198: 10–24.

LoVerde, P.T. 1998. Do antioxidants play a role in schistosome host-parasite interactions? Parasitology Today 14: 284–289.

Maizels, R.M., Balic, A., Gomez-Escobar, N., Nair, M., Taylor, M.D., and Allen, J.E. 2004. Helminth parasites-masters of regulation. Immunology Reviews 201: 89–116.

Medzhitov, R., and Janeway, C. Jr. 2000. Innate immune recognition: mechanisms and pathways. Immunological Reviews 173: 89–97.

Mei, H., and LoVerde, P.T. 1997. *Schistosoma mansoni*: the developmental regulation and immunolocalization of antioxidant enzymes. Experimental Parasitology 86: 69–78.

Miller, A.N., Raghavan, N., FitzGerald, P.C., Lewis, F.A., and Knight, M. 2001. Differential gene expression in haemocytes of the snail *Biomphalaria glabrata*: effects of *Schistosoma mansoni* infection. International Journal for Parasitology 31: 687–696.

Mitta, G., Galinier, R., Tisseyre, P., Girerd-Chambaz, Y., Guillou, F., Bouchut, A., and Coustau, C. 2005. Gene discovery and expression analysis of immune-relevant genes from *Biomphalaria glabrata* hemocytes. Developmental and Comparative Immunology 29: 393–407.

Mkoji, G.M., Smith, J.M., and Prichard, R.K. 1988a. Antioxidant systems in *Schistosoma mansoni*: correlation between susceptibility to oxidant killing and the levels of scavengers of hydrogen peroxide and oxygen free radicals. International Journal for Parasitology 18: 661–666.

Mkoji, G.M., Smith, J.M., and Prichard, R.K. 1988b. Antioxidant systems in *Schistosoma mansoni*: evidence for their role in protection of the adult worms against oxidant killing. International Journal for Parasitology 18: 667–673.

Monroy, F., Hertel, L.A., and Loker, E.S. 1992. Carbohydrate-binding plasma proteins from the gastropod *Biomphalaria glabrata*: strain specificity and the effects of trematode infection. Developmental and Comparative Immunology 16: 355–366.

Monroy, F.P., and Loker, E.S. 1993. Production of heterogeneous carbohydrate-binding proteins by the host snail *Biomphalaria glabrata* following exposure to *Echinostoma paraensei* and *Schistosoma mansoni*. Journal of Parasitology 79: 416–423.

Noda, S. 1992. Effects of excretory-secretory products of *Echinostoma paraensei* larvae on the hematopoietic organ of M-line *Biomphalaria glabrata* snails. Journal of Parasitology 78: 512–517.

Noda, S., and Loker, E.S. 1989a. Effect of infection with *Echinostoma paraensei* on the circulating hemocyte population of the host snail *Biomphalaria glabrata*. Parasitology 98: 35–41.

Noda, S., and Loker, E.S. 1989b. Phagocytic activity of hemocytes of M-line *Biomphalaria glabrata* snails: effect of exposure to the trematode *Echinostoma paraensei*. Journal of Parasitology 75: 261–269.

Nowak, T.S., and Loker, E.S. 2005. *Echinostoma paraensei*: differential gene transcription in the sporocyst stage. Experimental Parasitology 109: 94–105.

Nowak, T.S., Woodards, A.C., Jung, Y., Adema, C.M., and Loker, E.S. 2004. Identification of transcripts generated during the response of resistant *Biomphalaria glabrata* to *Schistosoma mansoni* infection using suppression subtractive hybridization. Journal of Parasitology 90: 1034–1340.

Perez-Sanchez, R., Ramajo-Hernandez, A., Ramajo-Martín, V., and Oleaga, A. 2006. Proteomic analysis of the tegument and excretory-secretory products of adult *Schistosoma bovis* worms. Proteomics 6: S226–S236.

Pham, L.N., Dionne, M.S., Shirasu-Hiza, M., and Schneider, D.S. 2007. A specific primed immune response in *Drosophila* is dependent on phagocytes. PLoS Pathogens 3: 26.

Raghavan, N., Miller, A.N., Gardner, M., FitzGerald, P.C., Kerlavage, A.R., Johnston, D.A., Lewis, F.A., and Knight, M. 2003. Comparative gene analysis of *Biomphalaria glabrata* hemocytes pre- and post-exposure to miracidia of *Schistosoma mansoni*. Molecular and Biochemical Parasitology 126: 181–191.

Roth, E. 1990. *Plasmodium falciparum* carbohydrate metabolism: a connection between host cell and parasite. Blood Cells 16: 453–460.

Sapp, K.K., and Loker, E.S. 2000a. Mechanisms underlying digenean-snail specificity: role of miracidial attachment and host plasma factors. Journal of Parasitology 86: 1012–1019.

Sapp, K.K., and Loker, E.S. 2000b. A comparative study of mechanisms underlying digenean-snail specificity: in vitro interactions between hemocytes and digenean larvae. Journal of Parasitology 86: 1020–1029.

Sapp, K.K., Meyer, K.A., and Loker, E.S. 1998. Intramolluscan development of the digenean *Echinostoma paraensei*: rapid production of a unique mother redia that adversely affects development of conspecific parasites. Invertebrate Biology 117: 20–28.

Schneider, O., and Zelck, U.E. 2001. Differential display analysis of hemocytes from schistosome-resistant and schistosome-susceptible intermediate hosts. Parasitology Research 87: 489–491.
Tielens, A.G., Van de Pas, F.A., Van den Heuvel, J.M., and Van den Bergh, S.G. 1991. The aerobic energy metabolism of *Schistosoma mansoni* miracidia. Molecular and Biochemical Parasitology 46: 181–184.
Tielens, A.G., Horemans, A.M., Dunnewijk, R., Van der Meer, P., and Van den Bergh, S.G. 1992. The facultative anaerobic energy metabolism of *Schistosoma mansoni* sporocysts. Molecular and Biochemical Parasitology 56: 49–57.
Van der Knaap, W.P., and Loker, E.S. 1990. Immune mechanisms in trematode-snail interactions. Parasitology Today 6: 175–182.
Vasta, G.R., Ahmed, H., and Odom, E.W. 2004. Structural and functional diversity of lectin repertoires in invertebrates, protochordates and ectothermic vertebrates. Current Opinion in Structural Biology 14: 617–630.
Vergote, D., Bouchut, A., Sautière, P.E., Roger, E., Galinier, R., Rognon, A., Coustau, C., Salzet, M., and Mitta, G. 2005. Characterisation of proteins differentially present in the plasma of *Biomphalaria glabrata* susceptible or resistant to *Echinostoma caproni*. International Journal for Parasitology 35: 215–224.
Walker, A.J. 2006. Do trematode parasites disrupt defence-cell signalling in their snail hosts?. Trends in Parasitology 22: 154–159.
Würzner, R. 1999. Evasion of pathogens by avoiding recognition or eradication by complement, in part via molecular mimicry. Molecular Immunology 36: 249–260.
Yoshino, T.P., and Granath, W.O. Jr. 1985. Surface antigens *of Biomphalaria glabrata* (Gastropoda) hemocytes: functional heterogeneity in cell subpopulations recognized by a monoclonal antibody. Journal of Invertebrate Pathology 45: 174–186.
Yoshino, T.P., and Laursen, J.R. 1995. Production of *Schistosoma mansoni* daughter sporocysts from mother sporocysts maintained in synxenic cultures with *Biomphalaria glabrata* embryonic (Bge) cells. Journal of Parasitology 81: 714–722.
Yoshino, T.P., and Vasta, G.R. 1996. Parasite-invertebrate host immune interaction. In Advances in comparative and environmental physiology, Vol. 24, pp. 125–167, Berlin: Springer.
Yoshino, T.P., Dinguirard, N., Kunert, J., and Hokke, C.H. in press. Molecular and functional characterization of a tandem-repeat galectin from the freshwater snail *Biomphalaria glabrata*, intermediate host of the human blood fluke *Schistosoma mansoni*. Gene 411: 46–58.
Zambrano-Villa, S., Rosales-Borjas, D., Carrero, J.C., and Ortiz-Ortiz, L. 2002. How protozoan parasites evade the immune response. Trends in Parasitology 18: 272–278.
Zhang, G., Schmidt, O., and Asgari, S. 2006. A calreticulin-like protein from endoparasitoid venom fluid is involved in host hemocyte inactivation. Developmental and Comparative Immunology 30: 756–764.
Zhang, S.-M., and Loker, E.S. 2003. The FREP gene family in the snail *Biomphalaria glabrata*: additional members, and evidence consistent with alternative splicing and FREP retrosequences. Fibrinogen-related proteins. Developmental and Comparative Immunology 27: 175–187.
Zhang, S.-M., Léonard, P.M., Adema, C.M., and Loker, E.S. 2001. Parasite responsive IgSF members in the snail Biomphalaria glabrata: characterization of novel genes with tandemly arranged IgSF domains and a fibrinogen domain. Immunogenetics 53: 684–694.
Zhang, S.-M., Adema, C.M., Kepler, T.B., and Loker, E.S. 2004. Diversification of Ig superfamily genes in an invertebrate. Science 305: 251–254
Zhang, S.-M., Zeng, Y., and Loker, E.S. 2007. Characterization of immune genes from the schistosome host snail *Biomphalaria glabrata* that encode peptidoglycan recognition proteins and gram-negative bacteria binding protein. Immunogenetics 59: 883–898.

Chapter 3
Echinostomes in the Second Intermediate Host

Shamus P. Keeler and Jane E. Huffman

Contents

3.1 Introduction ... 62
3.2 Echinostome Metacercariae ... 62
3.3 Encystment of Echinostome Metacercariae ... 62
3.4 Metacercarial Site Selection in Second Intermediate Hosts ... 63
3.5 Development of Echinostome Metacercariae in the Second Intermediate Host ... 65
3.6 Excystation of Echinostome Metacercariae ... 67
3.7 Invertebrate and Vertebrate Second Intermediate Hosts of Echinostomes ... 67
3.8 Behavioral and Morphological Changes in Invertebrate and Vertebrate Hosts Infected with Echinostome Metacercariae ... 73
3.9 Pathology of Metacercariae in Invertebrate and Vertebrate Hosts ... 75
3.10 Effects of Pesticides, Herbicides on Echinostome Metacercariae ... 77
3.11 Effects of Marinades of Echinostome Metacercariae ... 78
3.12 Concluding Remarks ... 79
References ... 80

Abstract Echinostomes have a wide range of both invertebrate and vertebrate second intermediate hosts. The most studied hosts are molluskan, particularly the gastropods and bivalves; next most studied hosts are the amphibians, particularly tadpoles and frogs. The encysted metacercaria or cyst is the dominant stage in the second intermediate host of echinostomes. This cyst is usually formed in or on tissues of the involved invertebrate or vertebrate second intermediate hosts. In gastropods the preferred site for encystment is usually the kidney or pericardium. In anurans the preferred site is often the kidney or the limb buds in echinostome-like forms such as *Ribieroia* sp. This chapter reviews the biology, development,

J.E. Huffman (✉)
Applied DNA Sciences, Northeast Wildlife DNA Laboratory, Fish & Wildlife Microbiology Laboratory East Stroudsburg University East Stroudsburg, PA 18301, USA
e-mail: jhuffman@po-box.esu.edu

S.P. Keeler
Southeastern Cooperative wildlife Disease Study (SCWDS), Department of Population Health, College of Veterinary Medicine, University of Georgia, Athens, GA 30602, USA

B. Fried and R. Toledo (eds.) *The Biology of Echinostomes*,
DOI: 10.1007/978-0-387-09577-6_3,

distinguishing morphological characteristics, site selection, behavioral changes in vertebrate and vertebrate hosts, pathology, and the effects of pesticides, herbicides, and marinades on the metacercariae of the Echinostomatidae.

3.1 Introduction

Echinostome trematodes target a diverse collection of second intermediate hosts both taxonomically (gastropod and bivalve mollusks, amphibian larvae, and fishes) and ethologically (lentic, pelagic, sessile). The nature and structure of the echinostome cysts, and their localization will vary (Huffman and Fried 1990; Fried 2001). The role of habitat and habitat structure on the distribution of echinostomes among and within intermediate host populations can play an important role in influencing the probability of contact between the parasite and potential host. Some intermediate hosts behave in ways that can increase or decrease the risk of infection (Mouritsen and Poulin 2003). Echinostomes can impact on host behavior and can be significant determinants of animal community structure and function. The metacercariae may also affect the general condition of the host. The pathogenicity of encysted echinostomes will depend on their localization in the host.

This chapter will address encystment, site selection, development of echinostome metacercariae, behavioral and morphological changes in the second intermediate host, and excystation of echinostome metacercariae. The effects of pesticides, herbicides, and marinades on echinostome metacercariae will be reviewed.

3.2 Echinostome Metacercariae

Histochemical and biochemical studies of echinostomes show considerable variability in the formation, ultrastructure, and biochemical composition of metacercarial cysts, reflecting diverse evolutionary adaptations. As indicated by Macy et al. (1968), the nature of the cyst wall also determines the conditions needed for excystment and may be important in determining final host specificity after ingestion. Although several biological functions (e.g., transport of nutrients or metabolites) have been attributed to the cyst wall, none, other than protection, has been clearly defined.

3.3 Encystment of Echinostome Metacercariae

Mechanisms of encystment either in vivo or in vitro for species of *Echinostoma* are not well understood. Cercariae of *Echinostoma* contain a complex array of tegumentary papillae (Fried and Fujino 1987) and a complicated neuromusculature (Humphries et al. 2000). Additionally, cystogenous glands are present throughout the cercarial body. Coordination of the sensory, nervous, and muscular systems with the cystogenous gland network is obviously involved in cercarial encystment.

Some echinostome species have been reported to encyst on snail mucus and *Himasthla rhigedana* encysts on vegetation (Adams and Martin 1963; Fried and Bennett 1979; Christensen et al.1980). Most echinostome cercariae require a second intermediate host for encystment. These hosts are usually gastropod and bivalve mollusks, amphibian larvae, and fishes. The site of encystment is variable. In gastropods, the cercariae have a predilection for the kidney-pericardial region; in bivalves the gills and gonads serve as a preferred site. In amphibians the kidney and the cloaca are preferred sites; and in fishes the skin, subtegument, and musculature are preferred (Marquardt and Demaree 1985; Huffman and Fried 1990). Species of *Echinochasmus* and *Petasiger* encyst within the gills of fish. Xiao et al. (2005) reported that *Echinochasmus liliputanus* cercariae encysted in vitro, in human gastric juice. This finding may explain the direct infection of final hosts by the cercariae.

For some species of 37-collar-spined *Echinostoma*, their cercariae successfully encyst and develop to metacercariae in vitro. Park et al. (2006a) cultured *Echinostoma cinetorchis* cercariae in 12 different media to study the formation of metacercariae. Gorbushin and Levakin (2005) evaluated trigger and toxic effects of *Mytilus edulis* (Bivalvia) hemolymph on encystment of cercariae *Himasthla elongata* obtained from infected *Littorina littorea* (Prosobranchia). The contact of *H. elongata* larvae with the whole hemolymph or mussel acellular plasma led to an intensive transformation of cercariae into metacercariae. In both tested media, the cercariae had to complete the encystment phase within 2 h; otherwise, the risk of injury to the larvae by humoral and cellular components of the mussel hemolymph would increase dramatically. The cercarial mortality after 24 h in the whole hemolymph was twice higher than in plasma (40 and 20%, respectively) and much higher than in the control medium (sea water). Both toxic and trigger effects of plasma were revealed to depend on concentration.

3.4 Metacercarial Site Selection in Second Intermediate Hosts

The site of encystment within gastropod hosts varies between species of echinostome. *Echinostoma trivolvis* mainly encysts within the kidneys, while *Echinostoma caproni* encysts within the heart, pericardium, and kidneys (Jeyarasasingam et al. 1972; Anderson and Fried 1987). *Echinoparyphium elegans* encyst within the pericardium and kidneys (Mouahid and Mone 1988). Morley et al. (2007) investigated the establishment and distribution of *Echinoparyphium recurvatum* metacercariae in the second intermediate host, *Lymnaea peregra*, at a temperature range of 5–29°C. The effect of temperature on encystment site choice (mantle cavity, kidney, pericardium) by metacercariae showed that the mantle cavity was the prime site of encystment, followed by the pericardium and the kidney. Temperatures at the lower and upper ranges (14 and 29°C), however, caused a significant reduction in encystment in the mantle cavity, but not in the pericardium or kidney.

Utilization of a single-species molluskan community of *L. peregra* by metacercariae of *E. recurvatum* over a summer (July–September) period in southeast England showed an increase in the mean number of cysts per host with host size and time of exposure. Aggregation resulting from host and habitat-related factors increased with host size and time of exposure. Encystment within the host was restricted to the peripheral organs in smaller juvenile snails, but as snails increased in size, metacercariae were distributed throughout the tissues (Morley et al. 2004b).

Laruelle et al. (2002) reported that zebra mussels (*Dreissena polymorpha*) served as the second intermediate host for trematodes in the family Echinostomatidae. Cysts were typically found embedded in epithelial tissues in the mantle cavity (including on gill surfaces) or in connective tissue just interior to the epithelial covering of the visceral mass, in regions near the digestive gland and gonads. Cysts were also occasionally recorded in the kidney, hemolymph sinuses, pericardial cavity, suprabranchial cavity, intestinal lumen, and gonads.

Many *Echinostoma* spp. have been reported to use larval anurans as intermediate hosts. The cercariae enter the tadpoles via the cloacal opening and form metacercarial cysts in the kidneys, pronephroi, and Wolffian ducts. Thiemann and Wassersug (2000b) examined the distribution of *Echinostoma* metacercariae in *Rana sylvatica* and *Rana clamitans* tadpoles; 200 individuals of each species were exposed to free-swimming cercariae. There was a significant left: right bias in the distribution of metacercariae within both *R. sylvatica* and *R. clamitans* tadpole hosts, with trematodes preferentially encysting in nephric structures on the right side. In *R. sylvatica* and *R. clamitans* 56.7 and 96.8% of the metacercariae, respectively, were on the tadpoles' right side. Asymmetry in the distribution of parasites followed the direction of the asymmetry in tadpole kidney size, but was much greater. Trematodes preferentially encysted in the head kidneys of *R. clamitans*, which regress at metamorphosis. The right head kidney was the most commonly infected structure in *R. clamitans* tadpoles, containing 72.7% of all cysts in that species. Despite the preference of trematodes to encyst in the head kidney, there was no correlation between the number of cysts in the right kidney and the number in the right head kidney. This suggests that limited space in the head kidney does not influence metacercarial formation in the kidney proper. The high frequency of unilateral encystment in both anurans, and in the head kidneys of *R. clamitans*, may be the result of a coevolved relationship that ultimately benefits both the host and the parasite by ensuring host survival (Thiemann and Wassersug, 2000b).

Light and scanning electron microscopy were used to examine the localization of echinostomatid metacercariae infecting the kidneys of leopard frogs, *Rana pipiens*; and green frogs, *R. clamitans*. Cysts occurred predominantly in the ventrolateral renal cortex, and at least some were confined to the lumen of the Bowman's capsules. Each vermiform metacercarial body was enclosed by a spherical cyst wall that had a uniform thickness. The wall was composed of a homogeneous material containing basic and keratin-like proteins, with sulfated acid mucopolysaccharides on the outer surface. Most cysts were enclosed by a fibrous capsule of host origin or were surrounded by an inflammatory focus (Martin and Conn 1990).

3.5 Development of Echinostome Metacercariae in the Second Intermediate Host

The mechanisms of cercarial encystation remain poorly understood. It is reasonable to speculate that cercarial encystation is mediated by chemoreceptors and the nervous system. Although there is no experimental evidence that any of the supposed digenean sense organs mediate sensory modalities, it is believed that ciliated papillae on the cercarial surface may act as chemoreceptors. Many studies have demonstrated that the ciliate papillae have both argentophilic characteristics, i.e., Ag1 binds selectively to the ciliated papillae, and cholinesterase activity. Silver staining, along with scanning electron microscopy, has revealed the presence of uniciliate and multiciliate papillae, which are undoubtedly sensory structures in some echinostome cercariae (Fried and Fujino 1987). Similar results were also obtained in *E. liliputanus* cercariae using scanning electron microscopy (Xiao et al. 2005).

The cysts of echinostome metacercaria are transparent and contain two to three cyst walls (Irwin and Fried 1990; Krejci and Fried 1994). Scanning electron microscopy studies have determined that *Echinoparyphium aconiatum* and *Echinostoma luisreyi* only have two cyst layers (Gorchilova and Kanev 1986; Maldonado et al. 2003). *E. trivolvis, E. echinatum, E. revolutum*, and *E. caproni* all have three metacercarial cyst layers (Gorchilova and Kanev 1986; Irwin and Fried 1990; Krejci and Fried 1994). Most echinostomes have an outer cyst, which may be smooth or rough in appearance and often has associated host-derived collagen fibers. The outer cyst layer of *E. trivolvis* was described to be more electron dense than underlying layers based on TEM observations (Gulka and Fried 1979). SEM and TEM observations of the outer cyst layer of *Himasthla quissetensis* found the layer to be homogeneous and derived from granules released from the tegument of the metacercaria (Laurie 1974). Species with three layers contain an amorphous middle layer (Gorchilova and Kanev 1986; Irwin and Fried 1990; Krejci and Fried 1994). Most species have a lamellated inner layer (Gorchilova and Kanev 1986; Irwin and Fried 1990; Krejci and Fried 1994). The inner layer of both *E. trivolvis* and *H. quissetensis* has been reported to have whorled or scrolled lamellar structures (Gulka and Fried 1979; Laurie 1974).

The cyst shape varies between the many species of echinostome from spherical as seen in *Echinoparyphium* and *Echinostoma* to oval and elliptical as seen in *Echinochasmus* and *Himasthla*, respectively (Laurie 1974; Krejci and Fried 1994; Kanev et al. 2000). The diameter of the cysts ranges from 100to 400 μm (Kanev et al. 2000). The diameter of the cyst of *E. caproni* is midsized about 150 μm, and the outer cyst wall is relatively smooth when compared to the coarser cyst wall of *E. trivolvis* (Krejci and Fried 1994).

Stein and Basch (1977) reported that cercariae of *Echinostoma paraensei* encysted normally in the presence of *Biomphalaria glabrata* embryo (Bge) cells in culture, partially in culture-conditioned medium, and not at all in fresh culture medium alone. At the ultrastructural level the cyst is composed of 2 well-defined regions. The outer cyst wall (OCW) is particulate to fibrous in nature, formed from secretory granules released from the cercarial tegument. Membranous scrolls or rodlets secreted from the subtegumental cystogenous gland cells are then added to this layer, forming the inner cyst wall (ICW). After 24 h, the cultured cyst is enclosed by a thin cellular capsule similar to that found around cysts in the snail host. The capsule also contains collagen fibers, not found elsewhere in Bge cell cultures.

A detailed study of the surface topography of the echinostome digenean *Mesorchis denticulatus* (Rudolphi 1802) Dietz 1909 was carried out using scanning electron microscopy by Køie (1987). The infective metacercariae, which occur in the gill filaments of euryhaline fishes, have conspicuous collar spines and smaller pointed spines, and occur on most of the body surface.

Metacercarial cysts of *E. revolutum* are infectious to the definitive host within 4 h of encystment, while other echinostomes, such as *Echinoparyphium*, have been reported to require up to 7 days of maturation (Donges 1969; Anderson and Fried 1987). Not much is known about morphological changes of metacercariae within the cyst. The excysted metacercariae of *Himasthla leptosoma* were observed by SEM and TEM (Irwin et al. 1984). The spines of the metacercarial collar were often withdrawn into pits (Irwin et al. 1984). Sensory papillae were observed around the oral sucker, but were not present around the ventral sucker (Irwin et al. 1984). Tegumental spines were observed and ranged from peg-like to palmate with an anterior-to-posterior distribution (Irwin et al. 1984). Fried and Reddy (2000) observed morphological changes of chemically excysted *E. caproni* metacercarial cysts that were maintained *in vitro*. Metacercaria of *E. caproni* have numerous excretory concretions in the protonephridial tubules, but the excretory concretions are absent in adults (Fried and Huffman 1996; Fried and Reddy 2000). Two days postexcystation (PE) most metacercariae were observed to contain 4 to 8 concretions, but by day 4 PE no concretions were present (Fried and Reddy 2000). Over the course of 2 days, PE collar and tegumentary spines became more pronounced (Fried and Reddy 2000). Protuberances and vacuolar blebbing were also observed in some metacercariae by day 4 PE, but these features are not typical of adult worms and may be the result of an inhospitable environment (Fried and Reddy 2000).

The potential influence of second intermediate host species on the infectivity of metacercarial cysts of *E. recurvatum* to the definitive host *Anas platyrhynchos* was examined experimentally by McCarthy (1999). *E. recurvatum* metacercarial cysts were obtained from the following experimentally infected second intermediate hosts 14 days postexposure to cercariae: *L. peregra*, *Physa fontinalis*, *L. stagnalis; Planorbis planorbis*; *Biomphalaria glabrata*; tadpoles of the amphibian *Rana temporaria*. Metacercarial cysts from each of these hosts were fed, in doses of 50 cysts per individual, to separate groups composed of between four and eight, 3-day-old *Anas platyrhynchos* ducklings. All *A. platyrhynchos* were necropsied 15 days postinfection, and the number, size, and reproductive status of *E. recurvatum* worms in the intestine were recorded. All worms recovered were gravid. McCarthy (1999) concluded that the species of second intermediate host utilized does not influence the infectivity of the metacercarial cyst of *E. recurvatum*, or the subsequent establishment and reproductive status of the parasite in *A. platyrhynchos*.

3.6 Excystation of Echinostome Metacercariae

The process of excystation from the metacercarial cyst varies between species of echinostomes. Fried (2000) reviewed the process of excystation for echinostome metacercariae. For some species, excystation is accomplished through digestion of the cyst wall by host-derived enzymes. For other species, excystation is accomplished through a combination of host-derived and parasite-derived enzymes (Sommerville and Rogers 1987). Scanning and transmission electron microscopy studies have been performed on *E. trivolvis* and *E. caproni* during *in vitro* excystation. The excystation process of both echinstomes was determined to be identical (Irwin and Fried 1990). The inner cyst layer undergoes fraying and fragmenting at a point on the inner layer containing lamellae and the metacercaria exits through the frayed area (Irwin and Fried 1990). The inner cyst layer of *H. leptosoma* contains regions lacking lamellae (Irwin et al. 1984). Electron microscopy studies of the excystation of *H. leptosoma* showed that the larva emerged from the inner cyst layer at these sites, and no fraying or fragmentation was observed (Irwin et al. 1984). Electron-dense globules were observed at the site of rupture in both *Echinostoma* species, and the authors speculated that these may be glandular secretions from the trematodes that may be involved in the disruption of the inner layer (Irwin and Fried 1990). The metacercaria enters the middle layer. Excystation is completed when the outer layer is ruptured and the larva emerges from the cyst (Irwin and Fried 1990). The outer layer of both *E. trivolvis* and *E. caproni* was observed to rupture at a single site, which allowed the larva to escape (Irwin and Fried 1990). The reader can find further information on the metacercarial excystment of echinostomes in Chap. 4 of this book.

3.7 Invertebrate and Vertebrate Second Intermediate Hosts of Echinostomes

Members of the family Echinostomatidae utilize a broad range of invertebrate and vertebrate organisms as second intermediate hosts. The hosts include multiple species of gastropods and bivalves, and numerous species of fish. Tables 3.1 and 3.2 summarize many of the reported hosts that harbor metacercariae of echinostomes.

The marine echinostomes of the genera *Acanthoparyphium* and *Himasthla* utilize multiple species of gastropods and bivalves as their second intermediate hosts. *Acanthoparyphium spinulosum* has also been reported to utilize polycheate worms (Bearup 1960; Martin and Adams 1961). Metacercariae of the genus *Echinoparyphium* have been reported in multiple species of gastropods, amphibian larvae, fish, and turtles. *Echinoparyphium megacirrus* was reported to utilize the ectoparasitic flatworm, *Temnocephala chilensis*, which is a common parasite of freshwater crustaceans, insects, and turtles in Australia (Viozzi et al. 2005). Species of the genus *Echinostoma* have been reported in a wide range of gastropods, amphibian larvae, and fish. The genus

Table 3.1 Invertebrate hosts of echinostome metacercariae

Echinostome Species	Host Species	Reference
Acanthoparyphium sp. n.	*Austrovenus stutchburyi* (marine bivalve)	Martorelli et al. (2006)
Acanthoparyphium paracharadrii	*Cerithium ornata* (marine gastropod)	Velazquez (1964a)
Acanthoparyphium spinulosum	*Salinator* spp. and *Pyrizus* spp. (freshwater gastropod) and species of polychaete worm; *Cerithidea californica* and *Crassostrea virginica* (marine gastropod)	Bearup (1960); Martin and Adams (1961); Little et al. (1966)
Acanthoparyphium tyosenense	*Mactra veneriformis*(marine bivalve)	Kim et al. (2004)
Curtuteria australis	*Austrovenus stutchburyi*(bivalve)	Allison (1979)
Curtuteria arguinae	*Cerastoderma edule*(bivalve)	Desclaux et al. (2006)
Echinoparyphium aconiatum	*Lymnaea stagnalis* (freshwater gastropod)	Faltýnková et al. (2007)
Echinoparyphium cinctum	*Lymnaea* spp. (freshwater gastropod)	Kanev et al. (1994)
Echinoparyphium dunni	*Lymnaea rubiginosa*, *Gyraulus convexiusculus*, *Indoplanorbis exustus*, and *Bellamya ingallsiana* (freshwater gastropods)	Lie and Umathevy 1965
Echinoparyphium echinatoides	*Dreissena polymorpha* (bivalve)	Molloy et al. (1997)
Echinoparyphium elegans	*Bulinus truncates*, *B. globosus*, *Physa acuta*, and *Biomphalaria glabrata* (freshwater gastropods)	Mouahid and Mone (1988); King and Van As (1996)
Echinoparyphium flexum	*Lymnaea* spp, *Helisoma* spp., and *Physa* spp. (freshwater gastropods)	Najarian (1953, 1954)
Echinoparyphium megacirrus	*Temnocephala chilensis* (flatworm)	Viozzi et al. (2005)
Echinoparyphium mordwilkoi	*Valvata piscinalis* (freshwater gastropod)	Kiseliene and Grabda-Kazubska (1990)
Echinoparyphium paraulum	*Dreissena polymorpha* (bivalve)	Molloy et al. (1997)
Echinoparyphium ralphaudyi	*Bulinus truncatus, B. glabrata,*and *B. alexandrina* (freshwater gastropods)	Lie et al. (1975)
Echinoparyphium recurvatum	*Radax auricularia* ; Lymnaea spp. (gastropod); *Dreissena polymorpha* (bivalve)	Sohn (1998); Evans and Gordon (1983); Molloy et al. (1997)
Echinoparyphium serratum	*Isidorella brazier* (freshwater gastropod)	Howell (1968)
Echinostoma caproni	*Bulinus*spp., *Biomphalaria* spp., *Physa* spp., *Lymnaea* spp., and *Helisoma* spp.; *Pisidium casertanum* and *Melania tuberculata* (freshwater gastropods)	Christensen et al. (1980); Richard and Brygoo (1978)

Echinostoma cinetorchis	*Hippeutis cantori; Radix auricularia coreana, Physa acuta, Segmentina hemisphaerula, Austropeplea ollula* and *Cipangopaludina* spp. (freshwater gastropods); *Corbicula fluminea* (bivalve)	Lee et al. (1990); Lee et al. (1988); Ahn et al. (1989); Chung and Jung (1999); Chung et al., (2001a, b)
Echinostoma echinatum (Echinostoma lindoense)	*Lymnaea* spp. , *Planorbarius* spp., *Planorbis* spp., *Anisus* ssp., *Gyraulus* spp, *Biomphalaria* spp. and *Viviparus* spp. (freshwater gastropods); *Corbicula* spp. (bivalve)	Carney et al. (1980); Kostadinova and Gibson (2000)
Echinostoma friedi	*Lymnaea peregra*, *L. corvus Gyraulus chinensis*, and *Physella acuta* (freshwater gastropods)	Toledo et al., (2000)
Echinostoma hortense	*Radix*s pp., *Physa* spp., *Planorbis* spp., and *Lymnaea* spp. (freshwater gastropods)	Chai and Lee (1990)
Echinostoma hystricosum	*Lymnaea rubiginosa, Cyraulus convexiusculus,*and *Indoplanorbis exustus* (freshwater gastropods)	Lie and Umathevy (1966)
Echinostoma ilocanum	*Gyraulus phrasadi*; *Pila conica*, *Viviparus javanicus* and *Lymnaea rubiginosa* brevis (freshwater gastropods)	Cross and Basaca-Sevilla 1981, 1986
Echinostoma luisreyi	*Physa*spp. and *Biomphalaria glabrata* (freshwater gastropods)	Maldonado et al. 2003
Echinostoma jurini	Mollusks, frogs, and freshwater turtles	Kanev et al. (1995b)
Echinostoma macrorchis	*Cipangopaludina malleata*, *C. japonica, Segmentina nitidella*, *Viviparus* spp., *Assiminea taiwanensis; Thiara granifera; Gyraulus* spp., *Segmentina hemisphaerula,Hippeutis cantor, Bulinus truncates, Physa*spp., and *Lymnaea* spp. (freshwater gastropods); *Corbicula fluminea* (bivalve)	Beaver et al. (1984); Lo (1995)
Echinostoma malayanum	*Lymnaea cumingiana* and *Pila luzonica* ; *Digoniostoma pulchella* (freshwater gastropods)	Monzon and Kitikoon (1989); Maji et al (1993)
Echinostoma melis	*Corbicula fluminea*and *Unis spinfoei* (bivalve); *Cipangopaludina chinensis, C. miyazagii,* and *Sinotia Quadrata* (freshwater gastropods)	Lu (1982)
Echinostoma miyagawai	*Planorbis planorbis*and *Anisus vortex* (freshwater gastropods)	Kostadinova et al. (2000)
Echinostoma paraensei	*Biomphalaria glabrata*(freshwater gastropod)	Stein and Basch (1977)
Echinostoma parvocirrus	*Biomphalaria glabrata*(freshwater gastropod)	Nassi and Dupouy (1988)

(continued)

Table 3.1 (continued)

Echinostome Species	Host Species	Reference
Echinostoma revolutum	*Corbicula flumineaand Unis spinfoei* (bivalve); *Physa* spp. and *Lymnaea* spp., *Cipangopaludina chinensis, C. miyazagii,* and *Sinotia Quadrata* (freshwater gastropods)	Lu (1982); Kanev (1994)
Echinostoma trivolvis	*Helisoma trivolvis;*pulmonate and prosobranch snails, mussels,and planarians	Schmidt and Fried (1996); Kanev et al. (1995a)
Euparyphium albuferensis	*Lymnaea*spp., *Gyraulus chinensis*, *Physa acuta* (freshwater gastropods)	Toledo et al. (1998)
Euparyphium albuferensis	*Lymnaea*spp. and *Physa* spp. (freshwater gastropods)	Esteban et al. 1997
Euparyphium paramurinum	*Viuipara angularis* (freshwater gastropod)	Velazquez (1964b)
Himasthla ambigua	*Tapes decussates* (bivalve)	Palombi (1934)
Himasthla avosettae	Marine mollusks	Diaz and Cremonte (2004)
Himasthla compacta	*Mya arenaria* and other Marine mollusks	Stunkard (1960)
Hypoderaeum conoideum	*Lymnaea* spp., *Physa* spp., and *Gyraulus chinensis*; *Planorbis planorbis* (freshwater gastropods)	Toledo et al. (1999); Yurlova (1987)
Himasthla continua	Marine mollusks	Diaz and Cremonte (2004)
Himasthla elongata	Unknown	Stunkard (1960)
Himasthla escamosa	Unknown	Diaz and Cremonte (2004)
Himasthla incisa	Unknown	Stunkard (1938); Diaz and Cremonte (2004)
Himasthla interupta	Marine mollusks	Diaz and Cremonte (2004)
Himasthla leptosoma	*Scrobicularia tenuis* (bivalve)	Stunkard (1960)
Himasthla limnodromi (alincia)	Marine mollusks	Didyk and Burt (1997)
Himasthla littorinae	Marine mollusks	Stunkard (1966)
Himasthla megacotyla	Unknown	Diaz and Cremonte (2004)
Himasthla militaris	Marine mollusks	Vanoverschelde and Vaes (1980)
Himasthla muehlensi	Marine mollusks	Stunkard (1960)
Himasthla multilecithosa	Marine mollusks	Mendheim (1940)
Himasthla piscicola	Snails or other invertebrates	Kohn and Fernanades (1981)

Himasthla quissetensis	*Mya* spp., *Mytilus* spp., and *Pectin* spp. (marine bivalves); *Crepidula* spp. (marine gastropod)	Stunkard (1938)
Himasthla rhigedana	Encyst on vegetation	Adams and Martin (1963)
Himasthla secunda	*Mytilus galloprovincialis* (bivalve)	Stunkard (1960)
Himasthla schachtachtinskoi	Marine mollusks	Diaz and Cremonte (2004)
Hypoderaeum conoideum	*Lymnaea* spp. (freshwater gastropod)	Mathias (1925)
Hypoderaeum cubanicum	*Lymnaea tumida*and *Planorbis planorbis* (freshwater gastropods)	Yurlova (1987)
Hypoderaeum dingeri	*Lymnaea rubiginosa*(freshwater gastropod)	Lie et al.(1973)
Paryphostomum sufrartyfex (Artyfechinostomum mehrail)	*Indoplanorbis exustus*(freshwater gastropod)	Roy and Tandon (1996)

Table 3.2 Vertebrate hosts of echinostome metacercariae

Echinostome Species	Host Species	Reference
Echinochasmus angustitestis	Fish	Yu et al. (1994)
Echinochasmus bagulai	*Aplocheilus panchax*, *Oryzias melastigma*, *Gambusia affinis*, and *Channa punctata* (fish)	Dhanumkumari et al. (1991)
Echinochasmus donaldsoni	*Notropis*spp, *Eucalia inconstans*, *Pfrille neogaeus*, and *Anzia calva* (fish)	Beaver (1941b)
Echinochasmus fujilanensis	Fish	Yu et al. (1994)
Echinochasmus liliputanus	Fish	Yu et al. (1994)
Echinochasmus japonicus	*Pseudorasbora parva Abbottina springeri*, *Gnathopogon coreanus*, and several other species of fish	Chai et al. (1985); Lin et al. (1985); Rhee et al. (1984)
Echinochasmus leopoldinae	*Cichlasoma* spp. (fish)	Scholz et al. (1996)
Echinochasmus macrocaudatus	*Astyanax fasciatus*	Ditrich et al. (1996)
Echinochasmus milvi	*Salmo gairdneri* (fish)	Uxmann and Hayduk (1964)
Echinochasmus perfoliatus	Characid fish	Dimitrov et al. (1998)
Echinoparyphium cinctum	Frogs and turtles	Kanev et al. (1994)
Echinoparyphium flexum	*Rana*spp, Hyla spp., and *Pseudacris* spp. (tadpoles)	Najarian (1953, 1954)
Echinoparyphium pseudorecurvatum	Tadpoles and fish	Kiseliene and Grabda-Kazubska (1990)
Echinostoma caproni	*Rana*spp. (tadpoles)	Richard and Brygoo (1978)
Echinostoma cinetorchis	*Misgurnus anguillicaudatus* (fish); *Rana nigromaculata* (tadpole)	Seo et al. (1984); Chung et al., (2001)
Echinostoma hortense	*Misgurnus anguillicaudatus* (fish)	Fan and Sun (1989)
Echinostoma macrorchis	*Rana*spp. and *Bufo bufo* (tadpoles); *Hynobius nigrescens* larvae (salamander)	Orchi (1931); Mori (1935); Lo (1995)
Echinostoma melis	Fish	Lu (1982)
Echinostoma revolutum	*Ameriurus rnelas* (Fish), *Rana* spp. tadpoles, and freshwater turtles	Lu (1982); Kanev (1994); Holland et al. (2006)
Echinostoma trivolvis	*Rana* spp. tadpoles; fishes, other amphibian larvae, and freshwater turtles	Martin and Conn (1990); Kanev et al. (1995a)
Episthmium caninum	Fish	Radomyos et al. (1991)
Hypoderaeum conoideum	Tadpoles	Toledo et al. 1999
Isthmiophora beaveri	Tadpoles	Beaver (1941a)
Isthmiophora melis	Tadpoles and fish	Kostadinova and Gibson (2002)
Petasiger grandivesicularis	*Puntius* spp. and other fish species	Kostadinova and Chipev (1992)
Petasiger nitidus	Fish	Beaver (1939a)
Petasiger variospinosus	Xenopus laevis (larval tadpoles) (natural host)	King and Van As (2000)
Stephanoprora aylacostoma	*Cnesterodon decemmaculatus* and *Poecilia reticulata* (fish)	Ostrowski de Nunez and Quintana (2007)
Stephanoprora denticulata	*Fundulus heteroclutus*(fish)	Schell (1985)
Stephanoprora uruguayense	*Cnesterodon decemmaculatus* (fish)	Ostrowski de Nunez (2007)

Hypoderaeum has been reported in gastropods and amphibian larvae. The genera *Eyparyphium* and *Paryphostomum* have only been reported in freshwater gastropods. Many genera have only been reported to utilize vertebrates as second intermediate hosts including *Isthmiophora* and *Petasiger*, which utilize amphibian larvae and fish, and *Stephanoprora*, which have only been reported in fish. Although the majority of the echinostomes encyst within a second intermediate host, some species have been reported to undergo ectopic encystment. *H. rhigedana* has been reported to encyst on vegetation (Adams and Martin 1963). Ectopic encystment has also been induced *in vitro* in many species of *Echinostoma* and *H. quissetensis* (Laurie 1974; Stein and Basch 1977; Fried and Bennett 1979).

The fingernail clam, *Pisidium coreanum*, is traditionally consumed raw as a so-called drug therapy by patients with bone fractures in Korea. Park et al. (2006a, b) confirmed the possibility of these clams serving as an experimental second intermediate host of *E. cinetorchis*; 20 of them were exposed to *E. cinetorchis* cercariae from experimentally infected *Segmentina hemisphaerula* that had been previously exposed to miracidia of *E. cinetorchis*; all exposed clams became infected. Metacercariae from clams at 14 days postinfection were fed to rats, and adult worms were recovered from the ileocecal regions. This suggests that *P. coreanum* can serve as a second intermediate host of *E. cinetorchis* (Park et al. 2006a, b).

Studies of oysters, *Crassostrea virginica*, collected from 10 sites in the Caloosahatchee River and Estuary, Florida, revealed a varied parasitic and symbiotic fauna that had never been reported in this area. *Echinostoma* sp. occurred at five (1% to 93%) sites and were found infecting the gonoducts of their hosts (Winstead et al.2004).

3.8 Behavioral and Morphological Changes in Invertebrate and Vertebrate Hosts Infected with Echinostome Metacercariae

One of the most widely described examples of behavioral manipulation by echinostome parasites of their second intermediate host involves the trematodes *Curtuteria australis* and *Acanthoparyphium* sp (Thomas and Poulin 1998; Babirat et al.2004). *Curtuteria australis* is a common trematode of the New Zealand cockle, *Austrovenus stutchburyi* (Allison 1979). Echinostome metacercariae accumulate in a cockle's foot, displacing host muscle tissue and preventing growth (Thomas and Poulin 1998). The disruption of the host muscle tissue prevents the cockles from burrowing into the sediment, which enhances predation by the parasite's definitive host oystercatchers (Thomas and Poulin 1998). This phenomenon has major implications for the benthic community, because surfaced cockles alter habitat properties such as sedimentation rates and availability of hard substrate, with consequences for other organisms. It was later determined that a second species of echinostome,

Acanthoparyphium sp., often coinfects the cockles and also utilizes the oystercatchers as their definitive host (Babirat et al.2004). The life cycle of both trematode species is enhanced by host manipulation, but the cockles are also subject to increased predation by species that are not suitable hosts for the parasites. To alter the cockles' behavior, the metacercariae need to encyst within the tip of the cockles' foot but fish will regularly nip the foot of stranded cockles (Mouritsen and Poulin 2003).

The patterns of association between parasites within a particular host are determined by a number of factors. One of these factors is whether or not infection by one parasite influences the probability of acquiring other parasite species. Leung and Poulin (2007a) investigated the pattern of association between various parasites of the New Zealand cockle *Austrovenus stutchburyi*. Hundreds of cockles were collected from one locality within Otago Harbour, New Zealand, and examined for trematode metacercariae and other symbionts. Two interspecific associations emerged from the study. First, the presence of the myicolid copepod *Pseudomyicola spinosus* was positively associated with higher infection intensity by echinostomes. The side effect of the copepod's activities within the cockle is suggested as the proximate mechanism that facilitates infection by echinostome cercariae, leading to a greater rate of accumulation of metacercariae in cockles harboring the copepod. Second, a positive association was also found between infection intensity of the metacercariae of foot-encysting echinostomes and that of gymnophallid metacercariae. This supports earlier findings and suggests that the gymnophallid is a hitch-hiker parasite because, in addition to the pattern of positive association, it (a) shares the same transmission route as the echinostomes, and (b) unlike the echinostomes, it is not capable of increasing the host's susceptibility to avian predation. Thus, both active hitch-hiking and incidental facilitation lead to nonrandom infection patterns in this parasite community (Leung and Poulin 2007a, b).

Echinostomes decrease amphibian performance in laboratory studies; however, the impact of echinostomes in nature has not been examined. Echinostomes have been reported in the kidneys of many North American anuran species (McAlpine and Burt 1998, Beasley et al. 2005). Echinostome cercariae penetrate a broad range of second intermediate hosts, including tadpoles, and subsequently migrate to the host's kidneys, where they encyst as metacercariae (Beaver 1937; Anderson and Fried 1987; Huffman and Fried 1990; Martin and Conn 1990; Fried et al.1997). Several laboratory studies have indicated that echinostomes slow growth and cause mortality in infected tadpoles (Fried et al.1997; Schotthoefer et al. 2003). Schotthoefer et al. (2003) demonstrated that the impact of echinostome infection on leopard frog tadpoles is dependent on the developmental stage. In another study, echinostome infection resulted in a significant decrease in the mean length and weight of the infected tadpoles (Fried et al.1997). Together, these studies indicate that echinostomes have a stage-dependent impact on growth and survival of leopard frog tadpoles in a laboratory setting. However, the impact of echinostomes on tadpoles in nature remains unknown.

3.9 Pathology of Metacercariae in Invertebrate and Vertebrate Hosts

Kuris and Warren (1980) reported that cercarial penetration and metacercarial encystment of *E. caproni* (referred to as *Echinostoma liei* in that study) in *B. glabrata* caused high mortality of juvenile snails (3–8 mm in diameter) after 4–6 days of continuous exposure to about 150 cercariae per snail per day. Larger snails (10–13 mm in diameter) withstood cercarial penetration longer, significant mortality appearing 16 days after exposure. The high rate of cercarial penetration needed to cause snail mortality suggests that echinostome penetration and encystment will not cause much mortality of snail intermediate hosts in nature unless the density of echinostome cercariae is high.

Encysted echinostome metacercariae infecting the gonoducts of oysters, beyond taking up space and eliciting localized hemocytic responses to necrotic worms, caused no significant damage to the host (Winstead et al. 2004).

Occasional signs of inflammation and hemocyte infiltration were observed in zebra mussels (*Dreissena polymorpha*) infected with echinostomatids, but hemocytic encapsulation was not seen (Laruelle et al. 2002). Bower et al. (1994) reported that most species of Echinostomatidae infecting marine mussels were relatively nonpathogenic, but such infections could, however, cause compression of adjacent tissues, reduce byssal production, or induce pearl formation.

Echinostome cercariae enter the tadpoles via the cloacal opening and form metacercarial cysts in the kidneys, pronephroi, mesonephroi, and Wolffian ducts (Thiemann and Wassersug 2000b; Schotthoefer et al. 2003). Figure 3.1 depicts an

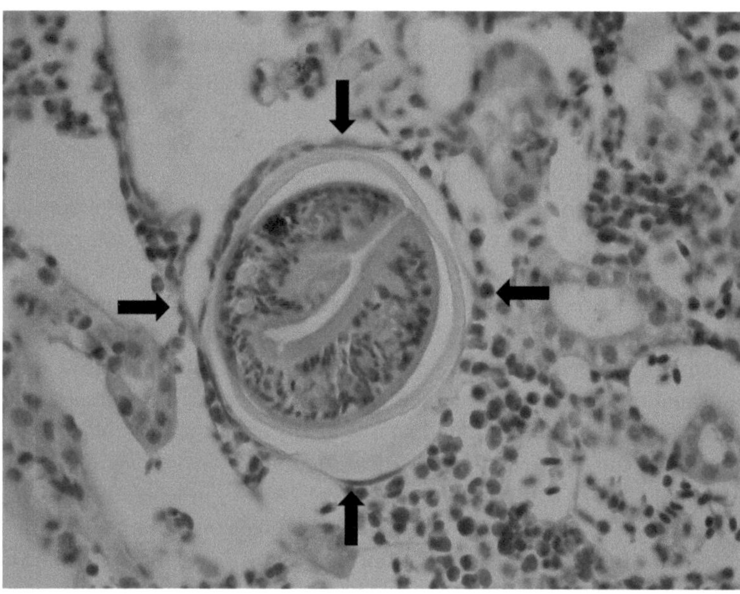

Fig. 3.1 *Echinostoma trivolvis* in the kidney of a naturally infected green frog (*Rana clamitans*)

encysted metacercaria of *E. trivolvis* in the kidney of a naturally infected green frog (*R. clamitans*). Experimental studies have shown that metacercarial infection causes edema, growth inhibition, and mortality of the tadpoles (Fried et al.1997; Schotthoefer et al. 2003).

In freshwater systems, environmental changes, such as eutrophication, are occurring that could impact the outcome of host–parasite interactions. Using tadpole infections with trematode cercariae as a host–parasite system, Belden (2006) examined growth, development, and maintenance of *E. trivolvis* infection levels in second intermediate host larval wood frogs (*R. sylvatica*) and postinfection impacts of eutrophication on *R. sylvatica* tadpoles infected to varying degrees with cercariae of this echinostome. Belden (2006) demonstrated that infection with *E. trivolvis* can impact *R. sylvatica* survivorship, at least above some threshold infection level, and that eutrophication may have minimal impacts on tadpole hosts once infection has occurred.

Johnson and Sutherland (2003) found that ponds in which anuran tadpoles were heavily infected with echinostomes showed a lower recruitment of tadpoles than ponds that had tadpoles with low echinostome infections. Several observations suggest that human-induced changes to aquatic ecosystems may increase the prevalence of echinostome infections or even change the dynamics of echinostome-tadpole interactions that may harm tadpole populations (Thiemann and Wassersug 2000a).

Skelly et al. (2006) have suggested that echinostome infection is an emerging disease in green frogs (*R. clamitans*) in urbanized environments. Echinostomes are widely distributed, and the infection intensity and prevalence in tadpoles is generally low, but this study demonstrated that echinostome infection prevalence in green frog tadpoles in some urban ponds can approach one, with mean infection intensities ranging from hundreds to thousands of echinostomes. Beasley et al. (2005) found that cricket frogs *Acris crepitans* populations experiencing high levels of echinostome infection had low rates of larval recruitment. A comparison of histological sections of renal tissue by Holland et al. (2006) from infected and uninfected green frog tadpoles revealed that echinostome-infected tissue often had fewer glomeruli. Histological sections from uninfected Gosner stage 33 tadpoles contained 10–20 glomeruli, while glomeruli were not apparent in similar sections from infected Gosner stage 33 tadpoles. Some metacercariae encysted within the renal ducts. The walls of ducts within infected renal tissue, especially those containing metacercarial cysts, consisted of a thick secretory columnar epithelium that was not observed in uninfected renal tissue. As observed by Martin and Conn (1990) in adult green frogs, an inflammatory response comprised mostly eosinophils and macrophages developed around some of the metacercariae in infected tissue. Holland et al. (2006) found no overall difference in the leukocyte population in infected versus uninfected histological sections.

Martin and Conn (1999) reported that the kidneys in leopard frogs (*R. pipiens*) and green frogs (*R. clamitans*) infected with echinostomatid metacercariae fibrosis were always focal. The degree of fibrosis varied between individual hosts and between different cysts within the same host. Some heavily encapsulated cysts were darkened and contained disintegrating worms. In heavily infected kidneys, confluence of fibrotic or inflammatory foci resulted in the displacement of functional renal tissue. These data suggest that infection by echinostomatids may impair renal function and that the host's response affects parasite viability (Martin and Conn 1990). The importance of echinostome infections as an emerging disease will be analyzed in Chap. 11 of this book.

3.10 Effects of Pesticides, Herbicides on Echinostome Metacercariae

Reddy et al. (2004) reported on the effects of copper sulfate toxicity on cercariae and metacercariae of *E. caproni* and *E. trivolvis*, and on the survival of *Biomphalaria glabrata* snails. [My paper]Studies were done on the effects of various concentrations of CuSO4 in artificial spring water (ASW) on the survival and infectivity of *E. caproni* and *E. trivolvis* cercariae. Solutions containing 1.0, 0.1, and 0.01% CuSO4 were 100% lethal within 2 h of exposure for both species. Time to 50% mortality in 0.001% CuSO4 was 8 h for *E. caproni* and 16 h for *E. trivolvis*; at 24 h, the controls showed 50 and 65% mortality, respectively. Treatment of cercariae of both species for 0.5 h in 0.001% CuSO4 had no effect on the ability of cercariae to form normal cysts in juvenile *B. glabrata* snails. However, treatment with 0.01% CuSO4 for 0.5 h caused a significant reduction in the ability of cercariae of both species to encyst in snails. Treatment of encysted metacercariae of both species in 0.001% CuSO4 for I h had no effect on subsequent excystation of these echinostomes in a trypsin-bile salts medium, whereas concentrations of 1.0, 0.1, and 0.01% CuSO4 and 1.0 and 0.1% CuSO4 decreased chemical excystation of *E. caproni* and *E. trivolvis* cysts, respectively. Survival studies on the effects of CuSO4 in Locke's solution on chemically excysted metacercariae of both species were also done. Excysted metacercariae of both species were killed by 2 h in either 0.1 or 0.01% CuSO4 in Locke's solution. However, time to 50% mortality for both species of excysted metacercariae in 0.001% CuSO4 was approximately 5 h. Time to 50% mortality for the controls was about 12 h. Survival of juvenile *B. glabrata* snails was also examined. All *B. glabrata* snails were dead by 6 h in 1 and 0.1% CuSO4 in ASW. *B. glabrata* snails showed 50% mortality by about 6 h in 0.01% CuSO4 and about 80% were still alive at 24 h in 0.001% CuSO4. All controls were alive at 24 h, at which time the experiment was terminated. Concentrations greater than 0.001% CuSO4 increased snail mortality, as well as that of the cercariae and excysted metacercariae of *E. caproni* and *E. trivolvis*. Their findings suggested that concentrations of copper sufficient to eliminate juvenile *B. glabratta* snails were also sufficient to kill the cercariae and excysted metacercariae of these digeneans, but not the encysted metacercariae, which may be protected by their cyst walls.

The toxicity of the antifouling biocides tributyltin (TBTO), copper, and Irgarol 1051 (irgarol) at a nominal concentration of 10 μg/l over a 30-day period was investigated against the viability of metacercarial cysts of *E. recurvatum* resident within the body of two common freshwater snails, *L. peregra* and *Physa fontinalis*. Reduced parasite viability was found under most exposures in both snail species. However, a greater effect of toxicant exposure was found in cysts within *P. fontinalis* compared to those in *L. peregra.* This was associated with an increased mortality of the host snail. Among all tested biocides, TBTO exposures induced the highest mortality to both the parasite and their hosts. These results suggest that parasite viability is interlinked with survival of the host snail (Morley et al.2004a, b).

Griggs and Belden (2007) examined the impact of two common herbicides, metolachlor and atrazine, on a host–parasite system consisting of the trematode, *E. trivolvis* and its two intermediate hosts, the snail *Helisoma trivolvis* (referred to as *Planorbella trivolvis* in the study), and larval *Rana* spp. tadpoles. Metolachlor and atrazine are two widely used agricultural herbicides that inhibit the growth of pre-emergent vegetation. Residues of these pesticides are commonly found in water bodies near agricultural areas. Griggs and Belden (2007) examined changes in survivorship when the echinostome cercariae were exposed to a low concentration (10 ppb: 15 ppb) and high concentration (85 ppb: 100 ppb) mixture of metolachlor and atrazine, respectively. These exposure levels were chosen to represent the higher end of herbicide levels that have been documented in aquatic systems. There was a significant decline in cercarial survivorship in the high concentration treatment at 14 h. Echinostomes, the second intermediate host tadpoles, or both echinostomes and the tadpoles, were exposed to the pesticide mixtures for a maximum of 10 h prior to infection, and subsequent tadpole infection levels were examined. The atrazine and metolachlor mixtures had no significant effects on the echinostome load, although newly shed cercariae were more likely than 10-h-old cercariae to infect tadpoles. Outdoor mesocosms were used to expose echinostomes, infected snail hosts, and *R. sylvatica* tadpoles to the pesticide mixtures. The pesticides had no significant effect on tadpole echinostome loads in the mesocosms. Griggs and Belden (2007) suggest that atrazine and metolachlor mixtures at the doses examined do not significantly alter the short-term dynamics of *E. trivolvis* infection in aquatic systems.

To examine the interactions of disease and pollution on amphibian populations, Budischak (2007) investigated the effect of infection on contaminant susceptibility in pickerel frog, *Rana palustris,* larvae. Toxicity tests were conducted to determine the effect of *E. trivolvis* infection (0, 10, or 30 cercaria) on the susceptibility of pickerel frog tadpoles to the widely used organophosphate insecticide, malathion. No difference in susceptibility to malathion was noted among echinostome treatments. The reverse interaction, the effect of pesticide exposure on susceptibility to echinostome infection, showed that tadpoles previously exposed to malathion as embryos suffered increased echinostome encystment rates compared to controls.

3.11 Effects of Marinades of Echinostome Metacercariae

Human echinostome infection results from ingestion of metacercariae that encyst in secondary intermediate hosts, usually freshwater snails, tadpoles, or fish. *E. malayanum* uses various species of gastropod mollusks for primary and secondary intermediate developmental stages (Garrison 1908; Waikagul 1991; Monzon et al. 1993; Radomyos et al.1994). Certain species of fish may also serve as secondary intermediate hosts (Monzon 1993). Several mollusks that may serve as primary and secondary intermediate hosts have been identified in the Philippines, including *Lymnaea* (*Bullastra*) *cumingiana*, *Radix quadrasi*, and *Physastra hungerfordiana* for *E. malayanum*, and *Pila luzonica* for *E. ilocanum* (Radomyos et al. 1998; Belizario et al. 2004).

In terms of eating habits, humans are infected with echinostomes by eating raw fish which are often dipped in a salt and vinegar mixture, known as *kinilaw*. Other methods of fish preparation are *tinola* (boiled), *ginataan* (stewed in coconut milk), and *sinugba* (charcoal-grilled). All echinostome-infected patients had a history of having eaten snails, *kuhol* and *kiambu-ay*, prepared raw with coconut milk and lime juice (*kinilaw*), especially when found in greater abundance during the rainy season. This suggests that various types of marinades and food preparations may not affect the viability of echinostome metacercaria (Belizario et al. 2007).

Wiwanitkit (2005) found that freshly killed freshwater fish (*Cyclocheilichthys armatus*), purchased from a local market in Thailand, were infected with large numbers of *Echinostoma* metacercariae. These infected fish were used to evaluate the effect of traditional food preparation on the viability of the metacercariae. The metacercariae in situ were evaluated using the following parameters: (1) left to dry at room temperature; (2) frozen; (3) refrigerated; (4) marinated in saline; and (5) marinated in 5% acetic acid solution. Degeneration of the metacercariae was slowed by cooling: degeneration of all metacercariae took approximately 10 h in the refrigerated or frozen fish, compared with 4 h in all other dishes left at room temperature and in the marinades.

Various physical and chemical factors were studied by Fried and Peoples (2007) to determine their effects on the viability of encysted metacercariae of *E. caproni*. Viability was equated with chemical excystation in an alkaline trypsin-bile salts (TB) medium. Of numerous marinades tested, the one that was most harmful to isolated and in situ cysts was vinegar. Concentrated solutions of NaCl and sucrose had no effect on the viability of isolated and in situ cysts, suggesting that their use in food preparations for mollusks would not be effective in killing echinostomatid cysts in tainted snail tissues.

3.12 Concluding Remarks

Information on the dynamics of the biology of echinostome metacercariae was presented. The review contributes to our understanding of the ecology and population dynamics between the echinostome metacercariae and the second intermediate host. The study of echinostomes has relevance to amphibian conservation and human health, as well as serving as important models for understanding how anthropogenic changes in the environment favor trematode transmission and lead to their negative effects on host populations. Prevention of human echinostome cases is dependent on eating habits, since raw or insufficiently cooked mollusks, fish, and amphibians are sources of infection for humans.

References

Adams, J.E., and Martin, W.E. 1963. Life cycle of *Himasthla rhigedana* Dietz, 1909 (Trematoda: Echinostomatidae). Transactions of the American Microscopical Society 82:1–6.

Ahn, Y.K., Ryangj, Y.S., Chai, J.Y., and Sohn, W.M. 1989. Cercarial shedding of *Echinostoma cinetorchis* and experimental infection of the cercariae to several kinds of snails. Korean Journal of Parasitology 27:203–211.

Allison, F.R. 1979. Life cycle of *Curtuteria australis* n. sp. (Digenea: Echinostomatidae: Himasthlinae), intestinal parasite of the South Island pied oystercatcher. New Zealand Journal of Zoology 6:13–20.

Anderson, J.W., and Fried, B. 1987. Experimental infection of *Physa heterostropha*, *Helisoma trivolvis*, and *Biomphalaria glabrata* (Gastropoda) with *Echinostoma revolutum* (Trematoda) cercariae. Journal of Parasitology 73:49–54.

Babirat, C., Mouritsen K.N., and Poulin, R. 2004. Equal partnership: Two trematode species, not one, manipulate the burrowing behavior of the New Zealand cockle, *Austrovenus stutchburyi*. Journal of Helminthology 78:195–199.

Bearup, A.J. 1960. Life history of *Acanthoparyphium spinulosum*. Johnson, 1917 (Trematoda: Echinostomatidae). Australian Journal of Zoology 8:217–225.

Beaver, P.C. 1937. Experimental studies on *Echinostoma revolutum* (Froelich) a fluke from birds and mammals. Illinois Biological Monographs 15:1–96.

Beaver, P.C. 1939. The morphology and life history of *Petasiger nitidus* Linton (Trematoda: Echinostomatidae). Journal of Parasitology 25:269–276.

Beaver, P.C. 1941a. Studies on the life history of *Euparyphium melis* (Trematoda: Echinostomatidae). Journal of Parasitology 27:35–44.

Beaver, P.C. 1941b. The life history of *Echinochasmus donaldsoni* n. sp., a trematode (Echinostomidae) from the pied-billed grebe. Journal of Parasitology 27:347–355.

Beaver, P.C., Jung, R.C., and Cupp, E.W. 1984. Clinical parasitology. Philadelphia: Lea & Febiger.

Beasley, V.R., Faeh, S.A., Wikoff, B., Eisold, J., Nichols, D., Cole, R., Schotthoefer, A.M., Staehle, C., Greenwell, M., and Brown, L.E. 2005. Risk factors and the decline of the cricket frog, *Acris crepitans*: evidence for involvement of herbicides, parasitism, and habitat modifications. In Amphibian declines: the conservation status of United States species, ed. M.J. Lannoo, pp. 75–87. Chicago: University of Chicago Press.

Belden, L.K. 2006. Impact of eutrophication on wood frog, *Rana sylvatica*, tadpoles infected with *Echinostoma trivolvis* cercariae. Canadian Journal of Zoology 84:1315–1321.

Belizario, V.Y., de Leon, W.U, Bersabe, M.J, Baird, J.K., and Bangs, M.J. 2004. A focus of human infection by *Haplorchis taichui* (Trematoda: Heterophyidae) in the southern Philippines. Journal of Parasitology 90:1165–1169.

Belizario V.Y., Geronilla, G.G., Anastacio, M.B.M., de Leon, W.U., Suba-an, A.P., Sebastian, A.C., and Bangs, M.J. 2007. *Echinostoma malayanum* infection, in the Philippines. Emerging Infectious Diseases 13:1130–1131.

Bower, S.M., McGladdery, S.F., and Price, I.M. 1994. Synopsis of infectious diseases and parasites of commercially exploited shellfish. Annual Review of Fish Diseases 4:1–199.

Budischak, S.A. 2007. The reciprocal influence of trematode parasites and malathion on developing pickerel frogs (*Rana palustris*). Master's Thesis Virginia Polytechnic Institute and State University. 67pp.

Carney, W.P., Sudomo, M., and Purnomo, A. 1980. Echinostomiasis: a disease that disappeared. Tropical and Geographic Medicine 32:101–105.

Chai, J.Y., and Lee, S.H. 1990. Intestinal trematodes of humans in Korea: *Metagonimus*, heterophyids and echinostomes. Korean Journal of Parasitolology 28:103–122.

Chai, J.Y., Hong, S.J., Sohn, D.W., Lee, S.H., and Seo, B.S. 1985. Metacercariae of *Echinochasmus japonicus* encysted in a fresh water fish, *Pseudorasbora parva*, and their development in experimental mice. Kisaengchunghak Chapchi 23:221–229.

Christensen, N. ø., Frandsen, F., and Roushdy, M.Z. 1980. The influence of environmental conditions and parasite-intermediate host-related factors on the transmission of *Echinostoma liei*. Zeitschrift für Parasitenkunde 63:47–63.

Chung, P.R., and Jung, Y. 1999. *Cipangopaludina chinensis malleata* (Gastropoda: Viviparidae): A new second molluscan intermediate host of a human intestinal fluke *Echinostoma cinetorchis* (Trematoda: Echinostomatidae) in Korea. Journal of Parasitology 85:963–964.

Chung P.R., Jung, Y., and Park, Y.K. 2001. *Segmentina hemisphaerula*: a new molluscan intermediate host for *Echinostoma cinetorchis* in Korea. Journal of Parasitology 87:1169–1171.

Chung, P.R., Jung, Y., Park, Y.K., Hwang, M.G., and Sohn, C.T. 2001. *Corbicula fluminea* (Bivalvia: Corbiculidae): a possible second molluscan intermediate host of *Echinostoma cinetorchis* (Trematoda: Echinostomatidae) in Korea. Korean Journal of Parasitology 39:329–332.

Cross, J.H., and Basaca-Sevilla, V. 1981. Intestinal parasitic infections in southeast Asia. Southeast Asian Journal of Tropical Medicine and Public Health 12:262–274.

Cross, J.H., and Basaca-Sevilla, V. 1986. Studies on *Echinostoma ilocanum* in the Philippines. Southeast Asian Journal of Tropical Medicine and Public Health 17:23–27.

Desclaux, C., Russell-Pinto, F., de Montaudouin, X., and Bachelet, G. 2006. First record and description of metacercariae of *Curtuteria arguinae* n. sp. (Digenea: Echinostomatidae), parasite of cockles *Cerastoderma edule* (Mollusca: Bivalvia) in Arcachon Bay, France. Journal of Parasitology 92:578–587.

Dhanumkumari, K., Hanumantha, R., and Shyamasundari, K. 1991. The life cycle of *Echinochasmus bagulai* (Trematoda: Echinostomatidae). International Journal for Parasitology 21:259–263.

Diaz, J.I., and Cremonte, F. 2004. *Himasthla escamosa* n. sp. (Digenea: Echinostomatidae) from the kelp gull, *Larus dominicanus* (Charadriiformes: Laridae), on the Patagonian Coast, Argentina. Journal of Parasitology 90:308–314.

Didyk, A.S., and Burt, M.D.B. 1997. *Himasthla limnodromi* n. sp. (Digenea: Echinostomatidae) from the short-billed dowitcher, *Limnodromus griseus* (Aves: Scolopacidae). Journal of Parasitology 83:1124–1127.

Dimitrov, V., Kanev, I., Bezproz-Vanich, V., and Radev, V. 1998. Argentophilic structures of the miracidium of *Echinochasmus perfoliatus* (Trematoda: Echinostomatidae). Parasite 5:185–188.

Ditrich, O., Scholz, T., and Vargas-Vazquez, J. 1996. The life cycle of *Echinochasmus bagulai* (Trematoda: Echinostomatidae). Systematic Parasitology 33:225–235.

Donges, J. 1969. Entwicklungs – und Lebensdauer von Metacercarien. Zeitschrift für Parasitenkunde 31:340–366.

Esteban, J.G., Toledo, R., Sanchez, L., and Muñoz-Antoli, C. 1997. Life-cycle of *Euparyphium albuferensis* n. sp. (Trematoda: Echinostomatidae) from rats in Spain. Systematic Parasitology 38:211–219.

Evans, N.A., and Gordon, D.M. 1983 Experimental studies on the transmission dynamics of the cercariae of *Echinoparyphium recurvatum* (Digenea: Echinostomatidae). Parasitology 87:167–174.

Faltýnková, A., Nasincova, V., and Kablaskova, L. 2007. Larval trematodes (Digenea) of the great pond snail, *Lymnaea stagnalis* (L.) (Gastropoda, Pulmonata), in Central Europe: a survey of species and key to their identification. Parasite 14:39–51.

Fan, S.Q., and Sun, M.F. 1989. *Echinostoma hortense*. Bulletin of Harbin Medical University 23:161.

Fried, B. 2000. Maintenance, Cultivation, and Excystation of Echinostomes. *In* Echinostomes as experimental models for biological research, eds. B. Fried and T.K. Graczyk, pp. 99–118. Dordrecht: Kluwer.

Fried, B. 2001. Biology of echinostomes except *Echinostoma*. Advances in Parasitology 49:163–215.

Fried, B., and Bennett, M.C. 1979. Studies on encystment of *Echinostoma revolutum* cercariae. Journal of Parasitology 65:38–40.

Fried, B., and Fujino, M.C. 1987. Argentophilic and scanning electron microscopic observation of the tegumentary papillae of *Echinostoma revolutum* (Trematoda) cercariae. Journal of Parasitology 73:1169–1174.

Fried, B., and Huffman, J.E. 1996. The biology of the intestinal trematode *Echinostoma caproni*. Advances in Parasitology 38:311–368.

Fried, B., and Peoples, R.C. 2007. Effects of various physical and chemical factors on excystation of the encysted metacercariae of *Echinostoma caproni*. Journal of Parasitology 93:1388–1391.

Fried, B., and Reddy, A. 2000. Postmetacercarial changes in *Echinostoma caproni* maintained in a defined medium plus calf serum. Korean Journal of Parasitology 38:173–175.

Fried, B., Pane, P.L., and Reddy, A. 1997. Experimental infection of *Rana pipiens* tadpoles with *Echinostoma trivolvis* cercariae. Parasitology Research 83:666–669.

Garrison, P.E. 1908. A new intestinal trematode of man. Philippine Journal of Science. B3:385–393.

Gorbushin, A., and Levakin, I. 2005. Encystment *in vitro* of the Cercariae *Himasthla elongata* (Trematoda: Echinostomatidae). Journal of Evolutionary Biochemistry and Physiology 41:428–436.

Gorchilova, L., and Kanev, I. 1986. Electron microscope studies of the metacercariae of *Echinostoma revolutum* (Forlich, 1802), *E. echinatum* (Zeder, 1803), and *Echinoparyphium aconiatum* (Dietz, 1909) (Trematoda: Echinostomatidae). Khelmintologiya 22:3–8.

Griggs J.L., and Belden, L.K. 2007. Effects of atrazine and metolachlor on the survivorship and infectivity of Echinostoma trivolvis trematode cercariae. Archives of Environmental Contamination and Toxicology 54:195–202.

Gulka, G.J., and Fried, B. 1979. Histochemical and ultrastructural studies on the metacercarial cysts of *Echinostoma revolutum* (Trematoda). International Journal for Parasitology 9:57–59.

Holland, M.P., Skelly, D.K., Kashgariam, M., Bolden, S.R., Harrison, L.M., and Cappello, M. 2006. Echinostome infection in green frogs (*Rana clamitans*) is stage and age dependent. Journal of Zoology 271:455–462.

Howell, M.J. 1968. The life cycle of *Echinoparyphium serratum* sp.nov. (Digenea: Echinostomatidae). Parasitology 58:573–582.

Huffman, J.E., and Fried, B. 1990. *Echinostoma* and echinostomiasis. Advances in Parasitology 29:215–269.

Humphries, J.E., Mousley, A., Maule, A.G., and Halton, D.W. 2000. Neuromusculature structure and functional correlates. *In* Echinostomes as experimental models for biological research, eds. B. Fried and T.K. Graczyk, p. 213–227. Dordrecht: Kluwer.

Irwin, S.W.B., and Fried, B.. 1990. Scanning and Transmission electron microscopic observations on metacercariae of *Echinostoma trivolvis* and *Echinostoma caproni* during in vitro excystation. Journal of Helminthology 57:79–83.

Irwin, S.W.B., Mckerr, G., Judge, B.C., and Moran, I. 1984. Studies on metacercarial excystment in *Himasthla leptosoma* (Trematoda: Echinostomatidae) and newly emerged metacercariae. International Journal for Parasitology 14:415–421.

Jeyarasasingam, U., Heyneman, D., Lim, H.K., and Mansour, N. 1972. Life cycle of a new echinostome from Egypt, *Echinostoma liei* sp. nov. (Trematoda: Echinostomatidae). Parasitology 65:203–222.

Johnson, P.T., and Sutherland, D.R. 2003. Amphibian deformities and *Ribeiroia* infection: an emerging helminthiasis. Trends in Parasitology 19:332–335.

Kanev, I. 1994. Life-cycle, delimitation and redescription of *Echinostoma revolutum* (Froelich, 1802) (Trematoda: Echinostomatidae). Systematic Parasitology 28:125–144.

Kanev, I., Radev, V., Vassilev, I., Dimitrov, V., and Minchella, D. 1994. The life cycle of *Echinoparyphium cinctum* (Rudolphi, 1803) (Trematoda: Echinostomatidae) with reexamination and identification of its allied species from Europe and Asia. Helminthologica 31:73–82.

Kanev, I., Fried, B., Dimitrov, V., and Radev, V. 1995a. Redescription of *Echinostoma trivolvis* (Cort, 1914) (Trematoda, Echinostomatidae) with a discussion on its identity. Systematic Parasitology 32:61–70.

Kanev, I., Dimitrov, V., Radev, V., and Fried, V. 1995b. Redescription of *Echinostoma jurini* (Skvortzov, 1924) with a discussion of its identity and characteristics. Annalen des Naturhistorischen 97B:37–53.

Kanev, I., Sterner, M., Radev, V., and Fried, B. 2000. An overview of the biology of echinostomes. *In* Echinostomes as experimental models for biological research, eds. B. Fried and T.K. Graczyk, pp. 1–29. Dordrecht: Kluwer.

Kim, Y., Yu, J., Chung, E., and Chung, P. 2004. *Acanthoparyphium tyosenense* (Digenea: Echinostomatidae): experimental confirmation of the cercaria and its complete life history in Korea. Journal of Parasitology 90:97–102.

King, P.H., and Van As, J.G. 1996. A description of the life stages of *Echinoparyphium elegans* (Trematoda: Echinostomatidae). South African Journal of Zoology 31:145–153.

King, P.H., and Van As, J.G. 2000. Morphology and life history of *Petasiger variospinosus* (Trematoda: Echinostomatidae) in the Free State, South Africa. Journal of Parasitology 86:312–318.

Kiseliene, V., and Grabda-Kazubska, B. 1990. *Echinoparyphium pseudorecurvatum* sp. N. (Trematoda: Echinostomatidae) and its life cycle. Acta Parasitologica Polonica 35:285–295.

Kohn A., and Fernandes, B.M.M. 1981. The adult from *Himasthla piscicola* Stunkard, 1960 and other trematodes from Brazil freshwater fishes. Journal of Helminthology 55:85–87.

Køie, M. 1987. Scanning electron microscopy of rediae, cercariae, metacercariae and adults of *Mesorchis denticulatus* (Rudolphi, 1802) (Trematoda, Echinostomatidae). Parasitology Research 73:50–56.

Kostadinova, A., and Chipev, N. 1992. Experimental data on the life cycle of *Petasiger grandivesicularis* Ishi, 1935 (Trematoda: Echinostomatidae). Systematic Parasitology 23:55–65.

Kostadinova, A., and Gibson, D.I. 2000. The systematics of the echinostomes.. *In* Echinostomes as experimental models for biological research, eds. B. Fried and T.K. Graczyk, p. 31–57. Dordrecht: Kluwer.

Kostadinova, A., and Gibson, D.I. 2002. *Isthmiophora* Lühe, 1909 and *Euparyphium* Dietz, 1909 (Digenea:Echinostomatidae) re-defined, with comments on their nominal species. Systematic Parasitology 52:205–217.

Kostadinova, A., Gibson, D.I., Biserkov, V., and Chipev, N. 2000. Revalidation of *Echinostoma miyaawi* Ishii 1932 (Digenea: Echinostomatidae) on the basis of the experimental completion of its life cycle. Systematic Parasitology 45:81–108.

Krejci, K.G., and Fried, B. 1994. Light and scanning electron microscopic observations of the eggs, daughter rediae, cercariae, and encysted metacercariae of *Echinostoma trivolvis* and *E. caproni*. Parasitology Research 80:42–47.

Kuris, A.M., and Warren, J. 1980. Echinostome cercarial penetration and metacercarial encystment as mortality factors for a second intermediate host, *Biomphalaria glabrata*. Journal of Parasitology 66:630–635.

Laruelle, F., Molloy, D.P., and Roitman, V.A. 2002. Histological analysis of trematodes in *Dreissena polymorpha*: their location, pathogenicity, and distinguishing morphological characteristics. Journal of Parasitology 88:856–863.

Laurie, J.S. 1974. *Himasthla quissetensis*: induced in vitro encystment of cercaria and ultrastructure of the cyst. Experimental Parasitology 35:350–362.

Lee, S.H., Chai, J.Y., Hong, S.T., and Sohn, W.M. 1990. Experimental life history of *Echinostoma cinetorchis*. Korean Journal of Parasitology 28:39–44.

Lee, J.K., Sohn, W.M., Hong, S.T., Hong, S.J., and Chai, J.Y., 1988. Metacercariae of *Echinostoma cinetorchis* encysted in the fresh water snail, *Hippeutis (Helicorbis) canrori,* and their development in rats and mice. Korean Journal of Parasitology 26:189–197.

Leung, T.L., and Poulin, R. 2007a. Interactions between parasites of the cockle *Austrovenus stutchburyi*: hitch-hikers, resident-cleaners, and habitat-facilitators. Parasitology 134:247–255.

Leung, T.L., and Poulin, R. 2007b. Recruitment rate of gymnophallid metacercariae in the New Zealand cockle *Austrovenus stutchburyi*: an experimental test of the hitch-hiking hypothesis. Parasitology Research 101:281–287.

Lie, K.J., and Umathevy, T. 1965. Studies on Echinostomatidae (Trematoda) in Malaya. X. The Life history of *Echinoparyphium dunni* sp. n. Journal of Parasitology 51:793–799.

Lie, K.J., and Umathevy, T. 1966. Studies on Echinostomatidae (Trematoda) in Malaya XI. The life history of *Echinstoma hystricosum* sp. n. Journal of Parasitology 52:449–453.

Lie, K.J., Nasemary, S., and Impand, P. 1973. Five echinostome species from Thailand. Southeast Asian Journal of Tropical Medicine and Public Health 4:96–101.

Lie, K.J., Heyneman, D., Jeyarasasingam, U., Mansour, N., Lee, H.F., Lee, H., and Kostanian, N. 1975. The life cycle of *Echinoparyphium ralphaudyi* sp. n. (Trematoda: Echinostomatidae). Journal of Parasitology 61:59–65.
Lin, J.X., Chen, Y.Z., Liang, C.Z., Yang, L.B., and Zhuang, H.J. 1985. Epidemiological investigation and experimental infection of *Echinochasmus japonicus*. Journal of Parasitology and Parasitic Diseases 3:89–91.
Little, J.W., Hopkins, S.H., and Schlicht, F.G. 1966. *Acanthoparyphium spinulosum* (Trematoda: Echinostomatidae) in Oysters at Port Isabel, Texas. Journal of Parasitology 52:663.
Lo, C.T. 1995. *Echinostoma macrorchis*: Life history, population dynamics of intramolluscan stages, and the first and second intermediate hosts. Journal of Parasitology 81:569–576.
Lu, S.C. 1982. Echinostomiasis in Taiwan. International Journal of Zoonoses 9:33–38.
Macy, R.W., Berntzen, A., and Benz, M. 1968. In vitro excystation of *Sphaeridiotrema globulus* metacercariae, structure of cyst and relationship to host specificity. Journal of Parasitology 54:28–38.
Maji, A.K., Bera, D.K., Manna, B., Nandy, A., Addy, M., and Bandyopadhyay, A.K. 1993. First record of human infection with *Echinostoma malayanum* in India. Transactions of the Royal Society of Tropical Medicine and Hygiene 87:673.
Maldonado, A., Vieira, G.O., and Lanfredi, R.M. 2003. *Echinostoma luisreyi* n. sp. (Playthelminthes: Digenea) by light and scanning electron microscopy. Journal of Parasitology 89:800–808.
Marquardt, W.C., and Demaree, R.S. 1985. Parasitology. Macmillan Publishing Company USA.
Martin, W.E., and Adams, J.E. 1961. Life cycle of *Acanthoparyphium spinulosum* Johnson, 1917 (Echinostomatidae: Trematode). Journal of Parasitology 47:777–782.
Martin, T.R., and Conn, D.B. 1990. The pathogenicity, localization, and cyst structure of echinostomatid metacercariae (Trematoda) infecting the kidneys of the frogs *Rana clamitans* and *Rana pipiens*. Journal of Parasitology 76:414–419.
Martorelli, S.R., Poulin, R., and Mouritsen, K.N. 2006. A new cercaria and metacercaria of *Acanthoparyphium* (Echinostomatidae) found in an intertidal snail *Zeacumantus subcarinatus* (Batillaridae) from New Zealand. Parasitology International 55:163–167.
Mathias, P. 1925. Recherches expCrimentales sur le cycle kvolutif de vuelques Trematodes. Bulliten Biological Frances et Belgique 59:1–123.
McAlpine, D.F., and Burt, D.B. 1998. Helminths of bullfrogs, *Rana catesbeiana*, green frogs, *R. clamitans*, and leopard frogs, *R. pipiens* in New Brunswick. Canadian Field Naturalist 112:50–68.
McCarthy, A.M. 1999. The influence of second intermediate host species on the infectivity of metacercarial cysts of *Echinoparyphium recurvatum*. Journal of Helminthology 73: 143–145.
Mendheim, H. 1940. Beiträge zur Systematik und Biologie der Familie Echinostomatidae. Nova Acta Leopoldina. Halle N.F. 8:489–588.
Molloy, D.P., Karatayev, A.Y., Burlakova, L.E., and Laruelle, F. 1997. Natural enemies of zebra mussels: predators, parasites and ecological competitors. Reviews in Fisheries Science 5: 27–97.
Monzon, R.B., and Kitikoon, V. 1989. *Lymnaea (Bullastra) cumingiana* Pfeiffer (Pulmonata: Lymnaeidae): second intermediate host of *Echinostoma malayanum* in the Philippines. Southeast Asian Journal of Tropical Medicine and Public Health 20:453–460.
Monzon, R.B., Kitikoon V., Thammapalerd, N., Temcharoen, P., Sornmani, S., and Viyanant, V. 1993. Ecological observations on *Lymnaea* (*Bullastra*) *cumingiana*. Southeast Asian Journal of Tropical Medicine and Public Health 24:563–569.
Mori, J. 1935. Experimental studies on the possibility of larval salamanders to act as the second intermediate host for echinostome cercariae. Tokyo Iji Shin Shi No 2929:1237–1244.
Morley, N.J., Adam, M.E., and Lewis, J.W. 2004a. The role of *Bithynia tentaculata* in the transmission of larval digeneans from a gravel pit in the Lower Thames Valley. Journal of Helminthology 78:129–135.
Morley, N.J., Lewis, J.W., and Adam, M.E. 2004b. Metacercarial utilization of a naturally infected single species (*Lymnaea peregra*) snail community by *Echinoparyphium recurvatum*. Journal of Helminthology 78:51–56.
Morley, N.J., Leung, K.M.Y., Morritt, D., and Crane, M. 2004c. Toxicity of anti-fouling biocides to encysted metacercariae of *Echinoparyphium recurvatum* (Digenea: Echinostomatidae) and their snail hosts Chemosphere 56:353–358.

Morley, N.J., Adam, M.E., and Lewis, J.W. 2007. The effects of temperature on the distribution and establishment of *Echinoparyphium recurvatum* metacercariae in *Lymnaea peregra*. Journal of Helminthology 81:311–315.

Mouahid, A., and Mone, H. 1988. *Echinoparyphium elegans* (Loos, 1899) (Digenea: Echinostomatidae): the life cycle and redescription of the adult with a revision of the 43-spined members of the genus *Echinoparyphium*. Systematic Parasitology 12:149–157.

Mouritsen, K.N., and Poulin, R. 2003. The risk of being at the top: foot-cropping in the New Zealand cockle Austrovenus stutchburyi. Journal of the Marine Biological Association of the United Kingdom 83:497–498.

Najarian, N.H. 1953. The life history of *Echinoparyphium flexum* (Linton 1892) (Dietz 1910) (Trematoda: Echinostomatidae). Science 117:564–565.

Najarian, N.H. 1954. Developmental stages in the life cycle of *Echinoparyphium flexum* (Linton 1892) (Dietz 1910) (Trematoda: Echinostomatidae). Journal of Morphology 84:165–197.

Nassi, H., and Dupouy, J. 1988. étude expérimentale du cycle biologique d'*Echinostoma parvocirrus* n. sp. (Trematoda: Echinostomatidae), parasite larvaire de Biomphalaria glabrata en Guadeloupe. Annales de Parasitologie Humaine et Comparee 63, 103–118.

Orchi, S. 1931. Studies on the development of *Echinostoma macrorchis*. Women's Medical Journal 1:12–32.

Ostrowski de Nunez, M. 2007. Life cycle of *Stephanoprora uruguayense* (Digenea: Echinostomatidae) in Argentina. Journal of Parasitology 93:1090–1096.

Ostrowski de Nunez, M., and Quintana, M.G. 2007. The life cycle of *Stephanoprora aylacostoma* n.sp. (Digenea: Echinostomatidae), parasite of the threatened snail *Aylacostoma chloroticum* (Prosobranchia: Thiaridae), in Argentina. Parasitology Research 102:647–655.

Palombi, A. 1934. Gli stadi larvali dei Trematodi del Golfo di Napoli. Pubblicazioni Stazione Zoologica Napoli 14:51–94.

Park, Y.K., Hwang, M.K., and Chung, P.R. 2006a. Encystment and metacercariae development of *Echinostoma cinetorchis* cercariae in an in vitro culture system. Journal of Parasitology 92:1010–1013.

Park, Y.K., Sohn, C.T., Park, G.M., Hwang, M.K., and Chung, P.R 2006b. Host specificity of *Pisidium coreanum* (Bivalvia: Sphaeriidae) to larval infection with a human intestinal fluke *Echinostoma cinetorchis* (Trematoda: Echinostomatidae) in Korea. Journal of Parasitology 92:1118–1120.

Radomyos, P., Charoenlarp, P., and Radomyos, B. 1991. Human *Episthmium caninum* (Digenea, Chinostornatidae) infection: report of two more cases. Journal of Tropical Medicine and Parasitology 14:48–50.

Radomyos, P., Radomyos, B., and Tangtrongchitr, A. 1994. Multi-infection with helminths in adults from northeast Thailand as determined by post-treatment fecal examination of adult worms. Tropical Medicine and Parasitology 45:133–135.

Radomyos, B., Wongsaroj, T., Wilairatana, P., Radomyos, P., Praevanich, R., and Meesomboon, V. 1998. Opisthorchiasis and intestinal fluke infections in northern Thailand. Southeast Asian Journal of Tropical Medicine and Public Health 29:123–127.

Reddy, A., Ponder, E.L., and Fried, B. 2004. Effects of copper sulfate toxicity on cercariae and metacercariae of *Echinostoma caproni* and *Echinostoma trivolvis* and on the survival of *Biomphalaria glabrata* snails. Journal of Parasitology 90:1332–1337.

Rhee J.K., Rim, M.H., Baek, B.K., and Lee, H.I. 1984. Survey on encysted cercariae of trematodes from fresh-water fishes in Tongjin riverside areas in Korea. Korean Journal of Parasitology 22:190–202.

Richard, J., and Brygoo, E.R. 1978. Life cycle of the trematode *Echinostoma caproni* Richard, 1964 (Echinostomatidae). Annales de Parasitologie Humaine et Comparee 53:265–275.

Roy, B., and Tandon, V. 1996. Effect of root-tuber extract of *Flemingia vestita*, a leguminous plant, on *Artyfechinostomum sufrartyfex* and *Fasciolopsis buski*: a scanning electron microscopy study. Parasitology Research 82:248–252.

Schell, S.C. 1985. Trematodes of North America, North of Mexico. Moscow, Idaho: University Press of Idaho.

Scholz, T., Ditrich, O., and Vargas-Vazquez, J. 1996. *Echinochasmus leopoldinae* n. sp. (Trematoda: Echinostomatidae) and data on its life-cycle. Systematic Parasitology 33:157–165.

Schotthoefer, A.M., Cole, R.A., and Beasley, V.R. 2003. Relationship of tadpole stage to location of echinostome cercariae encystment and the consequences for tadpole survival. Journal of Parasitology 89:475–482.

Schmidt, K.A., and Fried, B. 1996. Experimental infection of *Helisoma trivolvis* (Colorado Strain) snails with cercariae of *Echinostoma trivolvis*. International Journal for Parasitology 26:287–289.

Seo, B.S., Park, Y.H., Chai, J.Y., Hong, S.J., and Lee, S.H. 1984. Studies on intestinal trematodes in Korea. XIV. Infection status of loaches with metacercariae of *Echinostoma cinetorchis* and their development in albino rats. Korean Journal of Parasitology 22:181–189.

Skelly, D.K., Bolden, S.R., Holland, M.P., Freidenburg, L.K., Freidenfelds, N.A., and Malcolm, T.R. 2006. Urbanization and disease in amphibians. In Disease ecology: community structure and pathogen dynamics, eds. S.K. Collinge and C. Ray, pp. 153–167. New York: Oxford University Press.

Sohn, W.M. 1998. Life history of *Echinoparyphium recurvatum* (Trematoda: Echinostomatidae) in Korea. Korean Journal of Parasitology 36:91–98.

Sommerville, R.I., and Rogers, W.P. 1987. The nature and action of host signals. Advances in Parasitology 26:239–293.

Stein, P.C., and Basch, P.F. 1977. Metacercarial cyst formation in vitro of *Echinostoma paraensei*. Journal of Parasitology 63:1031–1040.

Stunkard, H.W. 1938. The morphology and life cycle of the trematode *Himasthla quissetensis* (Miller and Northrup, 1926). Biological Bulletin 75:145–164

Stunkard, H.W. 1960. Further studies on the trematode genus *Himasthla* with descriptions of *H. mcinthoshi* n. sp., *H. piscicola* n. sp., and stages in the life-history of *H. compacta* n. sp. Biological Bulletin 119:529–549.

Stunkard, H.W. 1966. The morphology and life history of the digenetic trematode, *Himasthla littorinae* sp. n. (Echinostomatidae). Journal of Parasitology 52:367–372.

Thiemann, G.W., and Wassersug, R.J. 2000a. Patterns and consequences of the behavioural responses to predators and parasites in *Rana* tadpoles. Biological Journal of the Linnean Society 71:513–528.

Thiemann, G.W., and Wassersug, R.J. 2000b. Biased distribution of trematode metacercariae in the nephric system of *Rana* tadpoles. Journal of Zoology 252:534–538.

Thomas, F., and Poulin, R. 1998. Manipulation of a mollusk by a trophically transmitted parasite: convergent evolution or phylogenetic inheritance? Parasitology 116:431–436.

Toledo, R., Muñoz-Antoli, C., Sanchez, L., Bayssade-Dufour, C., and Esteban, J.G. 1998. Cercarial chaetototaxy of *Euparyphium albuferensis* Esteban et al., 1997 (Trematoda: Echinostomatidae), with a review of some genera of the Echinostomatinae. Systematic Parasitology 39:35–44.

Toledo, R., Muñoz-Antoli, C., Perez, M., and Esteban, J.G. 1999. Encystment sites of metacercariae of *Hypoderaeum conoideum* (Trematoda: Echinostomatidae) in freshwater gastropods. Folia Parasitologica 46:157–158.

Toledo, R., Muñoz-Antoli, C., and Esteban, J.G. 2000. The life-cycle of *Echinostoma friedi* n. sp. (Trematoda: Echinostomatidae) in Spain and a discussion on the relationships within the *'revolutum'* group based on cercarial chaetotaxy. Systematic Parasitology 45:199–217.

Uxmann, J.R., and Hayduk, S.H. 1964. Larval *Echinochasmus* (Trematoda: Echinostomatidae) in rainbow trout, *Salmo gairdneri*. Journal of Parasitology 50:586.

Vanoverschelde, R., and Vaes, F. 1980. Studies on the lifecycle of *Himasthla militaris* (Trematoda: Echinostomatidae). Parasitology 81:609–617.

Velazquez, C.C. 1964a. Life history of *Acanthoparyphium paracharadrii* sp. n. (Trematoda: Echinostomatidae). Journal of Parasitology 50:261–265.

Velazquez, C.C. 1964b. Observations on the life cycle of *Euparyphium paramurinum* sp. n. (Trematoda: Echinostomatidae). Journal of Parasitology 50:106–111.

Viozzi, G., Flores, V., and Rauque, C. 2005. An ectosymbiotic flatworm, *Temnocephala chilensis*, as second intermediate host for *Echinoparyphium megacirrus* (Digenea: Echinostomatidae) in Patagonia (Argentina). Journal of Parasitology 91:229–231.

Waikagul, J. 1991. Intestinal fluke infections in Southeast Asia. Southeast Asian Journal of Tropical Medicine and Public Health 22:158–162.
Winstead, J.T., Volety, A.K., and Tolley, S.G. 2004. Parasitic and symbiotic fauna in oysters (*Crassostrea virginica*) collected from the Caloosahatchee River and estuary in Florida. Journal of Shellfisheries Research 23:831–840.
Wiwanitkit, V. 2005. Motility of intestinal fluke, *Echinostoma* spp. metacercariae in fish dishes prepared by different uncooked methods.The Internet Journal of Infectious Diseases 4.
Xiao, X., Wang, T., Zheng, X., Shen, G., and Tian, Z. 2005. In vivo and in vitro encystment of *Echinochasmus liliputanus* cercariae and biological activity of the metacercariae. Journal of Parasitology 91:492–498.
Yu, S., Hu, L., and Jiang, Z. 1994. Report of the first nationwide survey of the distribution of human parasites in China. 1. Regional distribution of parasite species. Chinese Journal of Parasitology and Parasitic Diseases 12:241–247.
Yurlova, N.I. 1987. Morphology and biology of the trematodes *Hypoderaeum cubanicum* n. comb. and *H. conoideum* (Echinostomatidae). Parazitologiya 21:35–42.

Chapter 4
Echinostomes in the Definitive Host: A Model for the Study of Host–Parasite Relationships

Rafael Toledo

Contents

4.1 Introduction 90
4.2 Host Specificity and Infectivity 90
4.3 Metacercarial Excystment 91
4.4 Establishment in the Definitive Host 93
4.5 Worm Distribution and Habitat Location 96
4.6 Growth and Development of Adult Echinostomes 97
4.7 Mating Behavior of Adult Echinostomes 98
 4.7.1 Chemical Attraction and Mate Finding 99
 4.7.2 Mating Behavior 100
4.8 Fecundity of Echinostomes in the Vertebrate Host 101
 4.8.1 Factors Affecting the Fecundity of Echinostomes 101
 4.8.2 Methods for Measuring the Fecundity of Echinostomes 103
4.9 Concluding Remarks 104
References 105

Abstract Echinostomes are hermaphroditic digeneans that parasitize, as adults, numerous vertebrate hosts of all classes. In this chapter, the main features of the development of adult echinostome in the definitive host are analyzed in detail. The stage of the echinostome in the definitive host undergoes a number of features such as metacercarial excystment, establishment, habitat location, adult development, mate, and release of eggs. These features are documented with an emphasis in their potential use as experimental models. The analysis is focused on members of the genus *Echinostoma*, although members of other genera of the Echinostomatidae are mentioned when they are relevant to the discussion. Although some of the nomenclature for echinostome species is disputed, the names used are those currently accepted names.

R. Toledo (✉)
Departamento de Biología Celular y Parasitología Facultad de Farmacia, Universidad de Valencia Av. Vicent Andrés Estellés s/n, 46100 Burjassot-Valencia, Spain
e-mail: Rafael.Toledo@uv.es

B. Fried and R. Toledo (eds.) *The Biology of Echinostomes*,
DOI: 10.1007/978-0-387-09577-6_4,

4.1 Introduction

Echinostomes are a rather heterogeneous group of cosmopolitan hermaphroditic digeneans that inhabit, as adults, the intestine of a great spectrum of vertebrate hosts, such as birds, mammals and, occasionally, reptiles, and fishes. These hosts become infected after ingestion of the second intermediate host harboring the encysted metacercariae (see Chap. 3). The stage of the echinostomes in the definitive host involves a number of processes such as metacercarial excystment, establishment, habitat location, adult development, mating, and release of eggs. In the present chapter, these features will be analyzed with emphasis on two aspects: their interest in general biology and the characteristics that convert the echinostomes into useful experimental models.

Echinostomes are an ideal model for the study of several aspects of the biology of intestinal helminths since they present a number of advantages such as their large worm size or the facility for the maintenance of their life cycles in the laboratory among other features. For these reasons, they have been used for decades as experimental models. Recently, a number of findings using echinostomes as models have shown that echinostomes may be of great importance for future developments in parasitology, particularly in regard to the study of intestinal helminths–vertebrate host relationships. Herein, these findings will be documented and related to previous knowledge in an attempt to establish a basis for the use of echinostomes as experimental models at this life cycle stage.

4.2 Host Specificity and Infectivity

Echinostomes are cosmopolitan parasites that infect a large number of different warm-blooded hosts, both in nature and the laboratory. The broad host specificity of echinostomes toward the definitive host is the result of phylogenetic, physiological, and ecological accommodations between the parasite and the host in an evolutionary dynamic process (Huffman and Fried 1990; Fried and Huffman 1996). For adult echinostomes, the main factor influencing the host specificity is host behavior, particularly the feeding habits of the vertebrate hosts (Huffman and Fried 1990; Fried and Huffman 1996; Graczyk and Fried 1998; Graczyk 2000). However, physiological factors, such as the characteristics of the digestive system or the host response, are also involved (Toledo and Fried 2005; Toledo et al. 2006a). Echinostomes have been reported in a wide range of species of birds, and domestic and wild mammals such as rodents, pigs, dogs, and foxes. Huffman (2000) provided a detailed list of the natural and experimental definitive hosts for selected species of *Echinostoma*.

However, the ability of a single echinostome species to infect a wide range of vertebrate hosts does not imply that the course of the infections is similar among the different host species. The course and the characteristics of an infection with a single species of echinostome are largely dependent on host-related factors

(Toledo and Fried 2005). This fact has allowed workers to define hosts of high and low compatibility for several echinostome species. The differences between both categories of host species are reflected mainly in worm establishment and growth, egg output, parasite fecundity, and survival of the worms. The comparative study of these features in different host species may be useful to analyze the factors on which depends the course of an intestinal helminth infection.

In highly compatible hosts, such as mice and hamsters, *Echinostoma caproni* infections become chronic, and the parasite can survive for up to 20 weeks postinfection (wpi) (Odaibo et al. 1988, 1989; Christensen et al. 1990; Hosier and Fried 1991; Toledo et al. 2004a; Muñoz-Antoli et al. 2007). In contrast, the worms are rapidly expelled in hosts of low compatibility such as rats (Hansen et al. 1991; Toledo et al. 2004a). Furthermore, significant differences in the egg output and growth of the adult worms have been observed in both types of hosts. In rats, *E. caproni* releases a significantly lower number of eggs than in hamsters and mice during the patent period (Toledo et al. 2004a; Muñoz-Antoli et al. 2007). Moreover, a significant greater worm growth has been observed in hamsters than in rats (Toledo et al. 2004a).

A similar situation has been described for other echinostome species such as *E. friedi* and *E. trivolvis*. Infections in rats with *E. friedi* are characterized by lower infection rate, worm survival, egg output, and growth than in hamsters (Muñoz-Antoli et al. 2004a). In the case of *E. trivolvis*, hamsters become 100% infected, and the infection is retained for at least 120 days postinfection (dpi), whereas the infection is expelled in mice at 2–4 wpi (Franco et al. 1986; Hosier and Fried 1986; Mabus et al. 1988).

Interestingly, host strain-related differences in the infectivity of *E. hortense* also have been observed. Lee et al. (2004) evaluated the susceptibility of five different strains of mice (C3H/HeN, BALB/c, C57BL6, FvB, and ICR) to this echinostome infection on the basis of the worm expulsion rate, worm size, and egg production during the first 8 wpi. It was observed by the authors that *E. hortense* is highly infective to ICR and C3H/HeN mice, but not to the other strains of mice. Based on the results obtained, the authors suggested that the susceptibility to *E. hortense* infection is dependent on the genetic and immunological background of the host.

In summary, the characteristics of the echinostome infections are dependent on the host species. The differences of the infection with an echinostome species in different host species are reflected in several parasitological features, though the origin appears to be on the genetic background of the host and the characteristics of the interactions in each host–parasite combination. The study of the intimate relationships of the echinostome with their vertebrate hosts may be of great utility in the context of the host–parasite relationships.

4.3 Metacercarial Excystment

Echinostome infections of the definitive hosts invariably begin with the excystation of the metacercariae in the digestive tract of its host. As mentioned the metacercarcariae of *Echinostoma* spp. are passively transported with the ingesta to the intestine, where the excystment process is initiated. There are probably more studies on excystment behavior than for any aspect of trematode behavior, and numerous reviews have been written on this topic (Lackie 1975; Sommerville and Rogers 1987; Sukhdeo and Mettrick 1987; Fried 1994; Irwin 1997). The signals that trigger excystation tend to be related to the physicochemical conditions within the intestine, and these may include pH, PCo_2 , enzymes, secretions, and bile (Fried 1994).

Several in vitro studies have provided valuable information on the metacercarial excystment of echinostomes. Although the details vary in terms of the echinostome species studied, the stages in the metacercarial excystment appear to be common for most of the echinostome species studied. Smoluk and Fried (1994) provided arbitrary stages during excystation of *E. trivolvis* cysts in an artificial medium. The following stages were considered:

1. Encysted stage with the organism quiescent and occupying most of the space within the inner cyst.
2. Activated stage with the organism rotating rapidly within the inner cyst.
3. Breached stage with the organism partially released from the inner cyst through an opening in the inner cyst wall.
4. Trapped stage with the organism trapped between the inner and outer cyst wall.

Several media and procedures have been used to study in vitro the excystation of metacercariae of echinostomatids (Howell 1968; Fried and Roth 1974; Irwin et al. 1984). Howell (1968) used a chemical excystation procedure for *Echinoparyphium serratum* including an acid pepsin pretreatment, followed by treatment in a 0.02-M sodium dithionite reductant solution. The percentage of excystation within 10 min was 75%. Other similar procedures have been also used (Kirschner and Bacha 1980; LeFlore and Bass 1982; Bass and Leflore 1984). Fried and Roth (1974) developed a medium to excyst *Parorchis acanthus* that was used thereafter for several species of Echinostomatidae. Similarly, Irwin et al. (1984) used a medium for excystation studies on the echinostomatid *Himasthla leptosoma*. This medium has since been used successfully on a number of nonechinostomatids (Irwin 1997). The composition, preparation, and use of all these media are described in detail in Fried (1994).

The use of these media has shown that excystation depends on a series of factors related to the metacercaria (intrinsic factors) and also to the procedure used for excystation (extrinsic factors). There is some evidence suggesting that metacercariae may play an active role in the excystation processes. The use of the medium of Irwin et al. (1984) results in intense metacercarial activity, and the larva emerges through the cyst wall after 20 min. The juvenile fluke emerges through a small area of the cyst that was devoid of a layer of lamellae present elsewhere toward the inner most surface. The appearance of the ruptured cyst walls indicated that they had been softened by the excystment medium. However, the intense activity of the organism during excystation and the possible release of metacercarial enzymatic factors also suggested an active role of the organism during excystation (Irwin et al. 1984). Moreover, Fried and Emili (1988) showed that excystation of *E. caproni* is higher than *E. trivolvis* using the medium of Fried and Roth, revealing a species-specific response to identical excystation conditions.

Regarding the extrinsic factors, excystation of echinostomes appears to be dependent on a synergistic effect of bile salts and trypsin, whereas acid pepsin pretreatment and reductants or special gas phases do not seem to be required (Fried and Emili 1988; Fried 1994). Metacercariae of *E. trivolvis* excyst using the Fried and Roth medium, but excystation does not occur in the absence of either the bile salts or the trypsin (Fried and Butler 1978). The alkaline pH and the high temperature were necessary for a high rate of excystation within a reasonable time (Fried and Emili 1988). Other authors have also shown that a number of factors such as acid pepsin treatment or the sodium dithionite reductant solution are not essential for excystment but increase the rate of excystment (Howell 1970; Mohandas and Nadakal 1978; Fried et al. 1997).

Moreover, there are a number of other factors that also may affect the excystation of echinostome metacercariae. Rossi et al. (2001) analyzed the effect of prolonged storage on the excystation of *E. caproni* metacercariae. Storage for 5 months in Locke's solution significantly diminished the metacercarial ability to excyst and to infect a vertebrate host. Cysts stored for more than 8 months did not excyst. Fried and Peoples (2007a) showed that excystation of *E. caproni* metacercariae was strongly affected by several physical and chemical factors such as vinegar treatment and boiling for 1 min. They also showed that concentration of potassium permanganate from 300 to 1,200 mg/l killed the cysts within 5 min.

In vivo excystment of echinostomes has been poorly studied. Fried and Kletkewicz (1987) studied the excystation in vivo of *E. trivolvis* in chicks. They fed domestic chicks large number of cysts and necropsied the hosts at intervals from 15 min to 24 h later. The authors also evaluated the viability of the cysts retained in the gizzard in an excystation medium and found that a sojourn in the gizzard for 3 h made the cysts nonviable. Thus, the long-term treatment of the cysts in an acid environment was detrimental. Moreover, they detected that intestinal emptying time influenced the rate of excystation in the host. A similar experiment was performed by Fried et al. (2001) using the combination *E. caproni*-ICR mice. Excystation of this echinostome occurred in the stomach and the anterior part of the small intestine. By 4 h the metacercariae had either excysted or were voided. Excysted metacercariae were widely scattered throughout the small intestine, with about 75% located in the duodenal-jejunum zone.

4.4 Establishment in the Definitive Host

After excystment, *Echinostoma* spp. remain as browsers in the intestine. However, ectopic parasitism in the bile duct, pancreas, gall bladder, and liver has been described in hamsters experimentally infected with *E. trivolvis* (Huffman et al. 1988). The infectivity and establishment of echinostomes vary greatly for each host–parasite combination, which makes it difficult to elaborate generalizations and to define rules. Furthermore, there are a number of factors such as infective doses, immunological status and age of the host, and coinfection with other helminths that also influence the course of the infection.

Several studies have analyzed the establishment of echinostomes in avian hosts (Beaver 1937; Senger 1954; Fried and Weaver 1969; Fried and Butler 1978; Fried and Alenick 1981; Fried 1984; Fried and Freeborne 1984; Fried et al. 1988; Kim and Fried 1989; Humphries et al. 1997). These studies have shown that some birds, such as domestic chicks, can serve as suitable experimental hosts for several echinostome species. Fried (1984) reported 100% infectivity with *E. trivolvis* cysts and a worm recovery of 25% in domestic chicks. Survival of adult worms in chicks was reported to be of at least 44 dpi. This is in contrast with other studies in which the infectivity of *E. trivolvis* in chicks was of 83–85% and the worm recovery of 15–22% (Simpkins and Fried 1999; Mullican et al. 2001). The infectivity and the worm recovery of *E. revolutum* in chicks seem to be less than those of *E. trivolvis*, ranging from 64 to 67% and 9 to 32%, respectively (Humpries et al. 1997; Mullican et al. 2001). The values for *E. caproni* in chicks are 100% for infectivity and 24% for worm recovery (Mullican et al. 2001).

The use of rodents as experimental hosts of *Echinostoma* spp. has served to increase our knowledge on these topics. As mentioned, several studies have allowed workers to define rodent species of high and low compatibility with each echinostome species (Toledo and Fried 2005). These categories of hosts are differentiated on the basis of the rates of establishment and the survival of the worms among other features. This fact is useful to study the factors determining the course of the infection.

The course of *E. caproni* infections has been studied in detail in several rodent species. Hamsters and mice are considered as highly compatible hosts, whereas rats and jirds display a low degree of compatibility with the parasite. The rates of infection of *E. caproni* in hamsters and mice are almost 100% independent of the infective doses employed (Odaibo et al. 1988, 1989; Isaacson et al. 1989; Christensen et al. 1990; Balfour et al. 2001; Toledo et al. 2004a; Fried and Peoples 2007b; Muñoz-Antoli et al. 2007). Moreover, the worm recovery over the course of the infection in both host species can be considered high. Christensen et al. (1990) studied the survival of *E. caproni* in hamsters during the first 13 wpi and reported that worm burden remained stable during the observation period. Toledo et al. (2004a) extended this observation up to 20 wpi. *E. caproni* infections in mice show a persistence that may reach 23 wpi (Odaibo et al. 1988, 1989; Fried and Peoples 2007c; Muñoz-Antoli et al. 2007). However, the worm burden is not uniform during this period. Muñoz-Antoli et al. (2007) observed a progressive decrease of the worm recovery from 5 wpi. In fact, Fried and Peoples (2007c) detected that only two of the five mice exposed to metacercariae of *E. caproni* remained infected at 23 wpi. Interestingly, no signs of atrophy were observed in the adult worms collected at 23 wpi.

Rats are considered as hosts of low compatibility with *E. caproni*. In a comparative study, Toledo et al. (2004a) showed that 100% of rats experimentally exposed to metacercariae of *E. caproni* became infected, though the weekly worm recovery was significantly lower than in hamsters. Moreover, all the worms were expelled from rats at 7–8 wpi (Hansen et al. 1991; Toledo et al. 2004a).

In the case of *E. trivolvis*, the compatibility with hamsters appears to be greater than with mice. Franco et al. (1986) reported 100% infection of hamsters with *E. trivolvis*, and the worm recovery averaged 38%. Mabus et al. (1988) showed that hamsters remained infected for 123 dpi. Although ICR and Swiss Webster mice can be infected with metacercariae of *E. trivolvis*, the worm recovery rapidly decreases (Hosier and Fried 1986). The parasite was eliminated at 4 wpi in both strains of mice (Hosier and Fried 1986).

A similar situation occurs with *E. friedi*. This echinostome species is able to infect different rodent species and, even, birds (Toledo et al. 2000). However, *E. friedi* shows a higher degree of compatibility with hamsters than with rats. The rate of infection in rats (48%) is significantly lower than in hamsters (100%) (Toledo et al. 2003, 2006b; Muñoz-Antoli et al. 2004a). Moreover, the infection is expelled at 4 wpi in rats, whereas the longevity of the worms in hamsters is of at least 12 wpi (Toledo et al. 2003). Interestingly, no differences in the worm recovery between both host species were observed during the first 4 wpi (Muñoz-Antoli et al. 2004a).

Recovery rates for *E. paraensei* were reported by Meece and Nollen (1996) to be 15% in mice. This low recovery rate appeared to be due to elimination of worms after 14 dpi. In contrast, worm recovery rates in hamsters were significantly higher (37%).

Factors causing the earlier expulsion of worms in hosts of low compatibility are difficult to elucidate. In this sense, differences in the local immune responses, including inflammatory responses, could be of great importance (Toledo et al. 2006a, 2006c). These aspects are reviewed in detail in Chap. 8. However, there are a number of other factors that also may influence the determination of the course of the echinostome infections in the definitive host.

Increased dosage levels influence several aspects of the worm development in the host intestine such as the parasite maturation, growth, or total weight gain (Huffman and Fried 1990; Fried and Huffman 1996). Regarding the effect of increased metacercarial doses on the rate of establishment, the available data are somewhat confusing. Several authors have indicated that primary worm establishment of *E. caproni* is infection-dose independent (Odaibo et al. 1988; Christensen et al. 1990; Balfour et al. 2001). In contrast, Yao et al. (1991) reported that the percentage of worm recovery of *E. caproni* in hamsters was significantly higher (63%) using 15 metacercariae/hamster than using 50 or 200 metacercariae/hamster (21 and 23%, respectively). The authors suggested that the negatively dose-dependent pattern could be related with the phenomenon of competition. However, this topic should be explored further.

Concurrent infections with other helminths may affect the rate of establishment of echinostomes in the definitive host. Iorio et al. (1991) demonstrated that concurrent infections of *E. caproni* and *E. trivolvis* in ICR mice increased the worm recovery rates of both echinostome species. The worm recovery of *E. caproni* increased from 38 to 73% in single and concurrent infections, respectively. The worm burden of *E. trivolvis* was of 6% in single infections versus 14% in concurrent infections. *E. paraensei* showed an increase in infectivity from 15 to 30% in concurrent infections with *E. caproni* in mice (Meece and Nollen 1996). In contrast, *E. caproni* recovery rates were much less in concurrent infections than in single infections (34% in concurrent versus 61% in single infections). Muñoz-Antoli et al. (2004b) reported that the recovery rate of the echinostomatid *Euparyphium albuferensis* was not affected by coinfection with *E. friedi*, whereas the recovery rate of this latter species increased from 8 to 17% in the concurrent infections. Furthermore, it has been shown that primary infections with the nematode *Nippostrongylus braziliensis* may have a dramatic effect on the establishment of *E. caproni* and *E. trivolvis* in mice (Fujino et al. 1996). All the *E. caproni* and *E. trivolvis* metacercariae used to challenge mice 8 dpi after primary infection with *N. braziliensis* were expelled. In contrast, considerable numbers of *E. caproni* (67%) and *E. trivolvis* (49%) remained in the intestine of hosts challenged on day 16 after primary infection with *N. braziliensis*. Interestingly, the expulsion of the worm coincided with a peak of goblet cells in mice infected with *N. braziliensis* (Fujino et al. 1996).

The nature of the changes induced on the worm recoveries due to concurrent infections is difficult to evaluate, though histopathological and immunological alterations induced by the coinfections appear to be involved. In this sense, it should be noted that dexamethasone treatment of C3H/HeN mice infected with *E. trivolvis* caused inhibition of worm expulsion (Fujino et al. 1997). However, these topics are analyzed in detail in Chap. 8.

4.5 Worm Distribution and Habitat Location

Relatively few studies have considered how echinostomes select their habitat within the vertebrate hosts. In fact, it is not known what mechanisms and host cues govern this orientation. However, it is well established that echinostomes have specific niches within the small intestine of their definitive hosts. Echinostomes are typically dispersed in the initial phases of the infection (Nollen 1996a, b) and, thereafter, they become confined to a small area of the intestine. Reviews by Huffman and Fried (1990) and Fried and Huffman (1996) report the distribution of *E. caproni* and *E. trivolvis* in the domestic chick. *E. trivolvis* occupies numerous sites in the intestine such as the ileum, rectum, cloaca, cecum, and the bursa of Fabricius. Mucha et al. (1990) reported that *E. trivolvis* in experimentally infected ducks (*Anas platyrhynchus*) was located in the cloaca and lower ileum. Humphries et al. (1997) and Fried et al. (1997) detected that *E. revolutum* occurred mainly in the posterior part of the intestine of chicks. Huffman (2000) examined the distribution of *E. caproni*, *E. trivolvis*, and *E. revolutum* in the intestine of domestic chickens. *E. caproni* was located in the midthird of the intestine, between the gizzard and the cloaca. Both *E. trivolvis* and *E. revolutum* were found more posteriorly, near to the cloaca, but typically *E. revolutum* was located more posteriorly than *E. trivolvis*.

The distribution of several species of echinostomes in the intestine of rodents also has been well established. *E. caproni* and *E. paraensei* show great specificity for the ileum and the duodenum, respectively (Odaibo et al. 1988, 1989; Yao et al. 1991; Meece and Nollen 1996; Huffman 2000). On the other hand, *E. trivolvis* can be found clustered, but spread along the entire intestinal tract (Hosier and Fried 1986; Huffman 2000). *E. friedi* occupies mainly the jejunum (Toledo et al. 2000).

Although this is the general pattern, several factors may alter the distribution of the worms. Firstly, the distribution may be subjected to changes over time and, secondarily, the coinfection with other echinostome species may affect the normal distribution of the worms.

The location of *E. paraensei* in mice and hamsters over time was studied by Meece and Nollen (1996). Adult worms were recovered throughout the entire length of the small intestine until 17 dpi. Thereafter, they were found clustered at the pylorus-duodenum surface with some worms extending into the stomach. After 20 dpi, the worms were again distributed throughout the small intestine.

Barus et al. (1974) demonstrated how in a double infection with *E. revolutum* and *Echinoparyphium recurvatum*, the final distribution of both fluke species in the intestine of the definitive host was the result of antagonistic interactions between both parasite species. Several worm distribution patterns have been observed in echinostome coinfections. Iorio et al. (1991) found that worm distribution was not affected in *E. caproni–E. trivolvis* coinfections in mice. Similarly, the normal distribution of *E. albuferensis* and *E. friedi* was not changed in coinfections in hamsters (Muñoz-Antoli et al. 2004b). This is in contrast with the situation described by Fried et al. (1997) for *E. revolutum* and *E. trivolvis* infections in chicks. In single infections, *E. revolutum* was found in the rectum, whereas *E. trivolvis* was in the lower ileum. In concurrent infections, both were found in the rectum.

4.6 Growth and Development of Adult Echinostomes

The growth and the development of preovigerous echinostomes have been studied in detail for *E. caproni* (Fried et al. 1988; Manger and Fried 1993; Fried and Huffman 1996). Excysted metacercariae are nonprogenetic and contain only genital anlage. Preovigerous adults from rodent or chicks show distinct testes from 2 to 3 dpi, an ovary distinct from the ootype by 4 dpi, and coiling of the uterus by day 5; the vitellaria are present by 6 dpi and the worms become ovigerous by 7–8 dpi. During development from the excysted metacercariae to ovigerous adult in rodents and chicks, the worm body area may increase some 80 times (Fried et al. 1988; Manger and Fried 1993; Fried and Huffman 1996).

The main features of the development of ovigerous adults over time have been studied for *E. caproni*, *E. trivolvis*, and *E. friedi* (Franco et al. 1988; Odaibo et al. 1988; Yao et al. 1991; Humphries et al. 1997; Toledo et al. 2003, 2004a; Muñoz-Antoli et al. 2004a, 2007; Stillson and Platt 2007). To this purpose, different methodologies and morphological markers have been used. In general, most of the morphometrical variables show a rapid increase during the first 2–3 wpi. Thereafter, the values continue the increase at a lower rate or even become stable.

Probably of particular interest are the works in which the growth of the echinostomes is studied in relation to other variables. In this context, echinostomes have served for studies on the influence of host species and the population density on the growth of digenetic trematodes.

Because most echinostomes can develop in different host species displaying different degrees of compatibility, they are good subjects for studies on the effect of host species on the development of digeneans. Toledo et al. (2004a) made a comparative study on the development of *E. caproni* in hosts of high (hamster) and low (rat) compatibility with the parasite. This study showed that the host species has a dramatic effect on most of the morphometrical features analyzed. The values were significantly greater in hamsters than in rats. Furthermore, time-host species interactions were detected for several variables, indicating that the kinetics of worm growth in each host species is different. The curves for all these variables showed a different pattern in hamsters and rats. Similar results were reported for *E. friedi* in the hamster (high compatible host) and the rat (low compatible host) (Muñoz-Antoli et al. 2004a).

As mentioned, echinostomes also have served for studies on the crowding effect in trematodes. Franco et al. (1988) studied the effect of crowding on adults of *E. trivolvis* in hamster. Their study reported that increased dosage levels and thus infrapopulation size influences several aspects of worm development, including a delay in maturation. Yao et al. (1991) found that *E. caproni* adults in golden hamsters fed a large number of cysts (200 cysts/hamster) were stunted compared with worms obtained from hosts only fed 15 cysts per host. Balfour et al. (2001) reported that *E. caproni* adult worms in mice infected with 100 metacercariae showed lower wet and dry weight than those collected from mice infected with 25 metacercariae.

Recently, Stillson and Platt (2007) made one of the most interesting approaches to the effect of population density or crowding on the morphometrical variability of *E. caproni* in mice. The hosts were infected with 25, 100, and 300 metacercariae each, and a total of 31 morphometrical variables (25 direct measurements and 6 ratios) were evaluated at 22 dpi. Univariate and multivariate statistical analyses revealed significant differences between worms from all three groups. A total of 27 characters showed significant intragroups differences, with the primary differences between worms from 25/100 versus 300 cysts infection. In general, there was an inverse relationship between inoculums size and worm size as described in previous studies. Reproductive structures were most sensitive to crowding effect. This fact had been previously reported (Yao et al. 1991; Balfour et al. 2001; Toledo et al. 2004a; Muñoz-Antoli et al. 2004a). Future studies on the effect of population size on the development of echinostomes may be of great interest for further understanding of host–parasite relationships in helminth infections.

4.7 Mating Behavior of Adult Echinostomes

Due to their availability and easy handling in the laboratory, echinostome species are favorite subjects for reproductive studies of digenetic trematodes. This is of great importance since reproductive strategies constitute a major factor shaping the evolution of organisms. Indeed, mating systems influence the genetic variability as well as the genetic structure of populations (Jarne 1995). In this context, hermaphroditic organisms, such as echinostomes, offer the possibility to analyze the evolution of several reproductive modes since a single individual can self- or crossfertilize. Echinostomes are hermaphroditic trematodes that undergo sexual reproduction in the intestine of their vertebrate host. The structure and the development of the reproductive system of echinostomes, together with processes such as oogenesis, vitellogenesis, spermiogenesis, and egg formation, have been reviewed in detail by Nollen (2000). The present section will focus on functional aspects of the reproduction of echinostomes such as chemoattraction of adults and mating behavior.

4.7.1 Chemical Attraction and Mate Finding

In some species of echinostomes adult worms are confined to a small area of the host's intestine, but in other species the adults may be spread out. Thus, interindividual communication is important to enhance pairings with other individuals for maintaining a viable species. In vitro studies showed that the worms find each other and that the nature must be chemical communication among the individuals (Fried 1986; Haseeb and Fried 1988). Several efforts have been made to investigate these chemical cues, and the excretory/secretory products appear to be involved in this process. Fried and Haseeb (1990) found that the compounds which stimulate the worm-to-worm attraction in excretory/secretory products were found to be lipophilic and the lipid fraction containing the free sterols. Trouve and Coustau (1998) investigated the differences in excretory/secretory products of three strains of *E. caproni* by electrophoresis. They found that although the three strains shared most of polypeptide bands, a few were strain specific. The authors suggested that this difference could be of great importance in evolutionary processes such as assortative mating or local host adaptation.

More information is needed on the nature of the attracting components and on the orientation mechanisms guiding the worms together, particularly considering that the available data are somewhat confusing. Using agar-containing petri dish cultures, Fried and Haseeb (1990) investigated intra- and interspecies chemoattraction between *E. caproni* and *E. trivolvis* adults. *E. trivolvis* showed significantly greater intraspecific attraction that *E. caproni*. Trouve and Coustau (1999) analyzed the chemical communication and mate attraction between two geographical isolates of *E. caproni* (Madagascar and Egypt) and another species of the same genus, *Echinostoma* sp. The results showed that echinostomes attracted each other and tended to pair under in vitro conditions, suggesting that this pairing can occur in the absence of external factors such as intestinal factors. However, echinostomes appear to lack a marked mate preference. The mate attraction was similar for any intra-, interisolate, or interspecific combination.

4.7.2 Mating Behavior

Echinostomes can either self- or crossinseminate depending on the situation. The species studied (*E. caproni*, *E. trivolvis*, and *E. paraensei*) self-inseminate in single infections with resulting viable eggs (Fried et al. 1990; Nollen 2000). In multiple infections, they cross- and self-inseminate (Nollen 2000). The act of copulation has been rarely directly observed, but the mating behavior was indirectly determined by sperm radiolabeling, transplantation of labeled worms, and subsequent autoradiography (Nollen 1997a, 2000). In general, the mating barriers in echinostomes appear to be weak, but individuals can identify partners of their own species. In concurrent infections, they show interspecies mating, which does not occur when individuals of the same species are available (Nollen 2000).

The mating patterns of echinostomes in multiple infections have been the subject of several investigations. The three species studied (*E. caproni*, *E. trivolvis*, and *E. paraensei*) showed an unrestricted pattern with cross- and self-insemination with other species present in the infection (Nollen 1990, 1993, 1996a, b). However, the percentage of cross- and self-insemination varied widely among the three species.

The cross- and self-insemination between different strains of *E. caproni* was studied by Trouve et al. (1996). Infections of mice with two individuals from two different isolates of *E. caproni* (originated from different geographic areas) showed that the parasite exhibit an unrestrictive mating pattern including both cross- and self-insemination. Moreover, in a mate-choice experiment in which mice were infected with three individuals, two of the same isolate and one of another isolate, it was found that both related individuals produced very few outcrossed offspring with the unrelated individual (Trouve et al. 1996). Although it was not possible to distinguish selfed from outcrossed offspring, this result suggests a marked preference of *E. caproni* between individuals of the same isolate. This prezygotic isolation seems to be followed by a postzygotic isolation characterized by hybrid breakdown. Indeed, it has been shown that the hybrids of the second and third generations display a significant lower fecundity compared to both parental isolates and to the F1 (i.e., hybrid breakdown) (Trouve et al. 1998).

Further investigation on this topic was made by Trouve et al. (1999). These authors showed that in mice infected with two individuals of the same isolate of *E. caproni* there were no differences between intra- and interisolate selfing rates. In triple infections, where two of the three individuals originated from the same isolate and the third from another isolate, a preferential outcrossing between individuals from the same isolate was observed. This latter fact may be useful to avoid hybrid breakdown. These results emphasize the important and synergistic roles of selfing, inbreeding depression, and hybrid breakdown in the evolution of echinostome reproductive strategies (Trouve et al. 1999).

Interspecies mating also has been studied in echinostomes. The first of these studies investigated the mating behavior between *E. caproni* and *E. paraensei* in concurrent infections in mice (Nollen 1996b). No interspecies mating was found, showing the strong mating barriers between these species. Nollen (1997b) looked at mating behavior between *E. caproni* and *E. trivolvis* in hamsters. Interspecies mating only was detected when *E. caproni* was the sperm donor. However, the percentage of interspecific mating was very low. Nollen (1999) analyzed the mating behavior of *E. trivolvis* and *E. paraensei* in concurrent infections in hamsters. Results of this study showed that interspecific mating occurred utilizing *E. trivolvis* as the sperm donor, though in a very low percentage. Further interspecific mating studies are needed since the possibility of producing hybrids between different echinostome species has not yet been pursued.

4.8 Fecundity of Echinostomes in the Vertebrate Host

Much of the interest in describing and modeling the population dynamics of helminths has concerned parasite fecundity in the definitive host. In this context, the easy handling of the echinostomes in the laboratory makes them excellent models for the study of the fecundity of helminths and the ability of different hosts to transmit a parasite species. Several studies have been conducted to develop quantitative methods for the measurement of these topics using echinostomes.

Quantitative approximations to the reproductive output of the echinostomes have been made on the basis of uterine egg counts (Christensen et al. 1990), eggs per gram of feces (Odaibo et al. 1988, 1989), and total amount of eggs in the feces of the host (Mahler et al. 1995; Muñoz-Antoli et al. 2004a, 2007; Toledo et al. 2003, 2004a). However, the analysis of reproduction on the basis of egg output constitutes only a partial measurement of the parasite fecundity. In this section, we analyze the factors that affect the fecundity of echinostomes and the methods developed to quantify the fecundity of these parasites in the definitive host.

4.8.1 Factors Affecting the Fecundity of Echinostomes

Adult worms of echinostomes may be ovigerous from 7 to 8 dpi. However, the first eggs in feces may appear later. The length of time for an echinostome species to begin the release of eggs in feces of the host generally ranges from 8 to 15 dpi (Huffman 2000). However, the prepatent period of an echinostome can be delayed depending on the host species. Muñoz-Antoli et al. (2004a) showed that the prepatent period of *E. friedi* is significantly longer in a host of low compatibility (rat) than in a highly compatible host (hamster).

During the patent period, the egg output of echinostomes depends on a number of factors such as echinostome species, population density, age of infection, and host species. The number of eggs produced is also characteristic of each echinostome species. However, this topic has been poorly studied. The egg production of *E. caproni* and *E. trivolvis* in hamsters was studied by Huffman (2000). In this study, two groups of hamsters were each infected with 25 metacercariae of either *E. caproni* or *E. trivolvis*. The results showed that *E. trivolvis* produced significantly more eggs than did *E. caproni*, with fecal egg counts from hamsters infected with *E. trivolvis*. Mahler et al. (1995) estimated the total number of eggs in the uterus of *E. caproni* in hamsters. The total number increased from around 600 at 15 dpi to over 1,700 by 50 dpi. Thus, it takes 2 days to fill the uterus. The peak of egg production would be approximately 855 per day (Mahler et al. 1995; Nollen 2000).

Population density strongly affects the egg production of echinostomes. In this sense, Christensen et al. (1990) detected that the number of eggs in the uterus of adult worms of *E. caproni* seems to be negatively infection dose dependent. In contrast, the number of eggs released in the feces is directly related to the infective dosage and, consequently, with the number of adult worms established in the intestine (Odaibo et al. 1988; Mahler et al. 1995).

The kinetics of egg release is similar for all the echinostome species studied. Egg release rapidly increases during the first weeks of infection to reach a maximum. Thereafter, the values become stable for a period and, finally, the egg release steadily declines (Odaibo et al. 1988, 1989; Christensen et al. 1990; Mahler et al. 1995; Toledo et al. 2003, 2004a; Muñoz-Antoli et al. 2004a, 2007). The decline of egg release has been often attributed to a decrease in the number of worms releasing eggs due to mortality. However, there is some evidence suggesting the involvement of other factors. The egg output of a parasite is the result of two separate influences: (1) number of reproducing parasites after the effect of host mortality; and (2) age-specific changes in the fecundity of the surviving worms (Whitfield et al. 1986). The results obtained by Toledo et al. (2003) suggest that the decline in egg output observed in *E. friedi* reflects functional changes in the adult worms that affect their fecundity. These changes could be related to worm age and/or the host response. Toledo et al. (2003) detected a progressive decline in egg output from 3 to 4 wpi, though neither worm mortality nor decline in the worm recovery was observed. Furthermore, this observation was also supported by the dynamics of viable eggs released, which does not depend on the number of surviving worms. The percentage of viable eggs of *E. friedi* produced steadily decreased in the last weeks of the study, though the worm recovery remained stable (Toledo et al. 2003).

Several studies have shown that the host species plays an important role in the egg output of echinostomes. In general, the egg output of an echinostome species is greater in a highly compatible host than in a host of low compatibility. Several studies illustrate this situation. Mahler et al. (1995) showed that the number of *E. caproni* eggs released is significantly greater in hamsters than in jirds. Toledo et al. (2004a) compared the number of eggs of *E. caproni* released in experimentally infected hamsters and rats during the course of the infection and concluded that the number of eggs released is time postinfection and host species dependent. The kinetics of egg release were similar in both host species, though the number of eggs released was significantly greater in hamsters. In the case of *E. caproni*, the lower number of eggs released in rats was correlated with a lower value of worm recovery in this host species. However, the results obtained with *E. friedi* suggest that the host species affects the functional processes of echinostome adult worms related to their capacity to produce eggs. Muñoz-Antoli et al. (2004a) found that the weekly worm recovery of *E. friedi* in hamsters and rats was similar during the first 4 wpi (period of survival of *E. friedi* in rat). Despite this fact, the number of eggs released weekly was significantly lower in rats than in hamsters indicating that the egg release of an echinostome species is host species dependent and is not only related to the number of adult worms established in each host species.

4.8.2 Methods for Measuring the Fecundity of Echinostomes

As shown earlier, the fecundity of echinostomes is the result of a complex set of interrelating factors. This makes it difficult to make estimations of the reproductive output of an echinostome in a host species and also for any other intestinal helminth. The approximations on the basis of egg counts may not reflect the dynamics of the reproductive output in trematode infections since other parameters should be taken into account (Whitfield et al. 1986). This suggests the need for further studies on the reproductive output and the standardization of the experimental procedures and criteria for elucidating the parasite fecundity. In this context, echinostomes may be of great utility.

In an attempt to complete and to standardize the studies on the parasite fecundity in echinostome infections, Mahler et al. (1995) defined the reproductive capacity of *E. caproni* in a particular host species considering the percentage of parasite establishment, the survival of adult worms, and the egg production rates. Using these parameters, the authors showed that the reproductive capacity of *E. caproni* in hamsters is significantly greater than in jirds. This method may be useful, but it cannot constitute a measure of the reproductive success since the total egg production is considered and only the output of viable eggs should be considered. In this context, Toledo et al. (2003) formulated a simple method to describe the population dynamics of *E. friedi* in hamsters from metacercariae to viable eggs. Using the concepts developed by Whitfield et al. (1986), Toledo et al. (2003) calculated the Reproductive Success of *E. friedi* defined as the total number of viable eggs produced by the cohort of adult worms infecting a host per metacercariae at which the host was exposed. This parameter takes into account all the variables that, according to Whitfield et al. (1986), influence the reproductive output of a digenean trematode in the definitive host. Furthermore, Toledo et al. (2003) developed the concept of Weekly Reproductive Success that may be defined as the total number of viable eggs produced each week per metacercariae at which the host was exposed. The results obtained showed that the reproductive success of *E. friedi* in hamsters is not constant over time attaining its maximum at 4 wpi. This variation seems to be related to changes in the egg output and the viability of the eggs produced over time. Moreover, the authors suggested that the transmission of *E. friedi* in hamsters is only viable from 3 to 8 wpi.

This methodology provides a framework for measuring the reproductive success of an echinostome throughout the phase of the life cycle from metacercaria to egg and its variations over the course of the infection. However, the parameters used by Toledo et al. (2003) only give a partial view of the ability of a host species to transmit an echinostome (or other helminth) to the next host and, thereby, to maintain its suprapopulation. Transmission success of an echinostome is also determined by other factors, i.e., the infectivity of the miracidia produced and its variations over time. In this sense, Toledo et al. (2004b) investigated the effect of aging of *E. friedi* adult worms on the miracidia yielded. Miracidia were obtained after hatching of eggs collected from adults of different ages. Miracidial infectivity, measured in terms of percentage of infection in *Lymnaea peregra*, was significantly influenced by the age of the adult worms from which the miracidia were derived. Infective miracidia only were obtained from adult worms in the age range of 4–9 wpi, with a maximal infectivity in the miracidia from adults of 8–9 wpi. This observation shows that the miracidial infectivity should be considered to measure the capacity of a host to transmit an echinostome species to the next host in its life cycle.

Using *E. friedi* as an experimental model, Toledo et al. (2006b) developed the concept of Experimental Transmission Success defined as the number of hosts B that became infected after exposure to a number of infective stages derived from a host A per unit of inoculation at which host A was exposed. In the case of echinostomes, this concept can be simply defined as the number of snail first intermediate hosts that become infected per unit of inoculation (metacercariae) at which the definitive host was exposed. Toledo et al. (2006b) calculated the Experimental Transmission Success of *E. friedi* in hamsters and rats experimentally infected. The results of this study showed that *E. friedi* is better adapted to successfully pass through this phase of its life cycle when using hamsters as final host than rats. The advantages of hamsters as final host are the result of a greater life span, egg output, and viable egg production which resulted in an Experimental Transmission Success of 91:1 with respect to rats. Moreover, Toledo et al. (2006b) calculated the Experimental Transmission Success of *E. friedi* for each week of the infection in both host species (Weekly Experimental Transmission Success). The results suggested that although the maturation of *E. friedi* in hamsters is slower, they are able to transmit *E. friedi* for a longer period than rats and with higher Experimental Transmission Success values. The application of Experimental Transmission Success allows for the quantification of rates under experimental conditions of transmission of a helminth in a definitive host and to estimate the contribution of a host species in the maintenance of a parasite population. Moreover, this method allows comparative studies between different hosts of a particular echinostome species to improve the efficacy in life cycle maintenance in the laboratory. The development of the Experimental Transmission Success concept constitutes an example of how echinostomes may serve as experimental models in the study of population dynamics of helminths.

4.9 Concluding Remarks

In the present chapter, the main features of the course of echinostome infections in the definitive host are analyzed. Although there are many aspects of these infections that should be explored further, the information presented indicates that echinostomes can be exploited further with regard to the study of relationships between vertebrate hosts and adult parasites. As shown earlier, the course of echinostome infections is dependent on host- and parasite-related factors. This fact, together with the easy maintenance of the echinostome life cycles in the laboratory, makes these digeneans excellent tools for the study of host- and parasite-dependent variables determining the course of intestinal helminth infections and their consequences in the other counterparts of the host–parasite combination. In this context, research with echinostomes holds great promise in the search for factors determining the course of intestinal helminth infections.

References

Balfour, C.D., Rossi, M., and Fried, B. 2001. Effects of a 100 metacercarial cyst inoculum on the host–parasite relationship of *Echinostoma caproni* and ICR mice. Journal of Helminthology 75: 321–324.

Barus, V., Moravec, F., and Rysavy, B. 1974. Antagonistic interaction between *Echinostoma revolutum* and *Echinoparyphium albuferensis* (Trematoda) in the definitive host. Folia Parasitologica 21: 155–159.

Bass, P.F., and LeFlore, W.B. 1984. *In vitro* excystment of the metacercaria of *Acanthoparyphium spinulosum* (Trematoda: Echinostomatidae). Proceedings of the Helminthological Society of Washington 51: 149–153.

Beaver, P.C. 1937. Experimental studies on *Echinostoma revolutum* (Froelich) a fluke from birds and mammals. Illinois Biological Monographs 15: 1–96.

Christensen, N. Ø., Simonsen, P.E., Odaibo, A.B., and Mahler, H. 1990. Establishment, survival and fecundity of *Echinostoma caproni* (Trematoda) infections in hamsters and jirds. Journal of the Helminthological Society of Washington 57: 104–107.

Franco, J., Huffman, J.E., and Fried, B. 1986. Infectivity, growth and development of *Echinostoma revolutum* (Digenea: Echinostomatidae) in the golden hamster, *Mesocricetus auratus*. Journal of Parasitology 72: 142–147.

Franco, J., Huffman, J.E., and Fried, B. 1988. The effects of crowding of adults of *Echinostoma revolutum* (Digenea: Echinostomatidae) in experimentally infected golden hamsters, *Mesocricetus auratus*. Journal of Parasitology 74: 240–243.

Fried, B. 1984. Infectivity, growth and development of *Echinostoma revolutum* (Trematoda) in the domestic chick. Journal of Helminthology 58: 241–244.

Fried, B. 1986. Chemical communication in hermaphroditic digenetic trematodes. Journal of Chemical Ecology 12: 1659–1677.

Fried, B. 1994. Metacercarial excystment of trematodes. Advances in Parasitology 33: 91–143.

Fried, B., and Alenick, D.S. 1981. Localization, length and reproduction in single- and multiple worm-infections of *Echinostoma revolutum* (Trematoda). Parasitology 82: 49–53.

Fried, B., and Butler, M.S. 1978. Infectivity, excystation and development on the chick chorioallantois of the metacercaria of *Echinostoma revolutum* (Trematoda). Journal of Parasitology 64: 175–177.

Fried, B., and Emili, S. 1988. Excystation in vitro of *Echinostoma liei* and *E. revolutum* (Trematoda) metacercariae. Journal of Parasitology 74: 98–102.

Fried, B., and Freeborne, N.E. 1984. Effects of *Echinostoma revolutum* (Trematoda) adults on various dimensions of the chicken intestine, and observations on worm crowding. Proceedings of the Helminthological Society of Washington 51: 297–300.

Fried, B., and Haseeb, M.A. 1990. Intraspecific and interspecific chemoattraction in *Echinostoma caproni* and *Echinostoma trivolvis* adults *in vitro*. Journal of the Helminthological Society of Washington 57: 72–73.

Fried, B., and Huffman, J.E. 1996. The biology of the intestinal trematode *Echinostoma caproni*. Advances in Parasitology 38: 311–368.

Fried, B., and Kletkewicz, K. 1987. Excystation of *Echinostoma revolutum* metacercariae in the domestic chick. Proceedings of the Helminthological Society of Washington 54: 267–268.

Fried, B., and Peoples, R.C. 2007a. Effects of various physical and chemical factors on excystation of the encysted metacercariae of *Echinostoma caproni*. Journal of Parasitology 93: 1388–1391.

Fried, B., and Peoples, R.C. 2007b. Effects of a 300-metacercarial cyst inoculum on worm recovery and crowding of *Echinostoma caproni* in Balb/c mice. Parasitology Research 101: 1701–1702.

Fried, B., and Peoples, R.C. 2007c. Longevity of *Echinostoma caproni* in Balb/c mice. Journal of Helminthology 82: 95–96.

Fried, B., and Roth, M. 1974. In vitro excystment of the metacercariae of *Parorchis acanthus*. Journal of Parasitology 60: 465.

Fried, B., and Weaver, L.J. 1969. Exposure of chicks to the metacercaria of *Echinostoma revolutum* (Trematoda). Proceedings of the Helminthological Society of Washington 36: 153–155.

Fried, B., Donovick, R.A., and Emili, S. 1988. Infectivity, growth and development of *Echinostoma liei* (Trematoda) in the domestic chick. International Journal for Parasitology 18: 413–414.

Fried, B., Huffman, J.E., and Weiss, P.M. 1990. Single and multiple worm infections of *Echinostoma caproni* (Trematoda) in the golden hamster. Journal of Helminthology 64: 75–78.

Fried, B., Schmidt, K.A., and Sorensen, R.E. 1997. *In vivo* and ectopic encystment of *Echinostoma revolutum* and chemical excystation of the metacercariae. Journal of Parasitology 83: 251–254.

Fried, B., Reddy, A., and Balfour, C.D. 2001. Excystation and distribution of metacercariae of *Echinostoma caproni* in ICR mice. Comparative Parasitology 68: 259–261.

Fujino, T., Yamada, M., Ichikawa, H., Fried, B., Arizono, N., and Tada, I. 1996. Rapid expulsion of the intestinal trematodes *Echinostoma trivolvis* and *E. caproni* from C3H/HeN mice after infection with *Nippostrongylus brasiliensis*. Parasitology Research 82: 577–579.

Fujino, T., Ichikawa, H., Fried, B., and Fukuda, K. 1997. The expulsion of *Echinostoma trivolvis*: worm kinetics and intestinal reactions in C3H/HeN mice treated with dexamethasone. Journal of Helminthology 71: 257–259.

Graczyk, T.K. 2000. Immunobiology and immunodiagnosis of echinostomiasis. In: Echinostomes as experimental models for biological research, eds. B. Fried and T.K. Graczyk, pp. 229–244. Dordrecht: Kluwer.

Graczyk, T.K., and Fried, B. 1998. Echinostomiasis: a common but forgotten food-borne disease. American Journal of Tropical Medicine and Hygiene 58: 501–504.

Hansen, K., Nielsen, J.W., Hindsbo, O., and Christensen, N. . 1991. *Echinostoma caproni* in rats: worm population dynamics and host blood eosinophilia during primary infections with 6, 25 and 50 metacercariae and resistance to secondary and superimposed infections. Parasitology Research 77: 686–690.

Haseeb, M.A., and Fried, B. 1988. Chemical communication in helminths. Advances in Parasitology 27: 169–207.

Hosier, D.W., and Fried, B. 1986. Infectivity, growth, and distribution of *Echinostoma revolutum* (Trematoda) in Swiss webster and ICR mouse. Proceedings of the Helminthological Society of Washington 53: 173–176.

Hosier, D.W., and Fried, B. 1991. Infectivity, growth, and distribution of *Echinostoma caproni* (Trematoda) in the ICR mouse. Journal of Parasitology 77: 640–642.

Howell, M.J. 1968. Excystment and in vitro cultivation of *Echinoparyphium serratum*. Parasitology 58: 583–597.

Howell, M.J. 1970. Excystment of the metacercariae of *Echinoparyphium serratum* (Trematoda: Echinostomatidae). Journal of Helminthology 44: 35–36.

Huffman, J.E. 2000. Echinostomes in veterinary and wildlife parasitology. In: Echinostomes as experimental models for biological research, eds. B. Fried and T.K. Graczyk, pp. 59–82. Dordrecht: Kluwer.

Huffman, J.E., and Fried, B. 1990. *Echinostoma* and echinostomiasis. Advances in Parasitology 29: 215–269

Huffman, J.E., Iglesias, D., and Fried, B. 1988. *Echinostoma revolutum*: pathology of extraintestinal infection in the golden hamster. International Journal for Parasitology 18: 873–874.

Humphries, J.E., Reddy, A., and Fried, B. 1997. Infectivity and growth of *Echinostoma revolutum* (Froelich, 1802) in the domestic chick. International Journal for Parasitology 27: 129–130.

Iorio, S.L., Fried, B., and Hosier, D.W. 1991. Concurrent infections of *Echinostoma caproni* and *Echinostoma trivolvis* in ICR mice. International Journal for Parasitology 21: 715–717.

Irwin, S.W.B. 1997. Excystation and cultivation of trematodes. In: Advances in trematode biology, eds. B. Fried and T.K. Graczyk, pp. 57–86. Boca Raton: CRC Press.

Irwin, S.W.B., McKerr, G., Judge, B.A., and Moran, I. 1984. Studies on metacercarial excystment in *Himasthla leptosoma* (Trematoda: Echinostomatidae) and newly emerged metacercariae. International Journal for Parasitology 14: 415–421.

Isaacson, A.C., Huffman, J.E., and Fried, B. 1989. Infectivity, growth, development and pathology of *Echinostoma caproni* (Trematoda) in the golden hamster. International Journal for Parasitology 19: 943–944.

Jarne, P. 1995. Mating system, bottlenecks and genetic polymorphism in hermaphroditic animals. Genetical Research 65: 193–207.

Kim, S., and Fried, B. 1989. Pathological effects of *Echinostoma caproni* (Trematoda) in the domestic chick. Journal of Helminthology 63: 227–230.

Kirschner, K., and Bacha, W.J. Jr. 1980. Excystation of *Himasthla quissetensis* (Trematoda: Echinostomatidae) metacercariae in vitro. Journal of Parasitology 66: 263–367.

Lackie, A.M. 1975. The activation of infective stages of endoparasites of vertebrates. Biological Reviews 50: 285–323.

Lee, K.J., Park, S.K., Im, J.A., Kim, S.K., Kim, G.H., Kim, G.Y., Yang, E.J., and Ryang, Y.S. 2004. Susceptibility of several strains of mice to *Echinostoma hortense* infection. The Korean Journal of Parasitology 42: 51–56.

LeFlore, W.B., and Bass, H.S. 1982. *In vitro* excystment of the metacercaria of *Himasthla rhigedana* (Trematoda: Echinostomatidae). Journal of Parasitology 68: 1177–1178.

Mabus, J., Huffman, J.E., and Fried, B. 1988. Humoral and cellular response to infection with *Echinostoma revolutum* in the golden hamster, *Mesocricetus auratus*. Journal of Helminthology 62: 127–132.

Mahler, H., Christensen, N. Ø., and Hindsbo, O. 1995. Studies on the reproductive capacity of *Echinostoma caproni* (Trematoda) in hamsters and jirds. International Journal for Parasitology 25: 705–710.

Manger, P.M. Jr., and Fried, B. 1993. Infectivity, growth and distribution of preovigerous adults of *Echinostoma caproni* in ICR mice. Journal of Helminthology 67: 158–160.

Meece, J.K., and Nollen, P.M. 1996. A comparison of the adult and miracidial stages of *Echinostoma paraensei* and *E. caproni*. International Journal for Parasitology 26: 37–43.

Mohandas, A., and Nadakal, A.M. 1978. *In vivo* development of *Echinostoma malayanum* Leiper, 1911, with notes on effects of population density, chemical composition and pathogenicity and *in vitro* excystment of the metacercaria (Trematoda: Echinostomatidae). Zeitschrift für Parasitenkunde 55: 139–155.

Mucha, K.H., Huffman, J.E., and Fried, B. 1990. Mallard ducklings (*Anas platyrhynchos*) experimentally infected with *Echinostoma trivolvis* (Digenea). Journal of Parasitology 76: 590–592.

Mullican, S.K., Huffman, J.E., and Fried, B. 2001. Infectivity and comparative pathology of *Echinostoma caproni, Echinostoma revolutum* and *Echinostoma trivolvis* (Trematoda) in the domestic chick. Comparative Parasitology 68: 256–259.

Muñoz-Antoli, C., Carpena, I., Espert, A., Esteban, J.G., and Toledo, R. 2004a. The effect of host species on the development of *Echinostoma friedi* (Trematoda: Echinostomatidae) adult worms. Revista Ibérica de Parasitología 64: 81–87.

Muñoz-Antoli, C., Trelis, M., Espert, A., Toledo, R., and Esteban, J.G. 2004b. Study of *Euparyphium albuferensis* and *Echinostoma friedi* (Trematoda: Echinostomatidae) antagonistic interactions in the experimental definitive host. Helminthologia 41: 81–84.

Muñoz-Antoli, C., Sotillo, J., Monteagudo, C., Fried, B., Marcilla, A., and Toledo, R. 2007. Development and pathology of *Echinostoma caproni* in experimentally infected mice. Journal of Parasitology 93: 854–859.

Nollen, P.M. 1990. *Echinostoma caproni*: mating behavior and the timing of development and movement of reproductive cells. Journal of Parasitology 76: 784–789.

Nollen, P.M. 1993. *Echinostoma trivolvis*: mating behavior of adults raised in hamsters Parasitology Research 79: 130–132.

Nollen, P.M. 1996a. The mating behavior of *Echinostoma paraensei* grown in mice. Journal of Helminthology 70: 43–45.

Nollen, P.M. 1996b. Mating behavior of *Echinostoma caproni* and *E. paraensei* in concurrent infections in mice. Journal of Helminthology 70: 133–136.

Nollen, P.M. 1997a. Reproductive physiology and behavior of digenetic trematodes. In: Advances in trematode biology, eds. B. Fried and T.K. Graczyk, pp. 117–147. Boca Raton: CRC Press.

Nollen, P.M. 1997b. Mating behavior of *Echinostoma caproni* and *E. trivolvis* in concurrent infections in hamsters. International Journal for Parasitology 27:71–75.

Nollen, P.M. 1999. Mating behavior of *Echinostoma trivolvis* and *E. paraensei* in concurrent infections in hamsters. Journal of Helminthology 73: 329–332.

Nollen, P.M. 2000. Reproductive physiology and behavior of echinostomes. In: Echinostomes as experimental models for biological research, eds. B. Fried and T.K. Graczyk, pp. 137–148, Dordrecht: Kluwer.

Odaibo, A.B., Christensen, N.ø., and Ukoli, F.M.A. 1988. Establishment, survival and fecundity in *Echinostoma caproni* (Trematoda) infections in NMRI mice. Proceedings of the Helminthological Society of Washington 55: 265–269.

Odaibo, A.B., Christensen, N.ø., and Ukoli, F.M.A. 1989. Further studies on the population regulation in *Echinostoma caproni* infections in NMRI mice. Proceedings of the Helminthological Society of Washington 56: 192–198.

Rossi, M.L., Balfour, C.D., and Fried, B. 2001. Excystation and infectivity of *Echinostoma caproni* metacercariae after prolonged storage. Comparative Parasitology 68: 126–129.

Senger, C.M. 1954. Notes on the growth, development and survival of two echinostome trematodes. Experimental Parasitology 6: 491–496.

Simpkins, H.L., and Fried, B. 1999. Effects of a diet deficient in vitamins A, D, and E on infectivity, growth and development of *Echinostoma caproni*. Parasitology Research 85: 873–875.

Smoluk, J.L., and Fried, B. 1994. Light microscopic observations of *Echinostoma trivolvis* (Trematoda) metacercariae during *in vitro* excystation. Parasitology Research 80: 435–438.

Sommerville, R.I., and Rogers, W.P. 1987. The nature and action of host signals. Advances in Parasitology 26: 239–293.

Stillson, L.L., and Platt, T.R. 2007. The crowding effect and morphometric variability in *Echinostoma caproni* (Digenea: Echinostomatidae) from ICR mice. Journal of Parasitology 93: 242–246.

Sukhdeo, M.V.K., and Mettrick, D.F. 1987. Parasite behavior: understanding platyhelminth responses. Advances in Parasitology 26: 73–144.

Toledo, R., and Fried, B. 2005. Echinostomes as experimental models in adult parasite–vertebrate host interactions. Trends in Parasitology 21: 251–254.

Toledo, R., Muñoz-Antoli, C., and Esteban, J.G. 2000. The life-cycle of *Echinostoma friedi* (Trematoda: Echinostomatidae) in Spain and a discussion on the relationships within the "*revolutum*" group based on cercarial chaetotaxy. Systematic Parasitology 45: 199–217.

Toledo, R., Espert, A., Carpena, I., Muñoz-Antoli, C., and Esteban, J.G. 2003. An experimental study of the reproductive success of *Echinostoma friedi* (Trematoda: Echinostomatidae) in the golden hamster. Parasitology 126: 433–441.

Toledo, R., Espert, A., Carpena, I., Muñoz-Antoli, C., Fried, B., and Esteban, J.G. 2004a. The comparative development of *Echinostoma caproni* (Trematoda: Echinostomatidae) adults in experimentally infected hamsters and rats. Parasitology Research 93: 439–444.

Toledo, R., Espert, A., Carpena, I., Trelis, M., Muñoz-Antoli, C., and Esteban, J.G. 2004b. *Echinostoma friedi*: the effect of age of adult worms on the infectivity of miracidia. Journal of Helminthology 78: 91–93.

Toledo, R., Esteban, J.G., and Fried, B. 2006a. Immunology and pathology of intestinal trematodes in their definitive hosts. Advances in Parasitology 63: 285–365.

Toledo, R., Carpena, I., Espert, A., Sotillo, J., Muñoz-Antoli, C., and Esteban, J.G. 2006b. A quantitative approach to the experimental transmission success of *Echinostoma friedi* (Trematoda: Echinostomatidae) in rats. Journal of Parasitology 92: 16–20.

Toledo, R., Monteagudo, C., Espert, A., Fried, B., Esteban, J.G., and Marcilla, A. 2006c. *Echinostoma caproni*: intestinal pathology in the golden hamster, a highly compatible host, and the Wistar rat, a less compatible host. Experimental Parasitology 112: 164–171.

Trouve, S., and Coustau, C. 1998. Differences in adult excretory–secretory products between geographical isolates of *Echinostoma caproni*. Journal of Parasitology 84:1062–1065.

Trouve, S., and Coustau, C. 1999. Chemical communication and mate attraction in echinostomes. International Journal for Parasitology 29: 1425–1432.

Trouve, S., Renaud, F., Durand, P., and Jourdane, J. 1996. Selfing and outcrossing in a parasitic hermaphrodite helminth (Trematoda, Echinostomatidae). Heredity 77: 1–8.

Trouve, S., Renaud, F., Durand, P., and Jourdane, J. 1998. Experimental evidence of hybrid breakdown between genetically distinct populations of *Echinostoma caproni*. Parasitology 117:133–135.

Trouve, S., Renaud, F., Durand, P., and Jourdane, J. 1999. Reproductive and mate choice strategies in the hermaphroditic flatworm *Echinostoma caproni*. The Journal of Heredity 90: 582–585.

Whitfield, P.J., Anderson, R.M., and Bundy, D.A.P. 1986. Host-specific components of the reproductive success of *Transversotrema patialense* (Digenea: Transversotrematidae). Parasitology 92: 683–698.

Yao, G., Huffman, J.E., and Fried, B. 1991. The effects of crowding on adults of *Echinostoma caproni* in experimentally infected golden hamsters. Journal of Helminthology 65: 248–254

Chapter 5
Maintenance, Cultivation, and Excystation of Echinostomes: 2000–2007

Bernard Fried and Robert C. Peoples

Contents

5.1 Introduction and Significance 112
5.2 Obtaining Material 116
5.3 Maintenance of Larval and Adult Echinostomes and the Intermediate and Definitive Hosts 119
5.4 In Ovo Cultivation 120
5.5 In Vitro Cultivation 120
5.6 Transplantation of Larval and Adult Echinostomes to New Sites 121
5.7 In Vivo and In Vitro Excystation of Metacercariae 121
5.8 In Vitro Encystment of Cercariae 123
5.9 Concluding Remarks 123
References 124

Abstract This review covers the salient studies on maintenance, cultivation, and excystation of echinostomes from 2000 to 2007. Some significant studies on these topics prior to 2000 are also included in the review. Ways to obtain echinostome material from the wild are discussed. The names of scientists and their laboratories where known species of echinostome life cycles are maintained are given. Information is given on the maintenance of larval and adult echinostomes and their intermediate and definitive hosts. The relatively sparse information from 2000 to 2007 on in ovo and in vitro cultivation and transplantation of larval and adult echinostomes to new sites is reviewed. Considerable new information on in vivo and in vitro excystation of echinostome metacercariae and the relatively new area of in vitro encystment of echinostome cercariae is included herein.

B. Fried (✉) and R.C. Peoples
Department of Biology, 111 Kunkel Hall, Lafayette College, Easton, PA 18042, USA
e-mail: friedb@lafayette.edu

B. Fried and R. Toledo (eds.) *The Biology of Echinostomes*,
DOI: 10.1007/978-0-387-09577-6_5,

5.1 Introduction and Significance

Fried (2000) reviewed the early literature on these topics. In this chapter we often refer the reader to that review. However, information of importance prior to 2000 on these topics is given in this chapter, so that dependence on the material in Fried (2000) is not absolutely necessary.

The need for consistent and reliable sources of living material is important for continued research on echinostomes and echinostomiasis. Often only a particular echinostome stage is available for research, and workers have used that stage. For instance, the rediae of *Echinostoma trivolvis* from naturally infected *Helisoma trivolvis* snails have been used to compare hemozoin pigment in the rediae with that of the malarial parasite *Plasmodium falciparum* (Pisciotta et al. 2005). The hemozoin pigment characteristics from both species were found to be very similar. Thus, a source of *E. trivolvis* material from naturally infected *H. trivolvis* snails served as a reliable supply for continuing research on the biology of this echinostome. Surveys indicate that this species of echinostome is a prevalent digenean in *H. trivolvis* snails in the USA, at least in the states of PA and NJ (Schmidt and Fried 1997; Klockars et al. 2007).

The optimal situation for continuing research on echinostomes is to raise in the laboratory the entire life cycle of species that are easy to maintain. Toledo and Fried (2005) obtained information from numerous parasitologists to determine how many echinostome species were maintained in laboratories worldwide. Table 5.1 from Toledo and Fried (2005) is reproduced herein (also as Table 5.1) with some modifications based on recent correspondence with most of the workers mentioned in the table. As of July 2007 of the ten species mentioned in the table, the most frequently maintained is *E. caproni*, and six of 13 research groups maintain this species. Next in frequency of maintenance is *E. paraensei* (maintained in three labs). A search of the citation literature from 2000 to 2007 in the ISI Web of Science made by us in July 2007 showed that most citations were available on *E. caproni* (98). Listed later are a number of widely used laboratory species and in parenthesis are the numbers of citations during 2000–2007 for these species: *E. trivolvis* (53); *E. paraensei* (36); *E. friedi* (14); *Echinoparyphium recurvatum* (8); *E. cinetorchis* (7); *Hypoderaeum conoideum* (6)

Although echinostome material from a number of sources is available, identification of these organisms to species (Fried and Toledo 2004) may be difficult; however, in most cases identification at least to the generic level may be made with the aid of the excellent recent key of Kostadinova (2005). One intent of this chapter is to make people aware of the sources of echinostome material and what type of work can be (and has been) done with such material. Table 5.2 provides a list of studies with useful information on work using echinostomes as experimental models in biology and chemistry from 2000 to 2007; the material compiled in the table is based on studies done by one of us (BF) in collaboration with numerous colleagues and students.

Table 5.1 Laboratories in which echinostome life cycles are currently maintained

Laboratory	Species
Fayez Bakry Department of Environmental Research and Medical Malacology, Theodore Bilharz Research Institute, Egypt (http://www.tbri.sci.eg/about/htm)	*Echinostoma caproni*
Christine Coustau Unite de Parasitologie Fonctionnelle, Universite de Perpignan, France (http://www.univ-perp.fr/see/rch/parasito/GB/perso/coustau.html)	*E. caproni*
Jillian Detwiler Department of Biological Sciences, Purdue University, USA (http://web.ics.purdue.edu/~jdetwile)	*Echinostoma revolutum*
Armand Kuris Department of Ecology, Evolution and Marine Biology, University of California, Santa Barbara, USA (http://www.lifesci.ucsb.edu/eemb/labs/kuris/index.html)	*E.caproni*
Bernard Fried Department of Biology, Lafayette College, USA (http://www.ufrj.br/)	*E. caproni*
Reinalda Marisa Lanfredi Laboratorio de Biologia de Helminthos Otto Wucherer, Universidade Federal do Rio de Janeiro, Brasil (http://www.ufrj.br/)	*Echinostoma paraensei*
Eric S. Loker Department of Biology, University of New Mexico, USA (http://biology.unm.edu/biology/esloker/)	*E. paraensei*
Arnaldo Maldonado, Jr. Laboratório de Biologia e Parasitologia de Mamíferos Silvestres Reservatórios, Instituto Oswaldo Cruz, Brasil (http://www.ioc.fiocruz.br/)	*E. paraensei*
Attef Saad Department of Zoology, Cairo University, Egypt (http://www.cu.edu.eg/)	*Echinoparyphium recurvatum*
Guillermo Salgado-Maldonado Istituto de Biologia, Universidad Nacional Autonoma de Mexico, Mexico (http://www.ibiologia.unam.mx/directorio/s/salgado_guillermo.htm)	*E. recurvatum*
Gregory J. Sandland Department of Biological Sciences, Purdue University, USA (http://www.purdue.edu/)	*Echinostoma revolutum*
Allen W. Shostak (http://www.biology.ualberta.ca/faculty/allen_shostak./)	*E. caproni*
Woon-Mok Sohn Department of Parasitology, Gyeongsang National University, Korea (http://www.gsnu.ac.kr/)	*Echinostoma cinetorchis; Echinostoma hortense*
Rafael Toledo, Carla Munoz-Antoli and J. Guillermo Esteban Departmento de Parasitologia, Universidad de Valencia, Spain (http://www.uv.es/~toledo/)	*E. caproni; Echinostoma friedi; Euparyphium albuferensis; Hypoderaeum conoideum*

All entries current as of December 2007. Information reprinted from Table 1 in Toledo and Fried (2005) with the permission of Elsevier. Some minor additions to this table were made

Table 5.2 Studies from Dr. Bernard Fried's laboratory that provide information on echinostome work from 1999 to 2007

Studies	Information	References
Echinostoma caproni infection and intestinal mucosa	Effects of infection on metal ions in the mouse gut	Layman et al. (1999)
TLC of lipids	Neutral lipids in cercariae of *E. trivolvis* and *Echinoparyphium sp.*	Muller et al. (1999)
Immunology	Mouse immune response to acute *E. caproni* infection	Brunet et al. (2000)
Nutrition	Effects of high-carbohydrate diet on growth of *E. caproni* in ICR mice	Daras et al. (2000)
Immunology	Expulsion of *E. trivolvis* from various hosts	Fujino et al. (2000a)
Ultrastructure of echinostome eggs	SEM and TEM on eggs of various species of echinostomes	Fujino et al. (2000b)
Postmortem worm survival	Survival and distribution of *E. caproni* in mouse gut after death of the host	Heim et al. (2000)
TLC of lipids	Neutral lipids in various larval stages of *E. caproni*	Marsit et al. (2000)
Cyst inoculum	Effects of various cyst inocula on mice infected with *E. caproni*	Balfour et al. (2001)
Excystation	In vivo excystation of *E. caproni* in mice	Fried et al. (2001)
Infectivity and pathology	Infectivity and comparative pathology of echinostomes in domestic chicks	Mullican et al. (2001)
Cyst storage	Effects of prolonged storage on metacercarial cysts of *E. caproni*	Rossi et al. (2001)
TLC of sugars	Analysis of sugars in *Biomphalaria glabrata* infected with *E. caproni*	Wagner et al. (2001)
Cercarial encystment	Encystment of *E. caproni*	Fried and LaTerra (2002)
Cercarial emergence	Emergence of *E. caproni* from *B. glabrata* under various conditions	Fried et al. (2002)
TLC of adult worms	Analysis of amino acids and sugars in *E. caproni*	Pachuski et al. (2002)
Heart rate in snails	Effects of larval *E. caproni* on heart rate of *B. glabrata*	Fried and Kim (2003)
Cercariae of *E. caproni*	Effects of temperature on in vitro encystment	Fried and Ponder (2003)
Sensory structures	Papillae patterns of *E. caproni* cercariae	Nakano et al. (2003)
TLC analysis	Free pool amino acids in larval *E. caproni*	Ponder et al. (2003a)
TLC analysis	Tonicity affects release of amino acids from adult *E. caproni*	Ponder et al. (2003b)
Immunology	Detection of coproantigens in *E. caproni*	Toledo et al. (2003)
TLC of pigments	Effects of echinostome infection on snail lutein and beta-carotene profiles in echinostomes	Evans et al. (2004)
Copper sulfate toxicity	Effects of copper sulfate treatment on in vitro encystment of *E. caproni* cercariae	Fried and Schneck (2004)
Treatment of metacercariae	Effects of tonicity, bile salts, and enzymes on *E. caproni* metacercariae	Fried et al. (2004)
Snail size	Effects of snail size on *E. caproni* cercarial infectivity	Ponder and Fried (2004a)
Glucose in culture	Glucose affects survival and linear movement of *E. caproni* cercariae	Ponder and Fried (2004b)

TLC of amino acids	Effects of *E. caproni* infection on amino acids in *B. glabrata*	Ponder et al. (2004)
Copper sulfate toxicity	Effects of copper sulfate treatment on larval *E. caproni*	Reddy et al. (2004)
Snail size	Effects of *B. glabrata* size on cercarial encystment of *E. caproni*	Schneck and Fried (2004)
Neutral lipid release	Effects of tonicity on release of neutral lipids in *E. caproni*	Schneck et al. (2004b)
TLC of neutral lipids	Lipids in snail-conditioned water and feces of *B. glabrata* infected with *E. caproni*	Schneck et al. (2004a)
Confocal microscopy	Neuromuscular studies in various stages of *E. caproni*	Sebelova et al. (2004)
Immunology	Characterization of somatic and excretory antigens of *E. caproni*	Toledo et al. (2004a)
Development of adult worms	Comparative development of *E. caproni* in experimental hosts	Toledo et al. (2004b)
Immunology	Kinetics of antigens in feces and serum of hamsters and rats infected with *E. caproni*	Toledo et al. (2004c)
Miracidial development	Age of adult *E. caproni* did not affect miracidial development	Fried and Bandstra (2005)
Tail loss	Tail loss in *E. caproni* cercariae during encystment; effects of copper sulfate treatment on tail loss	Fried and Schneck (2005)
Redial pigments	Hemozoin pigment in *E. trivolvis*	Pisciotta et al. (2005)
Growth of snails	Maintenance of *B. glabrata* for echinostome studies	Schneck and Fried (2005)
Immunology	Antigen and antibody kinetics in mice infected with *E. caproni*	Toledo et al. (2005)
Calcium carbonate in shells	*E. caproni* reduced calcium carbonate in infected *B. glabrata*	White et al. (2005)
TLC of lipids	*E. caproni* infection depleted snail lipids	Bandstra et al. (2006)
Intestinal pathology	*E. caproni* infection in compatible and non compatible hosts	Toledo et al. (2006)
Seasonal survey of larval trematodes	Isolation of *Helisoma trivolvis* snails and identification of larval trematodes	Klockars et al. (2007)
TLC of lipids	*E. caproni* infection altered fecal lipids in mice	Bandstra et al. (2007)
Cyst viability of *E. caproni*	Chemical excystation better determined of viability than light microscopy	Fried and Peoples (2007a)
Factors that kill cysts of *E. caproni*	Chemical and physical agents used to kill cysts	Fried and Peoples (2007b)
Worm burden and crowding	Effects of a 300-metacercarial cyst inoculum on worm recovery and crowding of E. caproni in Balb/C mice	Fried and Peoples (2007c)
Longevity	Longevity of *Echinostoma caproni* in Balb/C mice	Fried and Peoples (2007d)

5.2 Obtaining Material

Fried (2000) provided extensive information on obtaining preserved and living echinostome material from various commercial suppliers including Wards Biological Co., Carolina Biological Supply, Jones Biological Supply, and the Marine Biology Lab. The websites and/or catalogs of these and other suppliers should be consulted before attempting to get material from them.

The medically important snail *Biomphalaria glabrata* is a good experimental second intermediate host for cercarial encystment of many species of echinostomes. It is also a good experimental first intermediate host for some echinostome species including *E. caproni* and *E. paraensei*. This snail can be obtained for teaching and research purposes by qualified workers from Dr. Fred Lewis, Head, Schistosomiasis Laboratory, Biomedical Research Institute, 12111 Parklawn Drive, Rockville, MD 20852, USA.

In obtaining material from the wild, particularly from vertebrates and in some cases even invertebrates, care must be taken to follow the necessary country, state, and institutional guidelines for the care and handling of animal materials. Fried (2000) has discussed this issue in greater detail in his earlier review.

We have compiled Table 5.3 to help workers obtain material from the wild. The Table has 26 selected entries based on information compiled mainly from 2000 to 2007 from studies on eight genera of echinostomatids. The cosmopolitan nature of these entries is apparent based on the fact that the information was compiled from research done in more than 20 countries. The entries are based on selected studies in which echinostomes were obtained in the wild from naturally infected intermediate or definitive hosts. Most of the work examined freshwater echinostome life cycles, but some studies (see entries 19, 21, 26) were concerned with life cycles which occur in marine and brackish water. The marine forms have as second intermediate hosts, edible bivalves (cockles and mussels), which serve as potential vectors for transmitting echinostomes to humans who eat raw or undercooked mollusks tainted with metacercarial cysts. Such infections can cause human echinostomiasis.

Some entries in Table 5.3 show the value of collecting first or second intermediate hosts to initiate life cycles in the laboratory. For example, entry 1 mentions the use of naturally infected *Lymnaea peregra* snails in Spain to initiate the life cycle of *E. friedi* from cercariae and/or encysted metacercariae from the snail. Various laboratory animals including hamsters, rats, and chickens serve as experimental definitive hosts of this species to obtain adult worms. Other examples in the table in which life cycles were initiated based on naturally infected snails include entries 5, 7, 12, 19, 23, 24, and 26.

In many parts of the world, encysted metacercariae from naturally infected fish are used to initiate life cycles, or at least raise echinostomatid adults in vertebrate definitive hosts. Such examples from Table 5.3 include entries 2, 3, and 15. Not included in Table 5.3 are situations in which frogs (both tadpoles and adults) and snakes are infected with metacercarial stages. Cases of echinostomiasis in humans

Table 5.3 Relationships between echinostomes and their natural intermediate and definitive hosts in studies from more than 20 countries

Entry no.	Species	First intermediate host	Second intermediate host	Definitive host	Reference and (country/location)
	Echinostoma				
1	*E. friedi*	*Lymnaea peregra*	*L. peregra* and *Physella acuta*	*Rattus norvegicus (Norway rat)*	Toledo et al. (2000) (Spain)
2	*E. cinetorchis*	ND	Freshwater fishes	Rats, cats, and dogs	Chai and Lee (2002) (South Korea)
3	*E. hortense*	ND	Freshwater fishes	Rats, cats, and dogs	Chai and Lee (2002) (South Korea)
4	*E. revolutum*	ND	ND	Birds of prey in numerous families	Borgsteede et al. (2003) (Netherlands)
5	*E. luisreyi*	*Physa marmorata*	*P. marmorata*	ND	Maldonado et al. (2003) (Brazil)
6	*E. revolutum*	ND	ND	*Branta canadensis* (Canada goose)	Davis (2005) (New Zealand)
7	*E. trivolvis*	*Helisoma trivolvis*	*H. trivolvis*	ND	Klockars et al. (2007) (New Jersey, USA)
8	*E. malayanum*	ND	ND	Humans	Belizario et al. (2007) (Siargao Island, The Philippines)
	Echinoparyphium				
9	*E. agnatum*	ND	ND	Birds of prey in falconiformes and strigiformes	Borgsteede et al. (2003) (Netherlands)
10	*E. hydromyos*	ND	ND	*Rattus coiletti* (dusky rat) and *Hydromys chrysogaster* (water rat)	Mulder and Smales (2006) (Northern Territory, Australia)
11	*E. recurvatum*	ND	ND	*Aythya novaeseelandiae* (New Zealand scaup)	Davis (2006) (New Zealand)
12	*E. psuedorecurvatum*	*Lymnaea stagnalis*	Lymnaeids and other snails	ND	Faltynkova and Haas (2006) (S.E. Germany)
13	*E. aconiatum*	*Lymnaea stagnalis*	*L. stagnalis*	ND	Faltynkova et al. (2007) (Central Europe: i.e., Austria, Czech Rep., Germany, Poland)
14	*Echinoparyphium* sp.	*Lymnaea viatrix*	ND (exp: in *Biomphalaria glabrata*)	ND (exp: in domestic chicks)	Prepelitchi and de Nuñez (2007) (Patagonia, Argentina)

(continued)

Table 5.3 (continued)

Entry no.	Species	First intermediate host	Second intermediate host	Definitive host	Reference and (country/location)
	Echinochasmus				
15	*E. liliputanus*	ND	gills of goldfish	ND	Xiao et al. (2005) (Anhui Province, China)
16	*E. japonicus*	*Parafossarulus manouricues*	gills of *Pseudorasbora parva,* a freshwater fish	ND	Choi et al. (2006) (South Korea)
17	*Echinochasmus* sp.	ND	ND	*Chelydra serpentina* (snapping turtle)	Platt (2006) (Tennessee, USA)
18	*E. liliputanus*	ND	ND	Dogs	El Gayar (2007) (Egypt)
	Himasthla				
19	*H. elongata*	*Littorina littorea*	*Cerastoderma eduli* (cockles and edible bivalves)	ND	de Montaudouin et al. (2005) (France)
20	*H. elongata*	ND	*Mytilus eduli* (mussels)	ND	Nikolaev et al. (2006) (White Sea, Russia)
21	*Curtuteria arguinae*	ND	*Cerastoderma eduli* (cockles)	ND	Desclaux et al. (2006) (Arcachon Bay, France)
	Euparyphium				
22	*Euparyphium* sp.	ND	ND	*Procyon lotor* (Racoon)	Matoba et al. (2006) (most islands of Japan)
23	*E. albuferensis*	*Gyraulus chinensi*	*Gyraulus*, lymnaeids, and physids	*Rattus nonvegicus* and *R. rattus*	Esteban et al. (1997) (Spain)
	Other Genera				
24	*Hypoderaeun conoideum*	*Lymnaea peregra*	*Gyraulus*, lymnaeids, and physids	ND	Muñoz-Antoli et al. (2000) (Spain)
25	*Patagifer bilobus*	ND	ND	*Plegadis chihi* (white faced ibis)	Digiani (2000) (Buenos Aires Province, Argentina)
26	*Acanthoparyphium tyosenense*	*Lunatia fortans, a* marine gastropod	*Mactra veneriformis, a* marine bivalve	ND (exp: in *Larus crassiones*, a seagull)	Kim et al. (2004) (South Korea)

ND not defined, *Exp* experimental infection. Except for entry 23, all entries are from 2000 to 2007 eating raw frog and snake meat tainted with cysts have been reported (Fried et al. 2004a).

Mammals in the wild are a good source of initiating life cycles. Stray cats, dogs, and wild rats are often used to initiate various echinostome life cycles once the adult worms were obtained from these infected vertebrates. See entries 2, 3, 10, 18, 22, and 23 for such studies. Birds, particularly waterfowl, provide a good starting point for initiating echinostome life cycles. Davis (2005, 2006) pioneered such studies using Canada geese, *Branta canadensis*, as a starting point for maintaining the life cycle of *E. revolutum* and the New Zealand Scaup for initiating the life cycle of *Echinoparyphium recurvatum* (Table 5.3, entries 11 and 6).

Lastly, mention should be made of humans as a starting point for obtaining echinostome adults and eggs. A recent case of *E. malayanum* in humans in the Philippines provides a good example (see Belizario et al. 2007) as noted in entry 8 of Table 5.3. In this work, infected humans were purged with Dulcolax™, followed by treatment with praziquantel to allow the subject to expel echinostome eggs and adults into the stool. These stages were used for identification purposes in the Belizario et al. (2007) study. However, such samples may also provide a source of human material to initiate life cycles if suspected gastropod hosts are available to serve as experimental intermediate hosts.

5.3 Maintenance of Larval and Adult Echinostomes and the Intermediate and Definitive Hosts

Material relevant to this subject matter has been covered in detail by Fried (2000) and also recently reviewed by Toledo et al. (2007) in a section titled "Techniques for the maintenance of the life cycle in the laboratory." References from those sources should be consulted by workers who aim to maintain echinostomes in the laboratory.

In brief, echinostomes can be maintained in the appropriate first intermediate snail host following miracidial infection. Miracidia can be used to expose snails individually or en masse. For individual infections, 1–10 miracidia per snail may be used with 1–3 ml of natural or artificial spring water (ASW). These infections can be carried out conveniently in small Petri dishes or in multiwell chambers. Miracidia must be active and typically used in less than 6 h after hatching in order to optimize the host infection. Development of the intramolluskan stages will depend upon temperature and other factors, and cercarial sheds can be expected from 4 to 8 weeks postmiracidial infection. Cercariae can be obtained by isolating snails using standard procedures and then used to infect snails or other susceptible second intermediate hosts in multiwell chambers or fingerbowls. For many species of echinostomes, cysts are formed in the second intermediate host and are infective within 24 h postcercarial infection. Cysts can be removed from snails, tadpoles, or other second intermediate hosts to infect vertebrate hosts (Huffman and Fried 1990) or can be used for chemical excystation studies (Fried 1994).

Numerous vertebrate hosts serve as experimental definitive hosts; typically mice, hamsters, rats, domestic chicks, or ducklings are used. Trial and error are needed to determine optimal metacercarial dosage used to infect definitive hosts. Infection is by either per os or by gastric tube. The number of worms recovered does not always reflect the number of cysts given to the experimental host as noted in Fried and Peoples (2007c). Reasons for this situation are many, including operator error, viability of cysts, cyst age, and types and strains of hosts used.

Echinostome stages (eggs and cysts) can be stored at 4°C in saline for relatively long periods, enhancing the use of such material for teaching and research purposes. Eggs can be obtained from adult worms by teasing the uteri of worms and collecting and storing the eggs at 4°C in saline for up to 6 months (Idris and Fried 1996). Such eggs, transferred to ASW and maintained at 22–28°C, embryonate and produce miracidia that are infective to snails (Idris and Fried 1996). Eggs can also be obtained from the feces of definitive hosts as described in Toledo et al. (2007). Likewise, metacercarial cysts can be removed from snails or other second intermediate hosts and stored at 4°C for up to at least 6 months and still be infective to vertebrate hosts or capable of excystation (Rossi et al. 2001). The ability to store eggs and cysts of echinostomes for relatively long periods enhances the value of these organisms as research tools.

5.4 In Ovo Cultivation

The earlier literature was reviewed by Fried (2000). The most significant study on the topic was the cultivation of *E. caproni* on the chick chorioallantois from the excysted metacercaria to the ovigerous adult. The eggs contained miracidia, but attempts to infect snails with miracidia derived from chorioallantois-grown worms were not done. See Chien and Fried (1992) for details. In ovo cultivation studies on echinostomes or other digeneans have not been done during the 2000–2007 period covered in this review.

5.5 In Vitro Cultivation

The earlier work on in vitro cultivation of excysted metacercariae to young adults has been reviewed in Fried (2000). In spite of many attempts to cultivate echinostomes from the excysted metacercaria to the sexually mature adult, success in this area has been limited. Studies from 2000 to 2007 have been minimal. Irwin (1997) summarized the reasons for the lack of work in this area, including attempts to cultivate any digenean from the excysted metacercaria to the ovigerous adult. These reasons are as follow: inherent difficulties in such tasks, lack of consistent positive results leading to problems in publishing the work, and very labor-intensive studies.

Since the Fried (2000) review, there has been only one study on the development of an echinostome metacercaria in vitro. In that study, Fried and Reddy (2000) examined postmetacercarial changes in excysted metacercariae of *E. caproni* maintained in upright test tube cultures in the defined medium Mixture 199 plus 20% calf serum for 7 days at 41°C with a gas phase of air. Each culture was inoculated with 25 excysted metacercariae. Although some somatic changes were noted in metacercariae, and the organisms survived for up to 7 days, at which time the cultures were terminated, there was no indication of germinal development. Hence, as in earlier in vitro culture studies on echinostomes, the ability to stimulate genital development in vitro was not successful. Fried (2000) reviewed the earlier work on culturing echinostomes from the adult stage in either defined media or a defined medium plus natural products. Since that review, there have been no additional studies on the topic.

The relatively few studies on culturing the intramolluskan stages of echinostomes in vitro were reviewed by Fried (2000). There has been one significant study on that topic since that review by Gorbushin and Shaposhnikova (2002) using the rediae of *Himasthla elongata* obtained from the digestive gland of naturally infected marine snails *Littorina littorea*. The rediae were maintained in Leibovitz's L-15 medium at 14°C in atmospheric air. The cultured rediae were active and showed high synthetic activity in metabolic labeling experiments. Long-term cultivation showed 50% redial survival by 70 days postcultivation. The authors suggested that this culture method would be useful to obtain redial excretory–secretory products free of host-derived contaminants.

5.6 Transplantation of Larval and Adult Echinostomes to New Sites

The early studies on the transplantation of larval and adult echinostomes to new and altered sites have been reviewed by Fried (2000). This subject has not been explored since that review except for mention by Rossi et al. (2001) of transplanting chemically excysted metacercariae of *E. caproni* by the oral route into the intestines of ICR mice. The percentage of worm recovery from mice receiving excysted metacercariae from fresh cysts was 44% compared with 13% in mice receiving excysted metacercariae from cysts stored in Locke's solution for 5 months. Thus, cyst viability was determined by transplanting excysted metacercariae into the mouse gut.

5.7 In Vivo and In Vitro Excystation of Metacercariae

Fried (2000) reviewed the earlier literature on excystation of echinostomes and noted two significant works on excystation of digeneans, i.e., Fried (1994) and Irwin (1997). Both reviews provide considerable information on echinostomes as well as digeneans in general. Significant papers on this topic from 2000 to 2007 are related to work on *E. caproni* and all the in vitro excystation studies used the Fried and Roth (1974) alkaline trypsin-bile medium prepared in Earle's balanced salt solution. This medium is hereafter referred to as the TB medium.

Rossi et al. (2001) studied excystation of *E. caproni* metacercariae stored in Locke's 1:1 solution at 4°C for up to 12 months or maintained in *B. glabrata* for up to 3 months; comparisons were made to excystation in the TB medium of encysted metacercariae that were stored for 1–14 days postinfection (PI) and considered fresh cysts. Cysts stored for 8–12 months did not excyst, but excystation occurred in all the other cysts maintained in vitro and in vivo. Fried et al. (2001) studied in vivo excystation and distribution of encysted metacercariae of *E. caproni* fed to ICR mice. Each mouse was fed 400 metacercariae and necropsied at various intervals from 1 to 24 h PI. In vivo excystation of *E. caproni* occurred in the stomach and intestines of the mice; recovery of excysted metacercariae in these sites was seen mainly from 1 to 2 h PI and occurred less frequently after 3 h PI.

Fried et al. (2004b) excysted cysts of *E. caproni* in the TB medium and then studied the effects of various treatments on the excysted metacercariae; these treatments consisted of adding bile salts, digestive enzymes, and other nutrients to a Locke's 1:1 solution containing the excysted metacercariae. The results of the study suggested that the acidic pepsin environment in a host's stomach was detrimental to the survival of excysted metacercariae; however, prolonged survival in alkaline trypsin-bile salts probably facilitated the establishment of the metacercariae in the intestinal mucosa of the host.

Fried and Peoples (2007a) studied the viability of encysted metacercariae of E. caproni stored in Locke's 1:1 solution at 4°C for 1–24 week. Viability was judged by light microscopy based on morphological characteristics of the encysted metacercariae versus chemical excystation of the cysts in the TB medium. The percent viability was similar using both methods of assessment at 4-, 8-, and 16-week poststorage. At 1- and 24-week poststorage, viability was judged to be about two times greater based on excystation than using light microscopy. The authors concluded that light microscopy alone underestimated the viability of cysts and that determination of cyst viability was more accurate using assessment by chemical excystation than by light microscopy.

Fried and Peoples (2007b) observed the effects of various physical and chemical factors on the viability of encysted metacercariae of *E. caproni*. Viability was equated with chemical excystation in the TB medium. Cysts stored for 7 days at 28°C in a Locke's 1:1 solution showed 97% excystation. Isolated and in vivo formed cysts were killed by boiling for 1–3 min, but freezing at −10°C did not kill all isolated or in vivo formed cysts after 24 h. Concentrations of potassium permanganate ranging from 300 to 1,200 mg l^{-1} killed most isolated cysts within 5 min, but in vivo cysts survived these concentrations for 24 h. However, concentrated solutions of NaCl and sucrose had no effect on the viability of isolated and in vivo cysts. The killing effect of these agents could have economic implications, especially on the shellfish industry.

5.8 In Vitro Encystment of Cercariae

Studies on in vitro encystment of echinostome cercariae were not covered by Fried (2000). A number of species of echinostomes show some ability of their cercariae to encyst in vitro or ectopically on the surfaces of shells of aquatic invertebrates. A few earlier studies documented the experimental in vitro encystment of some cercariae of echinostomes (e.g., Fried and Bennett 1979). The implications of such in vitro encystment in the transmission of echinostomes as waterborne contaminants have been reviewed by Graczyk and Fried (2007). Of particular concern in transmission of water-borne echinostomatids are members of the genus *Echinochasmus*. Species of *Echinochasmus* are also transmitted to humans and domestic animals as waterborne parasites via the cercarial stage, when humans and animals inadvertently swallow such cercariae in tainted waters (Graczyk and Fried 2007). Some work from 2000 to 2007 on echinostome cercariae that encyst in vitro has been reported and key studies are listed below.

Fried and LaTerra (2002) observed encystation in vitro of *E. caproni* cercariae. Of the 15 media tested for the encystation of the in vitro cysts, only two were effective. They were a Locke's ASW (1:1) medium which yielded 67% encystment and a *B. glabrata* embryonic cell line medium which yielded 23% encystment. In vitro formed cysts treated in the TB medium were capable of excystation and also capable of infecting ICR mice.

Xiao et al. (2005) studied the in vivo and in vitro encystment of *Echinochasmus liliputanus* cercariae and the biological activity of the associated metacercariae. The cercariae encysted in vivo on the gills of goldfish, the second intermediate host. The cercariae also encysted in vitro in a Locke's solution containing artificial and human gastric juices. Full-strength Locke's was optimal for encystment within 24 h. Encystment of cercariae in vitro could be inhibited when the cercariae were treated with 1 μmol of silver nitrate. The authors suggested that these findings might be helpful in determining the mechanisms of the definitive hosts that allow for direct infection by the cercariae.

5.9 Concluding Remarks

This review has covered the salient literature on maintenance, cultivation, excystation, and in vitro encystment of echinostomes from 2000 to 2007. Significant earlier studies on these and related topics have also been mentioned. The review includes information on workers who are maintaining life cycles of known species of echinostomes in their laboratories. Also, an emphasis is placed on ways to obtain echinostome material from the wild as a way of initiating research on these digeneans. The period 2000–2007 saw little or no work on the in vitro or in ovo cultivation of echinostomes from the encysted metacercaria or adult stage. Some new work on chemical excystation of metacercarial cysts, cultivation of rediae, and in vitro encystment of the cercariae was done during 2000–2007 and covered in this review.

Useful information on echinostome research from the laboratory of one of us (B.F.) has been included to help new researchers plan future studies.

References

Balfour, C.D., Rossi, M., and Fried, B. 2001. Effects of a 100 metacercarial cyst inoculum on the host-parasite relationship of *Echinostoma caproni* and ICR mice. Journal of Helminthology 75: 321–324.

Bandstra, S.R., Fried, B., and Sherma, J. 2006. Effects of diet and larval trematode parasitism on lipids in snails as determined by thin-layer chromatography. Journal of Planar Chromatography and Modern TLC 19: 180–186.

Bandstra, S.R., Murray, K.E., Fried, B., and Sherma, J. 2007. High performance thin layer chromatographic analysis of neutral lipids in the feces of BALB/c mice infected with *Echinostoma caproni*. Journal of Liquid Chromatography and Related Technologies 30: 1437–1445.

Belizario, V.Y., Geronilla, G.G., Anastacio, M.B.M., de Leon, W.U., Subaan, A.P., Sebastian, A.C., and Bangs, M.J. 2007. *Echinostoma malayanum* infections, the Philippines. Emerging and Infectious Diseases 13: 1130–1131.

Borgsteede, F.H.M., Okulewicz, A., Zoun, P.E.F., and Okulewicz, J. 2003. The helminth fauna of birds of prey (Accipitriformes, Falconiformes and Strigiformes) in the Netherlands. Acta Parasitologica 48: 200–207.

Brunet, L.R., Joseph, S., Dunne, D.W., and Fried, B. 2000. Immune responses during the acute stages of infection with the intestinal trematode *Echinostoma caproni*. Parasitology 120: 565–571.

Chai, J.Y., and Lee, S.H. 2002. Food-borne intestinal trematode infections in the Republic of Korea. Parasitology International 51: 129–154.

Chien, W.Y., and Fried, B. 1992. Cultivation of excysted metacercariae of Echinostoma caproni to ovigerous adults in the allantois of the chick embryo. Journal of Parasitology 78: 1019–1023.

Choi, M.H., Kim, S.H., Chung, J.H., Jang, H.J., Eom, J.H., Chung, E.S., Sohn, W.M., Chai, J.Y., and Hongt, S.T. 2006. Morphological observations of *Echinochasmus japonicus* cercariae and the in vitro maintenance of its life cycle from cercariae to adults. Journal of Parasitology 92: 236–241.

Daras, M.R., Sisbarro, S., and Fried, B. 2000. Effects of a high-carbohydrate diet on growth of *Echinostoma caproni* in ICR mice. Comparative Parasitology 67: 241–243.

Davis, N.E. 2005. Endemic echinostome infections of candidate hosts. Journal of Helminthology 79: 127–132.

Davis, N.E. 2006. A survey of waterfowl for echinostomes and schistosomes from Lake Wanaka and the Waitaki River watershed, New Zealand. Journal of Helminthology 80: 33–40.

Desclaux, C., Russell-Pinto, F., de Montaudouin, X., and Brachalet, G. 2006. First record and description of Metacercariae of *Curtuteria arguinae* N. sp (Digenea: Echinostomatidae), parasite of cockles *Cerastoderma edule* (Mollusca: Bivalvia) in Arcachon Bay, France. Journal of Parasitology 92: 578–587.

Digiani, M.C. 2000. Digeneans and cestodes parasitic in the white-faced ibis *Plegadis chihi* (Aves: Threskiornithidae) from Argentina. Folia Parasitologica 47: 195–204.

El-Gayar, A.K. 2007. Studies on some trematode parasites of stray dogs in Egypt with a key to the identification of intestinal trematodes of dogs. Veterinary Parasitology 144: 360–365.

Esteban, J.G., Toledo, R., Sanchez, L., and Muñoz-Antoli, C. 1997. Life-cycle of *Euparyphium albuferensis* n. sp. (Trematoda: Echinostomatidae) from rats in Spain. Systematic Parasitology 38: 211–219.

Evans, R.T., Fried, B., and Sherma, J. 2004. Effects of diet and larval trematode parasitism on lutein and beta-carotene concentrations in planorbid snails as determined by quantitative high

performance reversed phase thin layer chromatography. Comparative Biochemistry and Physiology B-Biochemistry and Molecular Biology 137: 179–186.
Faltynkova, A., and Haas, W. 2006. Larval trematodes in freshwater molluscs from the Elbe to Danube rivers (Southeast Germany): before and today. Parasitology Research 99: 572–582.
Faltynkova, A., Nasincova, V., and Kablaskova, L. 2007. Larval trematodes (Digenea) of the great pond snail, *Lymnaea stagnalis* (L.), (Gastropoda, Pulmonata) in Central Europe: A survey of species and key to their identification. Parasite 14: 39–51.
Fried, B. 1994. Metacercarial excystment of trematodes. Advances in Parasitology 33: 91–144.
Fried, B. 2000. Maintenance, cultivation, and excystation of echinostomes. In Echinostomes as experimental models for biological research, eds. B. Fried and T.K. Graczyk, pp. 99–118. Dordrecht: Kluwer.
Fried, B., and Bandstra, S.R. 2005. Age of adult worms of Echinostoma caproni does not affect development of miracidia. Journal of Parasitology 91: 1220–1221.
Fried, B., and Bennett, M.C. 1979. Studies on encystment of *Echinostoma revolutum* cercariae. Journal of Parasitology 65: 38–40.
Fried, B., and Kim, Y. 2003. Effects of larval *Echinostoma caproni* and Schistosoma mansoni infection on the heart rate of *Biomphalaria glabrata* snails. Veliger 46: 182–183.
Fried, B., and LaTerra, R. 2002. In vitro and in vivo encystment of the cercariae of *Echinostoma caproni*. Journal of Parasitology 88: 1124–1129.
Fried, B., and Peoples, R.C. 2007a. Viability of the encysted metacercariae of *Echinostoma caproni* judged by light microscopy versus chemical excystation. Journal of Parasitology 93: 947.
Fried, B., and Peoples, R.C. 2007b. Effects of various physical and chemical factors on excystation of the encysted metacercariae of *Echinostoma caproni*. Journal of Parasitology 93: 1388–1391.
Fried, B., and Peoples, R.C. 2007c. Effects of a 300-metacercarial cyst inoculum on worm recovery and crowding of Echinostoma caproni in Balb/C mice. Parasitology Research 101: 1701–1702.
Fried, B., and Peoples, R.C. 2007d. Longevity of *Echinostoma caproni* in Balb/c mice. Journal of Helminthology 81: 95–96.
Fried, B., and Ponder, E.L. 2003. Effects of temperature on survival, infectivity and in vitro encystment of the cercariae of *Echinostoma caproni*. Journal of Helminthology 77: 235–238.
Fried, B., and Reddy, A. 2000. Postmetacercarial changes in *Echinostoma caproni* maintained in a defined medium plus calf serum. Korean Journal of Parasitology 38: 173–175.
Fried, B., and Roth, M. 1974. In vitro excystment of the metacercaria of *Parorchis acanthus*. Journal of Parasitology 60: 465.
Fried, B., and Schneck, J.L. 2004. Effects of copper sulphate on in vitro encystment of the cercariae of *Echinostoma caproni*. Journal of Helminthology 78: 371–372.
Fried, B., and Schneck, J.L. 2005. Cercarial tail loss in *Echinostoma caproni*: the influence of in vivo encystment and copper sulphate. Journal of Helminthology 79: 177–178.
Fried, B., and Toledo, R. 2004. Criteria for species determination in the 'revolutum' group of Echinostoma. Journal of Parasitology 94: 917.
Fried, B., Reddy, A., and Balfour, C.D. 2001. Excystation and distribution of metacercariae of *Echinostoma caproni* in ICR mice. Comparative Parasitology 68: 259–261.
Fried, B., LaTerra, R., and Kim, Y. 2002. Emergence of cercariae of *Echinostoma caproni* and Schistosoma mansoni from *Biomphalaria glabrata* under different laboratory conditions. Journal of Helminthology 76: 369–371.
Fried, B., Graczyk, T.K., and Tamang, L. 2004a. Food-borne intestinal trematodiasis in humans. Parasitology Research 93: 159–170.
Fried, B., Schneck, J.L., and Ponder, E.L. 2004b. Effects of tonicity, digestive enzymes and bile salts, and nutrient media on the survival of excysted metacercariae of *Echinostoma caproni*. Parasitology Research 93: 527–530.
Fujino, T., Shinohara, T., Fukuda, K., Ichikawa, H., Nakano, T., and Fried, B. 2000a. The expulsion of *Echinostoma trivolvis*: Worm kinetics and intestinal cytopathology in jirds, Meriones unguiculatus. Comparative Parasitology 67: 236–240.

Fujino, T., Nakano, T., Washioka, H., Tonosaki, A., Ichikawa, H., and Fried, B. 2000b. Comparative ultrastructure of eggs in *Echinostoma paraensei*, E. caproni, and E. trivolvis (Trematoda: Echinostomatidae). Parasitology Research 86: 427–430.

Gorbushin, A.M., and Shaposhnikova, T.G. 2002. In vitro culture of the avian echinostome *Himasthla elongata*: from redia to marita. Experimental Parasitology 101: 234–239.

Graczyk, T.K., and Fried, B. 2007. Human waterborne trematode and protozoan infections. Advances in Parasitology 64: 111–160.

Heim, J., Rossi, M., and Fried, B. 2000. Survival and distribution of *Echinostoma caproni* in the small intestine of ICR mice up to 36 hours after the death of the host. Journal of Parasitology 86: 415–416.

Huffman, J.E., and Fried, B. 1990. *Echinostoma* and echinostomiasis. Advances in Parasitology 29: 215–269.

Idris, N., and Fried, B. 1996. Development, hatching, and infectivity of *Echinostoma caproni* (Trematoda) eggs, and histologic and histochemical observations on the miracidia. Parasitology Research 82: 136–142.

Irwin, S.W.B. 1997. Excystation and cultivation of trematodes. In Advances in trematode biology, eds. B. Fried and T.K. Graczyk, pp. 57–85. New York: CRC Press.

Kim, Y.G., Yu, J.E., Chung, E.Y., and Chung, P.R. 2004. *Acanthoparyphium tyosenense* (Digenea: Echinostomatidae): Experimental confirmation of the cercaria and its complete life history in Korea. Journal of Parasitology 90: 97–102.

Klockars, J., Huffman, J., and Fried, B. 2007. Survey of seasonal trematode infections in *Helisoma trivolvis* (Gastropoda) from lentic ecosystems in New Jersey, USA. Comparative Parasitology 74: 75–80.

Kostadinova, A. 2005. Family Echinostomatidae Looss 1899. In Keys to the trematode, Vol. 2, eds. A. Jones, R.A. Brays, and D.I. Gibson, pp. 9–64 Wallingford: Commonwealth Agricultural Bureau International Publishing.

Layman, L.R., Mueller, T.J., Koehnlein, K.M., Fried, B., and Sherma, J. 1999. Effects of Echinostoma caproni infections on metallic ions in the intestinal mucosa of ICR mice. Journal of Helminthology 73: 367–368.

Maldonado, A., Vieira, G.O., and Lanfredi, R.M. 2003. *Echinostoma luisreyi* n. sp (Platyhelminthes: Digenea) by light and scanning electron microscopy. Journal of Parasitology 89: 800–808.

Marsit, C.J., Fried, B., and Sherma, J. 2000. Neutral lipids in cercariae, encysted metacercariae, and rediae of *Echinostoma caproni*. Journal of Helminthology 74: 365–367.

Matoba, Y., Yamada, D., Asano, M., Oku, Y., Kitaura, K., Yagi, K., Tenora, F., and Asakawa, M. 2006. Parasitic helminths from feral raccoons *(Procyon lotor)* in Japan. Helminthologia 43: 139–146.

de Montaudouin, X., Jensen, K.T., Desclaux, U., Wegeberg, A.M., and Sajus, M.C. 2005. Effect of intermediate host size *(Cerastoderma edule)* on infectivity of cercariae of *Himasthla quissetensis* (Echinostomatidae: Trematoda). Journal of the Marine Biological Association of the United Kingdom 85: 809–812.

Mulder, E. and Smales, L.R. 2006. Dusty rat, *Rattus colleti* (Muridae), a new host for *Echinopayphium hydromyos* (Trematoda: Digenea) in the Northern Territory, Australia. Comparative Parasitology 73: 289–290.

Muller, E.E., Simpkins, H., Fried, B., and Sherma, J. 1999. High performance thin-layer chromatographic analysis of neutral lipids in the marine snails *Ilyanassa obsoletus* and *Littorina littorea* infected with larval trematodes. Journal of Liquid Chromatography and Related Technologies 22: 1539–1545.

Mullican, S.K., Huffman, J.E., and Fried, B. 2001. Infectivity and comparative pathology of *Echinostoma caproni, Echinostoma revolutum*, and *Echinostoma trivolvis* (Trematoda) in the domestic chick. Comparative Parasitology 68: 256–259.

Muñoz-Antoli, C., Toledo, R., and Esteban, J.G. 2000. The life cycle and transmission dynamics of the larval stages of *Hypoderaeum conoideum*. Journal of Helminthology 74: 165–172.

Nakano, T., Fujino, T., Washioka, H., Tonosaki, A., Goto, K., and Fried, B. 2003. Tegumentary papillae of *Echinostoma caproni* cercariae (Trematoda: Echinostomatidae). Parasitology Research 89: 446–450.

Nikolaev, K.E., Sukhotin, A.A., and Galaktionov, K.A. 2006. Infection patterns in White Sea blue mussels *Mytilus edulis* of different age and size with metacercariae of Himasthla elongata (Echinostomatidae) and Cercaria parvicaudata (Renicolidae). Diseases of Aquatic Organisms 71: 51–58.

Pachuski, J., Wagner, S.D., Fried, B., and Sherma, J. 2002. Thin layer chromatographic analyses of amino acids and carbohydrates in adults of *Echinostoma caproni*. Comparative Parasitology 69: 202–205.

Pisciotta, J.M., Ponder, E.L., Fried, B., and Sullivan, D. 2005. Hemozoin formation in *Echinostoma trivolvis* rediae. International Journal for Parasitology 35: 1037–1042.

Platt, T.R. 2006. First report of *Echinochasmus* sp. from the snapping turtle (*Chelydra serpentina* L.) from Reelfoot Lake, Tennessee, USA. Comparative Parasitology 73: 161–164.

Ponder, E.L., and Fried, B. 2004a. Effects of glucose on survival, infectivity and linear movement of the cercariae of *Echinostoma caproni*. Journal of Helminthology 78: 185–187.

Ponder, E.L., and Fried, B. 2004b. Effects of snail size and diet on encystment of Echinostoma caproni cercariae in juvenile *Helisoma trivolvis* (Colorado strain) and observations on survival of infected snails. Journal of Parasitology 90: 422–424.

Ponder, E.L., Fried, B., and Sherma, J. 2003a. Effects of hypotonicity on amino acid release in adult *Echinostoma caproni* as determined by thin layer chromatography. Parasitology Research 89: 242–244.

Ponder, E.L., Fried, B., and Sherma, J. 2003b. Thin layer chromatographic analysis of free pool amino acids in cercariae, rediae, encysted metacercariae, and excysted metacerearіae of *Echinostoma caproni*. Journal of Liquid Chromatography and Related Technologies 26: 2697–2702.

Ponder, E.L., Fried, B., and Sherma, J. 2004. Free-pool amino acids in Biomphalaria glabrata infected with *Echinostoma caproni* as determined by thin-layer chromatography. Journal of Parasitology 90: 665–666.

Prepelitchi, L., and de Nuñez, M.O. 2007. Echinostomatid larval stages in *Lymnaea viatrix* (Gastropoda: Pulmonata) from Southwest Patagonia, Argentina. Journal of Parasitology 93: 323–327.

Reddy, A., Ponder, E.L., and Fried, B. 2004. Effects of copper sulfate toxicity on cercariae and metacercariae of *Echinostoma caproni* and Echinostoma trivolvis and on the survival of *Biomphalaria glabrata* snails. Journal of Parasitology 90: 1332–1337.

Rossi, M.L., Balfour, C.D., and Fried, B. 2001. Excystation and infectivity of *Echinostoma caproni* metacercariae after prolonged storage. Comparative Parasitology 68: 126–129.

Schmidt, K.A., and Fried, B. 1997. Prevalence of larval trematodes in Helisoma trivolvis (Gastropoda) from a farm pond in Northampton County, Pennsylvania with special emphasis on *Echinostoma trivolvis* cercariae. Journal of Helminthological Society of Washington 64: 157–159.

Schneck, J.L., and Fried, B. 2004. Effects of snail size on encystment of *Echinostoma caproni* in juvenile *Biomphalaria glabrata* (NMRI strain) and observations on the survival of infected snails. Journal of Helminthology 78: 277–279.

Schneck, J.L., and Fried, B. 2005. Growth of Biomphalaria glabrata (NMRI strain) and *Helisoma trivolvis* (Colorado strain) under laboratory conditions. American Malacological Bulletin 20: 71–73.

Schneck, J.L., Bandstra, S.R., Fried, B., and Sherma, J. 2004a. Thin layer chromatographic analysis of neutral lipids in snail conditioned water and feces of *Biomphalaria glabrata* infected with *Echinostoma caproni*. Journal of Liquid Chromatography and Related Technologies 27: 2039–2045.

Schneck, J.L., Fried, B., and Sherma, J. 2004b. Effects of tonicity on the release of neutral lipids in Echinostoma caproni adults and observations on lipids in excysted metacercariae. Parasitology Research 92: 285–288.

Sebelova, S., Stewart, M.T., Mousley, A., Fried, B., Marks, N.J., and Halton, D.W. 2004. The musculature and associated innervation of adult and intramolluscan stages of *Echinostoma caproni* (Trematoda) visualised by confocal microscopy. Parasitology Research 93: 196–206.

Toledo, R., and Fried, B. 2005. Echinostomes as experimental models for interactions between adult parasites and vertebrate hosts. Trends in Parasitology 21: 251–254.

Toledo, R., Muñoz-Antoli, C., and Esteban, J.G. 2000. The life-cycle of *Echinostoma friedi* n. sp (Trematoda: Echinostomatidae) in Spain and a discussion on the relationships within the 'revolutum' group based on cercarial chaetotaxy. Systematic Parasitology 45: 199–217.

Toledo, R., Espert, A.M., Muñoz-Antoli, C., Marcilla, A., Fried, B., and Esteban, J.G. 2003. Development of an antibody-based capture enzyme-linked immunosorbent assay for detecting *Echinostoma caproni* (Trematoda) in experimentally infected rats: Kinetics of coproantigen excretion. Journal of Parasitology 89: 1227–1231.

Toledo, R., Espert, A., Carpena, I., Muñoz-Antoli, C., Fried, B., and Esteban, J.G. 2004a. Immunological characterization of somatic and excretory–secretory antigens of Echinostoma caproni (Trematoda: Echinostomatidae) in experimentally infected rats. Comparative Parasitology 71: 42–48.

Toledo, R., Espert, A., Carpena, I., Muñoz-Antoli, C., Fried, B., and Esteban, J.G. 2004b. The comparative development of *Echinostoma caproni* (Trematoda: Echinostomatidae) adults in experimentally infected hamsters and rats. Parasitology Research 93: 439–444.

Toledo, R., Espert, A., Muñoz-Antoli, C., Marcilla, A., Fried, B., and Esteban, J.G. 2004c. Kinetics of *Echinostoma caproni* (Trematoda: Echinostomatidae) antigens in feces and serum of experimentally infected hamsters and rats. Journal of Parasitology 90: 752–758.

Toledo, R., Espert, A., Muñoz-Antoli, C., Marcilla, A., Fried, B., and Esteban, J.G. 2005. Kinetics of antibodies and antigens in serum of mice experimentally infected with *Echinostoma caproni* (Trematoda: Echinostomatidae). Journal of Parasitology 91: 978–980.

Toledo, R., Monteagudo, C., Espert, A., Fried, B., Esteban, J.G., and Marcilla, A. 2006. *Echinostoma caproni*: Intestinal pathology in the golden hamster, a highly compatible host, and the Wistar rat, a less compatible host. Experimental Parasitology 112: 164–171.

Toledo, R., Muñoz-Antoli, C., and Fried, B. 2007. The use of echinostomes to study host parasite relationships between larval trematodes and invertebrate and cold-blooded vertebrate hosts. Parasitology Research 100: 1177–1185.

Wagner, S.D., Kim, Y., Fried, B., and Sherma, J. 2001. Thin-layer chromatographic analysis of glucose, maltose, and raffinose in *Biomphalaria glabrata* (Gastropoda) infected with Echinostoma caproni (Trematoda). Journal of Planar Chromatography and Modern TLC 14: 459–461.

White, M.M., Chejlava, M., Fried, B., and Sherma, J. 2005. Effects of various larval digeneans on the calcium carbonate content of the shells of *Helisoma trivolvis*, Biomphalaria glabrata, and Physa sp. 2005. Parasitology Research 95: 252–255.

Xiao, X., Wang, T.P., Zheng, X.D., Shen, G.J., and Tian, Z.G. 2005. In vivo and in vitro encystment of *Echinochasmus liliputanus* cercariae and biological activity of the metacercariae. Journal of Parasitology 91: 492–498.

Chapter 6
Echinostomes in the Wild

Arnaldo Maldonado Jr. and Reinalda M. Lanfredi

Contents

6.1 Introduction 130
6.2 Major and Minor Species of Echinostomes in the Wild 130
6.2.1 Major Echinostome Species 131
6.2.2 Minor Echinostomes Species 134
6.3 Echinostomes as Experimental Models 134
6.3.1 Experimental Intermediate Hosts of Major Echinostome Species 135
6.3.2 Experimental Definitive Hosts of Major Echinostomes Species 137
6.4 Co-infection of Echinostomes and Other Helminths in the Wild 138
6.5 Pathological Aspects in Natural and Experimental Infections 138
6.6 Potential of Wild Echinostome Zoonoses 139
6.7 Impact of Parasitism on Wildlife Conservation 140
References 141

Abstract Echinostomes are closely related to wildlife and occasionally have been found to infect domestic animals. The low specificity of some species results in a large number of invertebrate and vertebrate hosts acting as natural hosts, and a large geographic distribution has been recorded. The zoonotic potential of echinostomes has been related to the ingestion of raw mollusks, fishes, and amphibians which are naturally found parasitized by these parasitic larvae. The identification of new natural hosts and the demonstration of their life cycles under experimental conditions have favored the understanding of biological variables involved in parasite transmission and may be useful predictors of human infection. The impact of echinostome parasitism in wildlife conservation needs more attention, since most of the known natural hosts are wild animals. The introduction or spread of echinostome species

A. Maldonado. Jr., and R.M. Lanfredi (✉)
Laboratório de Biologia de Helmintos Otto Wuchere, Programa de Biologia Celul, are Parasitologia, Instituto de Biofísica Carlos Chagas Filho, Universidade Federal do Rio de Janeiro, RJ, Rio de Janeiro 21941-902, Brazil

B. Fried and R. Toledo (eds.) *The Biology of Echinostomes*,

DOI: 10.1007/978-0-387-09577-6_6,

in new environments may imply a loss of biodiversity and increase the probability of new cases of zoonoses. Human behavior and the expansion of land use in a global context help to bring about changes in climate and induce new interfaces between humans and wildlife. Trade among nations and the exchange of new ideas and customs increase our concern about the spread and emergence of etiologic agents of animal and human diseases.

6.1 Introduction

The global biodiversity of animals in the wild is enormous. Each landscape encompasses many different habitats for a large number of animal species, and each species has its own parasite fauna. In the wild, most of the known echinostome species have been described mainly in wild birds, mammals, and reptiles and less frequently recorded in domestic animals. Echinostome infections are closely related to wildlife, and interest at this level is due to the importance of echinostomes as parasites of wild fauna (i.e., in the biodiversity of helminths) and also in their increasing significance in emerging and re-emerging zoonoses due to human activities, including a rapid increase of the human population, migration, and behavioral changes, among other factors.

Environmental changes have been associated with a major risk for wildlife, and the preservation of this fauna depends on the knowledge of the life history of each species in its natural environment in order to improve management to preserve the fauna without affecting the sustainable development of the planet.

Long-term studies of these parasites may allow scientists to gain a better comprehension of the transmission dynamics, both in the intermediate and definitive hosts in nature, their involvement in the risk for wildlife conservation, and, moreover, of the possibility of the spread of parasites to, thus, decreasing the food production of animals of economic importance. The aim of this chapter is to highlight some aspects of the biology of certain species of echinostomes in the wild and the experimental models developed with the purpose to gain a better understanding of the biology of this group of parasites.

6.2 Major and Minor Species of Echinostomes in the Wild

The Family Echinostomatidae comprises up to 91 genera currently considered as valid (Kostadinova 2005). Most of the known species of these taxa never have been studied after their original descriptions, and their life cycles in the wild remain unknown. Furthermore, in environments whose biodiversity is still poorly known, many other unknown echinostome species may exist.

6.2.1 *Major Echinostome Species*

Several genera of Echinostomatidae have biological, medical, veterinary, and experimental importance. From these genera, *Echinostoma* is the one genus with the largest number of species used as experimental models and whose biology has been well studied. Selective listings of *Echinostoma* species and their known natural intermediate and definitive hosts, and their geographical distributions are presented in Tables 6.1–6.5.

Table 6.1 Species of *Echinostoma* of wild animals in North America

Species	Definitive host	Intermediate hosts	Country	References
Echinostoma trivolvis	Hawks, owls		USA	Taft et al. (1993)
	Dendrocygna bicolor		USA	Forrester et al. (1994)
	Didelphis virginiana		USA	Alden (1995)
	Ondatra ziberthicus		USA	Zabiega (1996)
		Helisoma trivolvis	USA	Kanev et al. (1995)
		Rana catesbiana; *R. clamitans*; *R. pipiens*	Canada	McAlpine and Burt (1998)
Echinostoma revolutum		*Lymnaea elodes*	USA	Sorensen et al. (1997)

Table 6.2 Species of *Echinostoma* of wild animals in Central and South America

Species	Vertebrate hosts	First/second intermediate hosts	Country	References
Echinostoma parvocirrus	Birds	*Biomphalaria glabrata*	Guadeloupe	Nassi and Dupouy (1988)
Echinostoma paraensei	*Nectomys squamipes*		Brazil	Maldonado et al. (2001b)
	–	*Biomphalaria glabrata*	Brazil	Lie and Basch (1967)
Echinostoma. luisreyi	*Akodon* sp.	*Physa marmorata*	Brazil	Maldonado et al. (2003)
Echinostoma revolutum	*Cairina moschata*		Brazil	Lutz (1924)
	–		Brazil	Carvalho et al. (1975)
	–		Brazil	Travassos et al. (1968)
	Gallus gallus		Brazil	Mello (1933)
	Sus scrofa		Brazil	Kohn et al. (1972)
	Rattus norvegicus		Brazil	Barbosa (1991)
	Passer domesticus		Brazil	Brasil and Amato (1992)

Table 6.3 Species of *Echinostoma* of wild animals in Europe

Species	Vertebrate hosts	First/second intermediate hosts	Country	References
Echinostoma revolutum	*Anas platyrhynchus*		Yugoslavia	Kanev (1994)
	Buteo buteo		Czech Rep.	Sitko (1998)
	Turdos philomelos		Netherlands	Borgsteed et al. (2000)
	Ciconia ciconia		Germany	Schuster et al. (2002)
	Laurus dominicanus		New Zealand	Latham and Poulin (2002)
	Anser albitrons; A. tabalis; Branta Canadensis		Slovak Rep.	Macko et al. (2002)
	Accipiter gentilus; Aquila pomarina; Buteo buteo; B. lagopus; Circaetus gallicus; Falco subbuteo; Acthene noctua; Strix aluco; Tyto alba		Netherlands	Borgsteed et al. (2003)
		Lymnaea stagnalis	Finland	Niewiadomska et al. (1997)
		Lymnaea stagnalis L. perega	Filand	Vayrynen et al. (2000)
Echinostoma friedi	*Rattus norvegicus*	*Lymnaea peregra*	Spain	Toledo et al. (2000)
Echinostoma echinatum				Kostadinova and Gibson (2000)
Echinostoma miyagawai		*Planorbis planorbis; Anisus vortex*	Bulgaria	Kostadinova et al. (2000)

Table 6.4 Species of echinostomes of wild animals in Asia

Species	Vertebrate hosts	First/second intermediate hosts	Country	References
Echinostoma hortense	Rat domestic, dog	*Lymnaea pervia, Radix auricularia, Misgurnos unguiculatus*	Japan Korea	Asada (1926) Chai and Lee (2002)
Echinostoma echinatum			Korea	Chai and Lee (2002)
Echinostoma cinetorchis	Rat domestic	*Segmentina hemisphaerula, Misgurnos unguiculatus*	Japan Korea`	Ando and Osaki (1923) Chai and Lee (2002)
Echinochamus japonicus	Birds	Fish	Korea China	Tanabe (1926)
Acanthoparyphium tyosenense	Bird	*Mactra veneriformes*; *Solen grandis*; *Neverita bicolor*	Korea	Kim et al. (2004)

Table 6.5 Species of echinostomes attributed to zoonotic infections and main intermediate hosts

Species	Second intermediate host	Country	References
Echinostoma revolutum	Snails and clams	Indonesia	Carney (1991)
	Clams	Taiwan	Lu (1982)
Echinostoma hortense	Fresh water fish	Korea	Chai and Lee (2002)
	Misgurus unguillicaudatus, Burbatula toni	Japan	Miyamoto (1987)
Echinostoma japonicus	Fresh water fish	China	Carney (1991)
	Fresh water fish	Korea	Chai and Lee (2001)
	Fresh water fish	Japan	Rim (1982)
Echinostoma malayanum	*Lymnaea cumingiana*	Malaysia, Singapore	Rim (1982)
		Philippines	Monzon and Kitikoon (1989)
	Pila luzonica	Indonesia	
	Snails and clams		Carney (1991)
Echinostoma ilocanum	Snails and clams	Indonesia	Carney (1991)
	Snails and tadpoles	Thailand	Radomyos et al. (1982)
Echinostoma cinetorchis	Clam	Taiwan	Lu (1982)
	Fresh water fish	Korea	Chai et al. (2005)
	Fresh water fish	Japan	Rim (1982)
Echinostoma ilocanum	Snails and clams	Indonesia	Carney (1991)
		Philippines	Monzon and Kitikoon (1989)
		Thailand	Radomyos et al. (1982)
Echinostoma lindoense	*Corbicular* sp.	Indonesia	Cross (1984)
Echinoschamus. japonicus	Fresh water fish	Japan	Chai and Lee (2001)
Echinoparyphium recurvatum	Clam	Taiwan Indonesia	Lu (1982)
Achantoparyphium tyosenense	*Mactra veneriformis, Solen grandis, Neverita bicolor*	Korea	Chai and Lee (2001)

Echinostoma revolutum is the most cosmopolitan species found in natural infections and has been recorded from wild and domestic animals and humans. In addition, *Echinostoma hortense,* originally described in humans, and *E. cinetorchis*, *E. japonicus*, and *E. malayanum* are parasites of wild animals and may cause zoonoses (Sohn and Chain 2005; Chai and Lee 2002).

In addition to *Echinostoma,* some other genera of Echinostomatidae are also noteworthy for their medical, veterinary, and biological importance, i.e., *Echinoparyphium, Echinochamus, Himasthla, Hypoderaeum, Petasiger, Euparyphium, Stephanoprora, Isthmiophora, Acanthoparyphium*, and *Patagifer* (Fried 2001). Moreover, several species of these genera have significance as parasites of humans causing zoonoses, i.e., *Echinoparyphium recurvatum, Echinochasmus liliputanus* (commonly occurring in dogs and cats), *Himasthla quissetensis. H. rhigedana*, and

Hypoderaeum conoideum, usually found in natural infections in ducks, fowl, and rats (Fried 2001).

From these genera some species are commonly used as experimental models which have contributed to the knowledge of the biology and physiology of this group of Digenea, i.e., *Echinostoma paraensei, E. caproni, E. trivolvis, E. friedi,* and *Echinoparyphium recurvatum* (Fried and Graczyk 2004 and Toledo and Fried 2005).

6.2.2 *Minor Echinostomes Species*

From the 91 genera of Echinostomatidae considered valid by Kostadinova (2005), 85 include species parasitic in birds, mammals, and reptiles, with major importance in the biodiversity as parasites of wild animals. However, most of them have been recorded only once or a few times after their initial description. The knowledge of such biodiversity is important to improve diagnosis of parasites of wildlife and to assess the risk of such parasites in wild animal conservation and management, together with the aim to gain a better understanding of their potential role as a zoonotic disease.

6.3 Echinostomes as Experimental Models

The life cycle of echinostomatids includes two invertebrate hosts (first and second) and a vertebrate host (Fig. 6.1).(Maldonado et al. 2001a)

It is interesting to note that in most cases the echinostome species were described exclusively on the basis of one developmental stage isolated from a snail or other intermediate host species or from vertebrate hosts collected in the wild. Thus, only one of the hosts is known in such cases. In some cases, life cycles were experimentally established starting from one of the developmental stages collected in nature, using snails and vertebrates from the laboratory for the completion of the cycle. The life cycle of some species, such as *E. macrorchis* and *E. elegans,* was completed from materials collected in trapped wild birds as described by Lutz (1924). Thus, caution should be used with this approach since the finding of small samples of hosts infected with echinostomes in field surveys may lead to false interpretations in relation to the usual natural hosts of such species.

Some species of echinostomes have been used as experimental models with different aims, i.e., to clarify aspects of the biology of these platyhelminths, to analyze the parasite-vertebrate or invertebrate host relationships, including pathology, alterations of the snail biology and physiology, and to study coinfections in the vertebrate and invertebrate hosts, and biological control of different parasitosis, among others (Toledo and Fried 2005; Toledo et al. 2007).

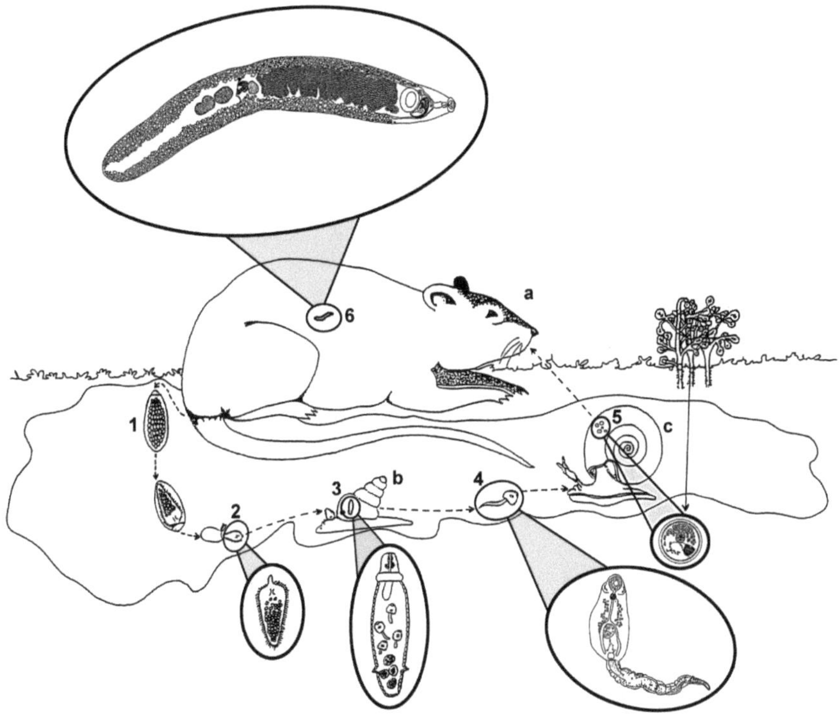

Fig. 6.1 Schematic representation of the life cycle of *Echinostoma* spp. in the wild

6.3.1 Experimental Intermediate Hosts of Major Echinostome Species

The maintenance of the life cycle of echinostomes requires the use of snail intermediate hosts. In this context, it is of interest to note that the specificity of echinostomes toward the first and second invertebrate hosts seems to be low since, in general, it is possible to infect more than one species of snail with one species of echinostome (Table 6.6).

Haas et al. (1995) analyzed the cercarial capacity of *Echinostoma revolutum*, *Pseudechinoparyphium echinatum*, and *Hypoderaeum conoideum* to recognize their respective snail second intermediate hosts and determined that each of these echinostome species was attracted by its specific intermediate hosts, mediated by the mucus compounds. This stimulus was inhibited by silver nitrate added to the water, since it binds to the cercarial receptors, inhibiting the capacity to find the host, snail, or fish, by the cercariae (Körner and Haas 1998). Under experimental

Table 6.6 Natural and experimental intermediate and definitive hosts of major *Echinostoma* species

	First intermediate host	Second intermediate host	Definitive host	References
E. paraensei	*Biophalaria glabrata, Physa marmorata, Lymnaea columella, B. tenagophila*	*Biomphalaria glabrata, Physa marmorata, Lymnaea columella, B. tenagophila*	Mouse, hamster, and *Rattus norvegicus Nectomys squamipes*	Lie and Basch, 1967 Maldonado et al., 2001
E. trivolvis	*Helisoma trivolvis Biophalaria glabrata, Helissoma trivolvis*	*Helissoma trivolvis-Biophalaria glabrata,*	Birds, marsupial Birds	Kanev et al., 1995 Fried et al., 1998
E. caproni	*Biomphalaria pfeifferi Biomphalaria spp*	*Molluscs and amphibians Biomphalaria spp,* Bulinus spp	*Falco newtoni* Hamster, mouse, birds	Richad and Brigood, 1978 Fried and Huffman, 1996
E. luizreyi	*Physa marmorata*	*Biomphalaria glabrata*	Hamster, mouse	Maldonado et al., 2003
E. friedi	*Lymnaea peregra* *L. corvus, Gyraulus chinenses*	*Lymnaea peregra* *L. corvus, Gyraulus chinenses, Physella acuta*	*Rattus norvegicus, hamster and chicken*	Toledo et al., 2000

conditions, echinostome cercariae are attracted, penetrate, and develop in a wide range of snails. When non-natural invertebrate host combinations were tested, random combinations of potential host species could produce minor biological adjustment (Maldonado et al. 2001b). This shows that the specificity of the echinostomes to the intermediate hosts may be adaptative. If the specific host is present, the efficiency is probably raised. Alternatively, if the specific host is not present larvae can locate, penetrate, and develop in an alternative snail species, although this will probably result in low cercarial release or metacercarial production.

Because of the zoonotic potential of echinostomes, it is important to determine the natural intermediate hosts to assess the risk of the ingestion of metacercariae harbored by raw mollusks, fishes, and/or amphibians, which in nature are found naturally infected. The experiments done by Park et al. (2006) demonstrated in a field research the capacity of the bivalve *Pisidium corneum* for the *E. cinetorchis* cercarial and metacercarial development in the field. This is of importance since this clam is usually consumed raw as a treatment for human bone fractures in Korea. Nevertheless, only mollusks are used as experimental intermediate hosts, mainly species of planorbids and lymnaeids.

The effect of echinostome infections on the survival and development of the snail, amphibian, and fish intermediate hosts has been investigated. It was demonstrated that tadpole *Rana pipiens* development can be modified by echinostome infection, and it was also observed that variations in the susceptibility to infection may occur in relation to the amphibian development stage (Schotthoefer et al. 2003).

Coinfection of snails with two different species of trematodes has been tested in an attempt to analyze the possibility to control schistomiasis using echinostomes. Sandland et al. (2007) demonstrated that attempts of coinfection with *Schistosoma mansoni* and *E. caproni* of *Biomphalaria glabrata* resulted in prevalences of infection of 23 and 63%, respectively, and the fecundity and egg production of the snail significantly decreased, demonstrating the antagonism of these trematode species in this host. This result corroborated the findings of Chai and Lee (2002) that also proposed the use of echinostomes as control agents for schistosomiasis.

6.3.2 Experimental Definitive Hosts of Major Echinostomes Species

Echinostoma revolutum has been reported from a large variety of definitive hosts, such as birds, pig, dog, rat, among others (Lutz 1924). This may serve as example of the wide range of hosts that an echinostome can use. However, this also may be caused by misidentification of the recovered helminths from natural hosts and, in any event, further investigations are required.

Among the mammals, mice and hamsters are the most used models as definitive hosts of echinostomatids, and among birds domestic chickens and ducks are the most commonly used (Table 6.6).

The life cycle of *E. paraensei* was reproduced in experimental conditions using the natural vertebrate host *Nectomys squamipes* reared in laboratory and sympatric snails *B. glabrata*, *Physa marmorata*, and *Lymnaea columella* (Fig. 6.1) (Maldonado et al. 2001b). Thereafter, it has been maintained in hamsters and the sympatric snail host, *B. glabrata*.

In experimental infections of BALC/c and C3H/HeN mice with *E. hortense*, the involvement of IL5 and immunoglobulin IgG1 in the induction of protective Th2 immunity and in worm expulsion from the intestine was observed (Cho et al. 2007). Pathological effects resulted from experimental infection by *E. caproni* in the golden hamster, and the Wistar rat are characterized by increased erosion of mucosal villi and elevated number of goblet cells in a low compatible host (rat) which could also be involved in the early expulsion of the parasite from this host (Toledo et al. 2006).

The identification of antigenic proteins recognized by humoral immune response in mice infected with *E. caproni* suggested an important role of the glycosylated forms of enolase which could be important in worm establishment on the host mucosal surface (Sotillo et al. 2008). The understanding of the factors which determine the fertility of adult worms during experimental infection by *E. caproni* was investigated (Fried and Bandstra 2005). The rate of survival and infectivity of *in vitro* encystment of the metacercariae of *E. caproni* was shown to be influenced by the temperature (Fried and Ponder 2003). The process involved in the excystation of *E. caproni* metacercariae was verified to be mediated by inorganic and organic molecules (Fried et al. 2004).

6.4 Co-infection of Echinostomes and Other Helminths in the Wild

The current knowledge of the biology of coinfection of echinostomes with other helminths in the wild is scarce, especially in relation to topics such as pathology, parasite loads in the vertebrate hosts, and worm fecundity.

A long-term survey focused on the role of the semiaquatic water rat *N. squamipes* in the transmission dynamics of *S. mansoni* revealed that concurrent infections with *E. paraensei* are usual in the field. However, no significant correlation was found between the intensity of infection between each pair of helminths. At the level of the helminth community structure, *E. paraensei* was classified as codominant exhibiting a highly aggregated pattern (Maldonado et al. 2006).

In experimental infections, it was demonstrated that *E. paraensei* can stimulate an antagonistic behavior toward *S. mansoni* resulting in a load decrease of *S. mansoni*. However, it should be considered that these trematodes occupy different niches within the host, and the mechanisms of interaction are probably indirect. In contrast, the increased persistence of the worm load of *E. paraensei* within hosts harboring *S. mansoni* suggests that adult schistosome infections are probably downregulating the immune response of the host, since the echinostome infection lasted for a longer time (Maldonado et al. 2001c).

Interaction between helminths during coinfections may be determined by direct influence of one species on another through secretory products, effects on the gut physiology, interspecific competition, or crowding effects, among others (Christensen et al. 1987). Furthermore, interactions would be mediated through the host's immune system (Lello et al. 2004). Simultaneous infections of two or more helminth species commonly occur in domestic and wild animals. This raises the possibility that one species may influence the transmission pattern or disease features caused by another (Noland et al. 2007).

Echinostomatids have specific niches within the small intestine of their definitive hosts. They are typically dispersed in the initial phase of infection (Nollen 1996). *E. paraensei* commonly occurs in the duodenum-jejunum. From our data, the distribution does not change when the host is challenged with *S. mansoni*. However, the host species may influence the *E. paraensei* distribution, as shown by the predilection of *E. paraensei* to occupy the pancreatic duct of *N. squamipes* when compared to *M. musculus* (Maldonado et al. 2001c). In addition, mice experimentally infected by *E. caproni* showed a decrease in natural resistance against *S. mansoni*, resulting in an increase of the *S. mansoni* load under experimental conditions (Christensen et al. 1985).

6.5 Pathological Aspects in Natural and Experimental Infections

Our knowledge of the pathology caused by echinostomatids in natural hosts is limited. However, as observed by Huffman (2000), the impact of echinostomes on wildlife health is significant, and these digeneans play an important role as causative agents of diseases in waterfowl and aquatic mammals. Most of the known cases are of migratory birds debilitated during their migrations. Moreover, the possibility of coinfections by other pathogens has not been considered.

Generally the level of pathogenicity evoked by the worms is low. Massive infection may induce a clinical picture of enteritis followed by hemorrhagic diarrhea, which could result in the death of animals (Huffman 2000). In the naturally infected water rodent, *N. squamipes* parasitized *with E. paraensei*, it was possible to observe transitory parasitism, which lasted no longer than a few weeks Animals generally harboring a few worms developed only minimal pathologic alteration (personal observation).

Some critical aspects of the pathology, such as multiple exposures to a low number of infective metacercariae or concurrent natural infection, require further studies. In fact, it would be convenient to use natural hosts raised in the laboratory for the study of pathological events. In this context it should be possible to get similar infection patterns or lesions as those observed in the wild.

Biological variability among isolates of *E. paraensei* showed distinct migratory behavior in the rodent *N. squamipes,* the natural host of this echinostome species (Maldonado et al. 2005), with temporal and spatial scale promoting differential kinetics in the evolution of disease (Maldonado et al. 2006). This makes evident the broad spectrum of the effects induced by this parasitism.

Histopathological studies in mice experimentally infected with *E. caproni* showed the modification of the intestinal wall involving modulation in the number of goblet cells, participation of mucosal neutrophils and mononuclear inflammatory cells in the mesentery. These features may contribute for the parasite establishment, enhancing the longevity of the infection and increasing the number of eggs released (Muñoz-Antoli et al. 2007).

However, intra- and interspecific variability in relation to parasite isolates and site of worm establishment needs to be taken into consideration (Maldonado et al. 2006).

6.6 Potential of Wild Echinostome Zoonoses

Echinostoma sp. eggs were found in intestinal material of a partially mummified body (dated 600–1200 years BC) in Brazil, indicating the possibility of the occurrence of this zoonoses in the pre-Colombian age in South American people (Sianto et al. 2005).

Human behavior associated with globalization, i.e., ease of travel has increased the risk of widespread diseases. This aspect was well summarized by Macpherson (2005): “Changing demographics and concomitant alterations to the environment, climate, technology, land use and changes in human behavior converge to favor the emergence and spread of parasitic zoonoses.”

In endemic countries, the traditional habit to ingest raw freshwater food together with the poverty, which pressures people to look for food in contaminated areas, has increased the risk of the transmission of echinostomiasis, among other food-borne diseases. Considering that humans may become infected and release eggs in the feces, people may act as reservoirs of such parasites. Since people can move from one place to another carrying the infection, environmental contamination of new habitats may occur, spreading the parasitism. In this case the lack of sanitation associated with ignorance, and the disregard of health authorities, increase the probability of infection to snails, elevating the transmission chances of echinostomiasis disseminated by humans.

Most of the human echinostomiasis have been diagnosed in Southeast Asia. The Philippines, India, Korea, and Japan are the countries with the highest prevalence rates (Chai and Lee 1990; Carney 1991; Chai et al. 2005). The human echinostomiasis is caused by a large number of echinostome species, always in relation to the ingestion of raw and/or undercooked snails or freshwater fishes. Among the species detected in human echinostomiasis, *E. malayanum*, *E. hortense*, and *E. revolutum* are major species with large geographical distributions (Table 6.5).

Parasitic diseases that usually occur in a restricted country or locality may be propagated to distant places by movements of people due to tourism, consequently acquiring new habits and massive migration, or merely by the popularization of dishes using raw or undercooked fishes, amphibians, and specially mollusks, together with the introduction of exotic animals.

The introduction of fish and shellfish to nonendemic countries as a result of international trade can result in the spread of food-borne trematode infections such as echinostomiasis (Dixon and Flohr 1997). Moreover, in these countries, wildlife may also harbor echinostomes representing potential reservoir of pathogens to man, infecting local consumers (with altered alimentary customs) or immigrants who migrate to such places.

Graczyk and Fried (1998) listed the following as the main symptoms of human echinostomiasis: abdominal pain, violent watery diarrhea, and anorexia. This was corroborated by the results of experimental human infections by the ingestion of raw metacercariae of *E. hortense* harvested from *Lymnaea japonica* that promoted variable symptoms in the volunteers such as abdominal pain and diarrhea (Miyamoto et al. 1984).

6.7 Impact of Parasitism on Wildlife Conservation

The literature on the impact of echinostomes in natural infections in the wild is very scarce. Although parasitism is common in the nature, for example, from 116 *Nectomys squamipes* necropsied, 7.8% were positive, in a two-year survey (Maldonado et al. 2006), no significant mortality caused by *E. paraensei,* or other etiological agents, was observed in the field or in laboratory conditions (personal observations in extensive field work).

Based on the literature, the specificity of echinostomes toward the vertebrate hosts is not high, and most echinostomes species are able to infect a variety of definitive hosts at least under experimental conditions. Thus, many vertebrates may be occasional hosts and reservoirs of species in nature. This may explain the fact that many vertebrate species have been observed only once or just a few times naturally infected, probably being only occasional hosts. This aspect raises the question if occasional parasitism can produce more dangerous lesions to a new host than to the natural host. As mentioned earlier, the impact of echinostomes on wildlife health is significant, and these parasites play an important role as causative agents of disease in waterfowl and aquatic mammals (Huffman 2000).

The spread of some parasitic diseases to new natural environments could imply in loss of biodiversity, especially among poorly adapted host species. On the other hand, emerging parasitic diseases in wildlife may be driven by ecological or human environmental changes (King et al. 2007). Recognition of parasites that may represent potential threat at the species level requires further investigation.

Restriction of animal movements by establishing ecological corridors or in zoo settings may exacerbate in some cases the intensity of infection with appreciable impact in life expectancy by the weakening of health.

Echinostomes can infect a large number of hosts in wildlife and more recently have been seen parasitizing domestic or peridomestic animals such as dogs in metropolitan areas. This probably is a result of the increase of the interface between wildlife and domestic animals or the reduction of the home range of wildlife.

Every year migratory birds land in the same localities to rest and to feed in their way to and from their natural locales. The area of these resting places, despite conservations laws, is becoming smaller, increasing the density of animals. Consequently, the environmental contamination and the risk of high intensities of infections weakening the birds' health are significantly increased. This fact also can create new niches for parasitism, that can be dangerous to the migratory birds and also widespread the infection to the local fauna and to humans.

Presently, climatic changes and other anthropogenic factors (agents of transformation of global dynamics) are receiving special attention, in order to create ways for sustaining the management of natural resources, with the principal aim of avoiding species extinction. The damage to natural environments as a consequence of demographic growth, fires, and deforestation pressure on the wildlife, induces them to move from their natural habitats and, in some cases, to share wild and urban interfaces when food is available.

The next decades will be crucial for life on earth, and nobody can predict with certainty the dynamics of parasitic diseases, including echinostomiasis, and how global environmental changes will affect each species.

References

Alden, K. J. 1995. Helminths of the opossum, *Didelphis virginiana*, in southern Illinois with a compilation of all helminths reported from this host in North America. Journal of the Helminthological Society of Washington 62: 197–208.

Ando, A., and Ozaki, Y. 1923. On four species of trematodes of genus *Echinostoma*. Dobuts Zasshi, 35: 108–119.

Asada, J. 1926. On a new echinostomatid trematode and life history. Transaction of Japanese Pathology. Society, 16: 194–199.

Barbosa, J. V. 1991. Infecção natural e formas evolutivas de *Echinostoma revolutum* (Froelich, 1802) em *Biomphalaria tenagophila* (Orbigny, 1835) e *Rattus norvergicus*. Susceptibilidade de *Mus musculus* ao parasitismo. Tese de Mestrado, Universidade Federal Rural do Rio de Janeiro, 58 pp.

Borgsteed, F. H. M., Okulewicz, A., and Okulewicz, J. 2000. A study of the helminth fauna of birds belonging to the Passeriformes in the Netherlands. Acta Parasitologica 45: 14–21.

Borgsteed, F. H. M., Okulewicz, A., Zoun, P. E. F., and Okulewicz, J. 2003. The helminth fauna of birds of prey (Accipitriformes, Falconiformes, and Strigiformes) in the Netherlands. Acta Parasitologica 48: 200–207.

Brasil, M. D., and Amato, S. B. 1992. Faunistic analysis of the helminths of sparrows (*Passer domesticus* L, 1758) captured in Campo Grande, Rio de Janeiro, R.J. Memórias do Instituto Oswaldo. Cruz 87: 43–48.

Carney, P. C. 1991. Echinostomiasis – A snail-borne intestinal trematode zoonoses. Southeast Asian Journal of Tropical Medicine and Public Health 22: 206–211.

Carvalho, O. S., Kawazoe, U., and Cuba, C. A. 1975. *Echinostoma revolutum* (Froelich, 1802) (Trematoda, Echinostomatidae) em pato doméstico de Minas Gerais. Atas da Sociedade de Biologia do Rio de Janeiro 17: 73–75.

Chai, J. Y., and Lee, S. H. 1990. Intestinal trematodes of human in Korea: *Metagonimus*, Heterophyids and Echinostomes. Korean Journal of Parasitology 28: 103–122.

Chai, J. Y., and Lee, S. H. 2002. Food-borne trematode infection in the Republic of Korea. Parasitology International 51: 129–154.

Chai, J. Y., Han, E. T., Park, Y. K., Guk, S. M., and Lee, S. H. 2001. *Acanthoparyphium tyosenense*: the discovery of human infection and identification of its source. Journal of Parasitology 87: 794–800.

Chai, J. Y., Darwin Murrel, K. and Lymbery, A. J. 2005. Fish-borne parasitic zoonoses: status issues. International Journal for Parasitology 35: 1233–1254.

Cho, Y. K., Ryang, Y. S., Kim, I. S., Park, S. K., Im, J. A., and Lee, K. J. 2007. Differential immune profiles following experimental *Echinostoma hortense* infection in BALB/c and C3H/HeN mice. Parasitology Research 100: 1053–1061.

Christensen, N. È., Knudsen, J., Fagbemi, B., and Nassen, P. 1985. Impairment of primary expulsion of *Echinostoma revolutum* in mice concurrently infected with *Schistosoma mansoni*. Journal of Helminthology 59: 333–335.

Christensen, N. È., Nansen, P., Fagbemi, B. O., and Monrad, J. 1987. Heterologous antagonistic and synergistic interactions between helminths and between protozoans in concurrent experimental infection of mammalian hosts. Parasitology Research 73: 387–410.

Cross, J. H. 1984. Changing patterns of some trematode infections in Asia. Azneimittelforshung 34: 1224–1226.

Dixon, B. R., and Flohr, R. B. 1997. Fish and Shellfish-borne trematode infection in Canada. Southeast Asian Journal Tropical Medicine Public Health 28: 58–64.

Forrester, D. J., Kinsella, J. M., Mertins, J. W., Price, R. D., and Turnbull, R. E. 1994. Parasitic helminths and arthropods of fulvous whistling ducks (*Dendrocygna bicolor*) in southern Florida. Journal of the Helminthological Society of Washington 61: 84–88.

Fried, B., and Huffman, J. E. 1996. The Biology of the Intestinal Trematode Echinostoma caproni. Advances in Parasitology 38: 311–368.

Fried, B., Frazer, B. A., and Kanev, I. 1998. Comparative observations on cercariae and metacercariae of *Echinostoma trivolvis* and *Echinoparyphium* sp. Journal of Parasitology 84(3): 623–626.

Fried, B. 2001. Biology of echinostomes except *Echinostoma*. Advances in Parasitology 49: 163–210.

Fried, B., and Bandstra, S. R. 2005. Age of adult of *Echinostoma caproni* does not affect development of miracidia. Journal of Parasitology 91: 1220–1221.

Fried, B., and Graczyk, T. K. 2004. Recent advances in the biology of *Echinostoma* species in the "*revolutum*"group. Advances in Parasitology 58: 139–195.
Fried, B., and Ponder, E. L. 2003. Effects of temperature on survival, infectivity and in vitro encystment of the cercariae of *Echinostoma caproni*. Journal of Helminthology 77: 235–238.
Fried, B., Schneck, J. L., and Ponder, E. L. 2004. Effects of tonicity, digestive enzymes and bile salts, and nutrient media on the survival of excysted metacercaria of *Echinostoma caproni*. Parasitology Research 93: 1–4.
Graczyk, T. K., and Fried, B. 1998. Echinostomiasis: a common but forgotten food-borne disease. American Journal Tropical Medicine and Hygiene 58: 501–504.
Haas, W., Köner, M., Hutterrer, E., Wegner, M., and Habert, B. 1995. Finding and recognition of the intermediate hosts by 3 species of echinostomes cercariae. Parasitology 110: 133–142.
Huffman, J. E. 2000. Echinostomes in Veterinary in wild life parasitology. In Echinostomes as experimental models for biological research, eds. B. Fried and T.K. Graczyk, pp. 59–82. Dordrecht: Kluwer Academic Publishers.
Kanev, I. 1994. Life-cycle, delimitation and redescription of *Echinostoma revolutum* (Froelich, 1802) (Trematoda: Echinostomatidae). Systematic Parasitology 28: 125–144.
Kanev, I., Fried, B., Dimitrov, B., and Radev, V. 1995. Redescription of *Echinostoma trivolvis* (Cort, 1914) (Trematoda, Echinostomatidae) with a discussion on its identity. Systematic Parasitology 32: 62–70.
King, K. C., McLaughlin, J. D., Gendron, A. D., Pauli, B. D., Giroux, I., Rondeau, B., Boily, M., Juneau, P., and Marcogliese, D. J. 2007. Impacts of agriculture on the parasite communities of northern leopard frogs (*Rana pipiens*) in southern Quebec, Canada. Parasitology 134: 2063–2080.
Kim, Y. G., Yu, J. E., Chung, E. Y., and Chung, P. R. 2004. *Acanthoparyphium tyosenense* (Digenea: Echinostomatidae): experimental confirmation of the cercaria and its complete life history in Korea. Journal of Parasitology 90(1): 97–102.
Kohn, A., Buhrnhein, U., and Costa, H. M. A. 1972. Infestação natural de *Echinostoma revolutum* (Froelich, 1802) (Trematoda, Echinostomatidae) em porco doméstico. Atas da Sociedade de Biologia do Rio de Janeiro 10: 5–6.
Körner, M., and Haas, W. 1998. Chemo-orientation of echinostome cercariae towards their snail host aminoacids signal low host specificity. International Journal for Parasitology 28: 511–506.
Kostadinova, A. 2005. Family Echinostomatidae Looss, 1899. In Keys to the Trematoda, Vol. 2, eds. A. Jones, R.A. Bray, and D.I. Gibson, pp. 9–64, London: CABI Publishing and The Natural History Museum.
Kostadinova, A., and Gibson, D. I. 2000. The systematics of the echinostomes. In Echinostomes as experimental models for biological research, eds. B. Fried and T.K. Graczyk, pp. 31–57, Dordrecht: Kluwer Academic Publishers.
Kostadinova, A., Gibson, D. I., Biserkov, V., and Chipev, N. 2000. Re-validation of *Echinostoma miyagawai* Ishii, 1932 (Digenea: Echinostomatidae) on the basis of experimental completion of its life-cycle. Systematic Parasitology 45: 81–108.
Latham, A. D. M., and Poulin, R. 2002. New records of gastrointestinal helminths from the southern black-backed gull *(Larus dominicanus)* in New Zealand. New Zealand Journal of Zoology 29: 253–257.
Lello, J., Boag, B., Fenton, A., Stevenson, I. R., and Hudson, P. J. 2004. Competition and mutualism among the gut helminths of a mammalian host, Nature 428: 840–844.
Lie, J. K., and Basch, P. F. 1967. The life history of *Echinostoma paraensei* sp. n. (Trematoda: Echinostomatidae). Journal of Parasitology 53: 1192–1199.
Lu, S. C. 1982. Echinostomiasis in Taiwan. International. Journal of Zoonoses 9: 33–38
Lutz, A. 1924. Estudos sobre a evolução dos Endotrematodes brasileiros. Memórias do Instituto Oswaldo Cruz 17: 55–57.
Macko, J. K., Hanzelova, V., and Mackva, A. 2002. Contribution to the helminths of wild geese in the Slovak Republic. Helminthologia 39: 159–163.

Macpherson, C. N. 2005. Human behavior and the epidemiology of parasitic zoonoses. International Journal for Parasitology 35: 1319–1331.

Maldonado, A., Jr., Loker, E. S., Morgan, J. A. T., Rey, L., and Lanfredi, R. M. 2001a. Description of the adult worms of a new Brazilian isolate of *Echinostoma paraensei* (Platyhelmintes: Digenea) from its natural vertebrate host *Nectomys squamipes* by light and scanning electron microscopy and molecular analysis. Parasitology Research 87: 840–848.

Maldonado, A., Jr., Vieira, G. O., Garcia, J. S., Rey, L., and Lanfredi, R. M. 2001b. Biological aspects of a new isolate of *Echinostoma paraensei* (Trematoda: Echinostomatidae): susceptibility of sympatric snails and the natural vertebrate host. Parasitolology Research 87: 853–859.

Maldonado, A., Jr., Vieira, G. O., and Lanfredi, R. M. 2003. *Echinostoma luisreyi* n. sp. (Platyhelminths: Digenea) by light and scanning electron microscopy. Journal of Parasitology 89: 800–808.

Maldonado, A., Jr., Vieira, G. O., Garcia, J. S., Rey, L., and Lanfredi, R. M. 2005 Changes on *Schistosoma mansoni* (Digenea: Schistosomatidae) worm load in *Nectomys squamipes* (Rodentia: Sigmodontinae) concurrently infected with *Echinostoma paraensei* (Digenea: Echinostomatidae). Memórias do Instituto Oswaldo Cruz 96: 193–198.

Maldonado, A., Jr., Zeitone, B. K., Amado, L. A., Rosa, I. F., Machado-Silva, J. R., and Lanfredi, R. M. 2005. Biological variation between two Brazilian geographical isolates of *Echinostoma paraensei*. Journal of Helminthological 79: 345–351.

Maldonado, A., Jr., Gentile, R., Fernandes-Moraes, C. C., D'Andrea, P. S., Lanfredi, R. M., and Rey, L. 2006. Helminth communities of *Nectomys squamipes* naturally infected by the exotic trematode *Schistosoma mansoni* in southeastern Brasil. Journal of Helminthology 80: 369–375.

McAlpine, D. F., and Burt, M. D. B. 1998. Helminths of bullfrogs, *Rana catesbiana*, green frogs, *R. clamitans*, and leopard frogs, *R. pipiens* I New Brunswick. Canadian Field Naturalist 112: 50–68.

Mello, A. 1993. Da ocorrência de *Echinostoma revolutum* (Froelich, 1802) em *Gallus domesticus* no Brasil. Archivos da Escola Superior de Agricultura Medica Veterinária do Rio de Janeiro 10: 172.

Miyamoto, K. 1987. Studies on the zoonoses in Hokkaido, Japan. 8. The prevalence of metacercariae of *Echinostoma hortense* Asada, 1926 in Loaches. Hokkaido Igaku Zasshi 62: 890–894.

Miyamoto, K., Nakao, M., Ohnishi, K., and Inoaka, T. 1984. Studies on zoonoses in Hokkaido, Japan. 6. Experimental human Echinostomiasis. Hokkaido Journal of Medical Sciences 59: 696–700. et al, 198.

Mozoon, R. B., and Kitikoon, V. 1989. *Lymnaea (Bullastra) cumingiana* Pfeifer (Pulomonata: Lymnaeidae) second intermediate host of *Echinostoma malayanum* in the Philippines. Southeast Asian J. Trop Méd Public Health 22: 16–22.

Muñoz-Antoli, C., Sotillo, J., Monteagudo, C., Fried, B., Marcilla, A., and Toledo, R. 2007. Development and pathology of *Echinostoma caproni* in experimentally infected mice. Journal of Parasitology 93: 854–859.

Nassi, H., and Dupouy, J. 1988. étude expérimentale du cicle biologique d *Echinostoma parvocirrus* n. sp. (Trematoda: Echinostomatidae), parasite larvaire de Biomphalaria glabrata em Guadeloupe. Annales de Parasitologie Humaine et Comparee 63: 103–118.

Niewiadomska, K., Valtonen, E. T., and Siddall, R. 1997. Cercariae from *Lymnae stagnalis* in Lake Kuuhankavesi (Central Finland). Acta Parasitologica 42: 132–137.

Noland, G. S., Graczyk, T. K., Fried, B., and Kumar, N. 2007. Enhanced malaria parasite transmission from helminth co-infected mice. American Journal of Tropical Medicine and Hygiene 76: 1052–1056.

Nollen, P. M. 1999. Mating behaviour of *Echinostoma trivolvis* and *E. paraensei* in concurrent infections in hamsters. Journal of Helminthology 73: 329–332.

Park, Y. K., Sohn, C. T., Park, G. M., Hwang, M. K., and Chung, P. R. 2006. Host specificity of *Pisidium coreanum* (Bivalvia: Sphaeriidae) to larval infection with a human intestinal fluke *Echinostoma cinetorchis* (Trematoda: Echinostomatidae) in Korea. Journal of Parasitology 92: 1118–1220.

Radomyos, P., Bunnag, D., and Harinasuta, T. 1982. *Echinostoma ilocanum* (Garrison, 1908) Odhner, 1911 infection in man in Taliland. Southeast Asia Journal Tropical Medicine Public Health 13: 265–269.
Rashed, A. A. 2002. Biological studies on the snail intermediate host of schistosomiasis with a special emphasis on using larval echinostomes as biocontrol agent against larval schistosomes and snails. Journal of the Egyptian Society of Parasitology 32: 775–784.
Richards, J., and Brygoo, E. R. 1978. Life cycle of the trematode Echinostoma caproni Richard, 1964 (Echinostomatoidea) Annales de Parasitologie Humaine et Comparée. 53(3): 265–75.
Rim, H. L. 1982. Intestinal trematodiasis. Hillyer, G.V., Hopla, C.E. (eds.). CRC Handbook Series in Zoonoses. Boca Raton, Florida: CRC Press, Inc., 53–69.
Sandland, G. J., Rodgers, J. K., and Minchella, D. J. 2007. Interespecific antagonism and virulence in hosts exposed to two parasite species. Journal of Invertebrate Pathology 96: 43–47.
Schotthoefer, A. M., Cole, R. A., and Beasley, V. R. 2003. Relationship of tadpole stage to location of echinostome cercariae encystment and the consequences for tadpole survival. Journal of Parasitology 89: 475–482.
Schuster, R., Schaffer, T., and Shimalov, V. 2002. The helminth fauna of indigenous white storks *(Ciconia ciconia)*. Berliner und Munchener Tierarzliche Wochenschrift 115: 435–439.
Sianto, L., Reinhard, K. J., Chame, M., Chaves, S., Mendonça, S., Gonçalves, M. L., Fernandes, A., Ferreira, L. F., and Araújo, A. 2005. The finding of Echinostoma (Trematoda: Digenea) and hookworm eggs in coprolites collected from a Brazilian mummified body dated 600–1,200 years before present. Journal of Parasitology 9: 972–975.
Sitko, J. 1998. Trematodes of birds of prey (Falconiformes) in Czech Republic. Helminthologia 35: 131–146.
Sohn, W. M., and Chain, J. Y. 2005. Infection status with helminthes in feral cats purchased from a market in Busan, Republic of Korea. Korean Journal of Parasitology 43: 93–100.
Sorensen, R. E., Kanev, I., Fried, B., and Minchella, D. J. 1997. The occurrence and identification of *Echinostoma revolutum* from North American *Lymnea elodes* snails. Journal of Parasitology 83: 169–170.
Sotillo, J., Valero, L., Sánchez del Pino, M. M., Fried, B., Esteban, J. G., Marcilla, A., and Toledo, R. 2008. Identification of antigen proteins from *Echinostoma caproni* (Trematoda) recognized by mouse immunoglobulin, M, A and G using an immunoproteomic approach. Parasite Immunology 30: 271–279.
Suhardono Roberts, J. A., and Coperman, D. B. 2006. Biological control of *Fasciola gigantica* with *Echinostoma revolutum*. Veterinary Parasitology 31: 166–170.
Taft, S. J., Suchow, K., and VanHorn, M. 1993. Helminths from some Minnesota and Wisconsin raptors. Journal of Helminthological Society of Washington 60: 260–263.
Toledo, R., and Fried, B. 2005. Echinostomes as experimental models for interactions between adult parasites and vertebrate hosts. Trends in Parasitology 21: 251–254.
Toledo, R., Muñoz-Antoli, C., and Esteban, J. G. 2000. The life-cycle of *Echinostoma friedi* n. sp. (Trematoda: Echinostomatidae) in Spain and a discussion on the relationships within the *"revolutum"* group based on cercarial chaetotaxy. Systematic Parasitology 45: 199–217.
Toledo, R., Monteagudo, C., Espert, A., Fried, B., Esteban, J. G., and Marcilla, A. 2006. *Echinostoma caproni*: intestinal pathology in the golden hamster, a highly compatible host, and the wistar rat, a less compatible host. Experimental Parasitology 112: 164–171.
Toledo, R., Muñoz-Antolí, C., and Fried, B. 2007. The use of echinostomes to study host-parasite relationships between larval trematodes and invertebrate and cold-blooded vertebrate hosts. Parasitology Research 100: 1177–1185.
Travassos, L., Freitas, J. F., and Kohn, A. 1968. Encontro de *Echinostoma revolutum* (Froelich, 1802) em falo de pato doméstico. Atas da Sociedade de Biologia do Rio de Janeiro 11: 215–216.
Vayrynen, T., Siddall, R., Valtonen, E. T., and Taskinen, J. 2000. Patterns of trematode parasitism in Lymnaeid snails from northern and central Finland. Annales Zoologici Fennici 37: 189–1999.
Zabiega, M. H. 1996. Helminths of mink, *Mustela vison*, and muskrats, *Ondatra zibethicus*, in southern Illinois. Journal of the Helminthological Society of Washington 63: 246–250.

Chapter 7
Echinostomes in Humans

Jong-Yil Chai

Contents

7.1 Introduction ... 148
7.2 Taxonomy, Biology and Epidemiology of Echinostomes Infecting Humans ... 150
7.2.1 *Acanthoparyphium tyosenense* Yamaguti, 1939 ... 152
7.2.2 *Artyfechinostomum malayanum* (Leiper, 1911) Mendheim, 1943 ... 154
7.2.3 *Artyfechinostomum oraoni* Bandyopadhyay et al., Manna and Nandy, 1989 ... 154
7.2.4 *Echinochasmus fujianensis* Cheng et al., 1992 ... 155
7.2.5 *Echinochasmus japonicus* Tanabe, 1926 ... 155
7.2.6 *Echinochasmus jiufoensis* Yu and Mott, 1994 ... 156
7.2.7 *Echinochasmus liliputanus* (Looss, 1896) Odhner, 1910 ... 156
7.2.8 *Echinochasmus perfoliatus* (Ratz, 1908) Dietz, 1910 ... 157
7.2.9 *Echinoparyphium recurvatum* (Von Linstow, 1873) Diez, 1909 ... 158
7.2.10 *Echinostoma angustitestis* Wang, 1977 ... 160
7.2.11 *Echinostoma cinetorchis* Ando and Ozaki, 1923 ... 160
7.2.12 *Echinostoma echinatum* (Zeder, 1803) Rudolphi, 1809 ... 160
7.2.13 *Echinostoma hortense* Asada, 1926 ... 162
7.2.14 *Echinostoma ilocanum* (Garrison, 1908) Odhner, 1911 ... 164
7.2.15 *Echinostoma macrorchis* Ando and Ozaki, 1923 ... 165
7.2.16 *Echinostoma revolutum* (Froelich, 1802) Looss, 1899 ... 165
7.2.17 *Episthmium caninum* (Verma, 1935) Yamaguti, 1958 ... 166
7.2.18 *Himasthla muehlensi* Vogel, 1933 ... 166
7.2.19 *Hypoderaeum conoideum* (Block, 1872) Diez, 1909 ... 166
7.2.20 *Isthmiophora melis* (Schrank, 1788) Lühe, 1909 ... 167
7.3 Pathology and Symptomatology of Echinostomes Infecting Humans ... 167
7.3.1 Pathogenicity of Echinostomes ... 167
7.3.2 Pathological Findings in the Intestine of Experimental Animals ... 168
7.3.3 Symptoms and Signs of Human Echinostomiasis ... 168

J.-Y. Chai (✉)
Department of Parasitology and Tropical Medicine, Seoul National University College of Medicine and Institute of Endemic Diseases, Seoul National University Medical Research Center, Seoul 110-799, Korea
e-mail: cjy@snu.ac.kr

B. Fried and R. Toledo (eds.) *The Biology of Echinostomes*,
DOI: 10.1007/978-0-387-09577-6_7, © Springer Science+Business Media, LLC 2009

7.3.4 Endoscopic Findings in *E. hortense* Infection 171
7.4 Immunology of Human Echinostomiasis 172
7.4.1 Host Resistance to Echinostome Infection 172
7.4.2 Role of Goblet Cells and Mast Cells 172
7.4.3 Role of Other Effector Cells and Antibodies 173
7.4.4 Cytokines and Th1/Th2 Immune Responses 173
7.5 Diagnosis, Treatment, and Control of Echinostomes Infecting Humans 174
7.5.1 Diagnosis of Human Echinostomiasis 174
7.5.2 Treatment and Control of Echinostomiasis in Humans 175
References 176

Abstract A total of 20 species belonging to nine genera of Echinostomatidae are known to cause human infections around the world. They constitute an important group of food-borne trematodes of public health importance chiefly in the Southeast Asia and Far East. The largest genus is *Echinostoma* (seven species), followed by *Echinochasmus* (five species) and others (1–2 species for each genus), i.e., *Acanthoparyphium, Artyfechinostomum, Echinoparyphium, Episthmium, Himasthla, Hypoderaeum,* and *Isthmiophora*. Various types of food animals, including freshwater fish, brackish water fish, freshwater snails, brackish water snails (gastropods and bivalves), and amphibians are the sources of human infection. The pathogenicity and host-parasite relationships have been studied extensively in several species, including *Echinostoma hortense* (human-infecting species) and *Echinostoma trivolvis* and *Echinostoma caproni* (animal-infecting species). Mechanical damages by the worms and toxic effects by their metabolites seem to be important factors in the pathogenesis of human echinostomiasis. Mucosal ulcerations and bleeding at the upper part of the duodenum or at the distal part of the stomach are occasionally encountered in gastroduodenal endoscopy of human *E. hortense* infections. However, clinical importance of each echinostome species, as well as mucosal defense mechanisms of different hosts, is yet poorly understood. Diagnosis of echinostomiasis can be done by recovery of eggs in the feces, but specific diagnosis is difficult because of morphological similarity of eggs. Praziquantel has been proved to be effective against echinostomiasis. Epidemiological surveys and detection of human infections are required for better understanding of the geographical distribution and endemicity of each echinostome species.

7.1 Introduction

Flukes of the family Echinostomatidae (= echinostomes) are morphologically characterized by the presence of a head collar with collar spines around the oral sucker, and the number and arrangement of collar spines is an important key for taxonomic purposes (Toledo et al. 2006). They consist of 50 genera, 355 species, and six subspecies (Yamaguti 1958). They are predominantly parasites of birds, and also of mammals, including humans (Yamaguti 1958). Among them, until 1964, a total of 13 species were known to cause human infections (Yamashita 1964). However, at present, the number of human-infecting echinostomes has increased to a total of 20–21 species (Chai 2007); in this chapter 20 species are listed (Table 7.1). They consist of three groups of genera (Beaver et al. 1984); those with collar spines arranged in two rows, interrupted ventrally but not dorsally (*Echinostoma, Acanthoparyphium, Hypoderaeum, Artyfechinostomum*, *Echinoparyphium*, and *Isthmiophora*), collar spines arranged in a single row, interrupted ventrally (*Himasthla*), collar spines arranged in a single or alternative row, interrupted both ventrally and dorsally (*Echinochasmus* and *Episthmium*). The human-infecting echinostomes are distributed mainly in the Southeast Asia and Far East (Rim 1982; Chai 2007).

Table 7.1 List of species and morphological characters of echinostomes infecting humans

Species	Worm size (length: mm)	No. of collar spines	References
Acanthoparyphium tyosenense	2.5–3.1	23	Chai et al. (2001)
Artyfechinostomum malayanum	5.0–10.0	43	Belizario et al. (2007)
Artyfechinostomum oraoni	5.5–13.9	41	Bandyopadhyay and Nandy (1986)
Echinochasmus fujianensis	1.5–2.0	24	Cheng et al. (1992a)
Echinochasmus japonicus	0.7–1.0	24	Seo et al. (1985a)
Echinochasmus jiufoensis	0.5–0.8	24	Liang and Ke (1988)
Echinochasmus liliputanus	0.5–0.8	24	Witenberg (1932)
Echinochasmus perfoliatus	0.5–12.0	24	Rim (1982)
Echinoparyphium recurvatum	2.8–4.0	43–45	Rim (1982)
Echinostoma angustitestis	5.8–6.1	41	Cheng et al. (1992b)
Echinostoma cinetorchis	10.8–12.6	36–38 (mostly 37)	Seo et al. (1980)
Echinostoma echinatum	13.0–15.0	37	Rim (1982) as E. lindoense
Echinostoma hortense	9.0–12.6	27–28	Seo et al. (1985b)
Echinostoma ilocanum	2.5–6.5	49–51	Rim (1982)
Echinostoma macrorchis	4.6–7.8	43–47 (mostly 45)	Rim (1982)
Echinostoma revolutum	6.5–30.0	37	Kanev (1994)
Episthmium caninum	0.8–1.2	24	Radomyos et al. (1985)
Himasthla muehlensi	11.0–17.7	32	Rim (1982)
Hypoderaeum conoideum	5.0–12.0	47–53 (mostly 49)	Rim (1982)
Isthmiophora melis	4.6–7.4	27	Kostadinova and Gibson (2002)

Echinostomes have a 3-host life cycle; namely, the first intermediate host (aquatic snails), the second intermediate host (snails, clams, fishes, amphibian, and reptiles), and the definitive host (fishes, reptiles, birds, and mammals) (Toledo et al. 2006; Chai 2007). The definitive host is mainly infected by consuming the second intermediate hosts harboring the metacercariae. When infected in the definitive host, including humans, the main habitat of the flukes is the small intestine, and probably less frequently the pyloric area of the stomach. The clinical symptoms in humans are mostly gastrointestinal, including abdominal pain, diarrhea, weakness, and weight loss. In this chapter, the author focused on the taxonomy, biology, epidemiology, geographical distribution, pathology, symptomatology, diagnosis, treatment, prevention, and control of human-infecting echinostomes.

7.2 Taxonomy, Biology and Epidemiology of Echinostomes Infecting Humans

For taxonomic differentiation of echinostome flukes, the number, shape, arrangement, and relative size of collar spines, the position of the cirrus sac and structure of the seminal vesicle, the development of the pars prostatica, the size and armament of the cirrus, the position of the ovary and testes, the location and structure of vitellaria, the nature of the tegumental armament, the presence of a uroproct, and the body shape, in terms of length/width ratios are important morphologic characters at the generic as well as specific levels (Kostadinova and Gibson 2000; Maldonado et al. 2005). In addition, biological characteristics, including cercarial morphology (Kanev et al. 1995) and chaetotaxy (Kostadinova 1999; Toledo et al. 2000), and genetic characteristics, including gene sequences (Morgan and Blair 1995; Fujino et al. 1995b; Sorensen et al. 1998; Kostadinova et al. 2003), are also used for species identification.

The echinostome eggs are immature when laid, but they mature after leaving the host, and hatch in about 3 weeks in the environment (Rim 1982). Miracidia enter the snail host, where they develop into mother rediae and in turn, daughter rediae, and cercariae (Beaver et al. 1984). The cercariae have well-developed tails and usually bear collar spines around the oral sucker similar to that of the adults (Rim 1982). The mature cercariae may directly encyst within the body of their snail host, or escape from the snail, and enter and encyst in other snails of the same or different species (Rim 1982). They may also encyst in bivalves, insects, tadpoles and frogs, fishes, or on vegetation (Rim 1982). The encysted metacercariae are round or oval, and show two branches of the excretory bladder filled with coarse granules and a head collar with collar spines (Rim 1982).

Humans or animals are infected through ingestion of metacercariae encysted in the second intermediate host. Eating raw snails, clams, fishes, or vegetation harboring metacercariae is the main practical mode of infection in humans (Table 7.2). However, it has been postulated that humans can also be infected through drinking untreated water containing echinostome cercariae, which could become encysted when exposed to the human gastric juice (Xiao et al. 2005).

Table 7.2 Possible source of human or animal infections with echinostomes

Parasite species	Source of human or animal infections
Fish-borne	
Echinochasmus fujianensis	Freshwater fish, *Pseudorasbora parva, Cyprinus carpio*
Echinochasmus japonicus	Freshwater fish, *Pseudorasbora parva, Hypomesus olidus, Gnathopogon strigatus*
Echinochasmus jiufoensis	Unknown
Echinochasmus liliputanus	Freshwater fish, *Pseudorasbora parva*, goldfish
Echinochasmus perfoliatus	Freshwater fish, *Carassius* sp.
Echinostoma angustitestis	Freshwater fish
Echinostoma cinetorchis	Freshwater fish, *Misgurnus anguillicaudatus*
Echinostoma hortense	Freshwater fish, *Misgurnus anguillicaudatus, Misgurnus mizolepis, Odontobutis obscura interrupta, Moroco oxycephalus, Coreoperca kawamebari, Squalidus coreanus*
Episthmium caninum	Freshwater fish
Snail-borne	
Acanthoparyphium tyosenense	Bivalve, *Mactra veneriformis, Solen grandis, gastropod, Neverita bicolor*
Artyfechinostomum malayanum	Snail, *Digoniostoma pulchella, large snail, Pila scutata, Lymnaea (Bullastra) cumingiana*
Echinoparyphium recurvatum	Freshwater snail, *Planorbis planorbis, Lymnaea* sp., *Lymnaea stagnalis*
Echinostoma cinetorchis	Freshwater snail, *Radix auricularia coreanus, Physa acuta, Cipangopaludina chinensis malleata*
Echinostoma echinatum	Mussel, *Corbicula lindoensis, Corbicula sucplanta, Idiopoma javanica, freshwater snail, Biomphalaria glabrata*
Echinostoma ilocanum	Large snail, *Pila conica, Viviparus javanicus*
Echinostoma macrorchis	Large snail, *Cipangopaludina malleata, Cipangopaludina japonica, Segmentina nitiella, Viviparus malleatus*
Echinostoma revolutum	Snail or clam, *Corbicula producta*
Himasthla muehlensi	Clams, *Venus mercenaria*, bivalve mollusk, *Mytilus, Mya* spp.
Hypoderaeum conoideum	Snail *Lymnaea stagnalis, Lymnaea limosa, Lymnaea tumidae, Planorbis planorbis*
Amphibia-borne	
Echinoparyphium recurvatum	Tadpole and frog of *Rana temporaria*
Echinostoma macrorchis	Frog of *Rana* sp.
Echinostoma revolutum	Tadpole
Hypoderaeum conoideum	Tadpole
Isthmiophora melis	Tadpole

The history of discovery, morphology of taxonomic importance, life cycle, kinds of intermediate and definitive hosts, epidemiology, and geographical distribution (Table 7.3) of each human-infecting echinostome is as follows.

Table 7.3 Geographical distribution of echinostomes infecting humans

Species	Distribution
Acanthoparyphium tyosenense	Republic of Korea, Japan
Artyfechinostomum malayanum	Malaysia, Singapore, Thailand, Indonesia, India, the Philippines
Artyfechinostomum oraoni	India
Echinochasmus fujianensis	China
Echinochasmus japonicus	Japan, Republic of Korea, China
Echinochasmus jiufoensis	China
Echinochasmus liliputanus	Egypt, Syria, Palestine, China
Echinochasmus perfoliatus	Hungary, Italy, Romania, Russia, Japan, China, Taiwan, Denmark
Echinoparyphium recurvatum	Cosmopolitan, especially Taiwan, Indonesia, Egypt
Echinostoma angustitestis	China
Echinostoma cinetorchis	Japan, Republic of Korea, China
Echinostoma echinatum	European countries (esp. Germany), Asia (Indonesia), South America (Brazil)
Echinostoma hortense	Japan, Republic of Korea, China
Echinostoma ilocanum	The Philippines, Indonesia, China, Thailand, India
Echinostoma macrorchis	Japan
Echinostoma revolutum	Asia, Europe, Africa, Australia, New Zealand, North and South America
Episthmium caninum	India, Thailand
Himasthla muehlensi	USA (?)
Hypoderaeum conoideum	Europe (including Spain), Russia (Siberia), Japan, Thailand
Isthmiophora melis	Europe (including Rumania), Russia, China, Taiwan, North America

7.2.1 *Acanthoparyphium tyosenense* Yamaguti, 1939

The first description of this echinostome was based on worms obtained from the small intestines of the duck *Melanitta fusca stejnegeri* and *Melanitta nigra americana* caught in the Republic of Korea (Yamaguti 1939). It has a characteristic head collar with a total of 23 collar spines arranged in a single row, without ventral corner spines, a long cirrus sac reaching beyond the posterior margin of the acetabulum, and vitellaria extending to the level of the cirrus sac or Mehlis' gland (Fig. 7.1) (Chai et al. 2001). The marine megagastropods *Lunatia fortuni* and *Glassaulax didyma* have been verified to be the first intermediate hosts in the Republic of Korea (Kim et al. 2004). The marine gastropods, *Tympanotonus microptera, Cerithidea cingulata,* and *Cerithidea largillierti,* shed *Cercaria yamagutii* in Japan (Ito 1957), and these cercariae are now considered to have been those of *A. tyosenense* (Kim et al. 2004). The metacercariae were detected in various species of brackish water mollusks, i.e., four species of bivalves, *Mactra veneriformis, Solen grandis, Solen strictus*, and *Ruditapes philippinarum*, and a species of gastropod *Neverita bicolor* (Chai et al. 2001; Kim et al. 2004). The adult flukes were experimentally

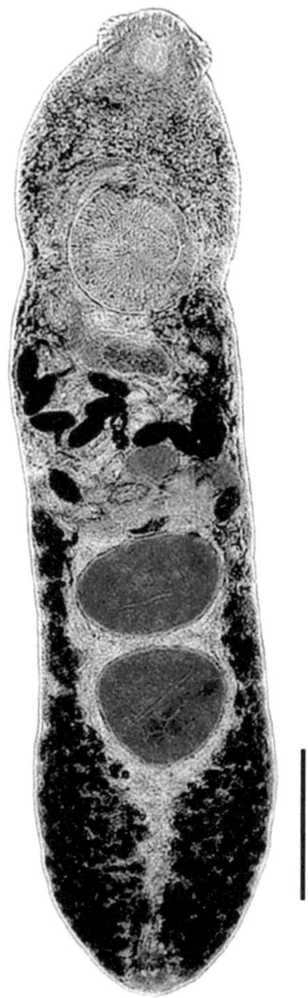

Fig. 7.1 *Acanthoparyphium tyosenense* adult from an experimentally infected chick, day 15 postinfection. Acetocarmine stain. Scale bar = 0.4 mm

obtained from domestic chicks (Chai et al. 2001; Han et al. 2003), and sea gulls *Larus crassiostris* (Kim et al. 2004). In chicks, the worms grew quickly to become ovigerous adults in 5 days and survived at least up to 38 days, chiefly in the jejunum (Han et al. 2003). The number of uterine eggs reached a peak during day 10 and day 15 after infection (Han et al. 2003). Human infections were discovered in ten patients residing in two coastal villages in Chollabuk-do (Province), Republic of Korea (Chai et al. 2001). The patients recalled that they had eaten improperly cooked marine bivalves and gastropods. This echinostome has been reported from the Republic of Korea and Japan (Kim et al. 2004).

7.2.2 *Artyfechinostomum malayanum* (Leiper, 1911) Mendheim, 1943

(syn. *Echinostoma malayanum* Leiper, 1911, *Euparyphium malayanum* Odhner, 1913, *Artyfechinostomum sufrartyfex* Lane, 1915, *Artyfechinostomum mehrai* Faruqui, 1930, *Paryphostomum sufrartyfex* Bhalerao, 1931; *Isthmiophora malayana* Yamaguti, 1958)

Artyfechinostomum malayanum (under the name *Echinostoma malayanum*) was first described from humans in Malaysia, and then from people in Singapore, Thailand, Indonesia, India, and the Philippines (Beaver et al. 1984; Maji et al. 1993; Radomyos et al. 1998; Yu and Mott 1994; Belizario et al. 2007). It has a small head collar, with a total of 43 (39–45) collar spines arranged in two alternating rows at the dorsal side, two deeply lobed testes, and a large and long cirrus sac reaching beyond the ventral sucker (Premvati and Pande 1974). Premvati and Pande (1974) and Kostadinova and Gibson (2002) described that *Echinostoma malayanum* Leiper 1911 better fits the generic features of *Artyfechinostomum* Lane 1915 and thus *E. malayanum* should be renamed as *A. malayanum.* A related species, *A. sufrartyfex,* was found in an Assamese girl in India (Beaver et al. 1984). Later, *A. sufrartyfex* was found in pigs (Ahluwalia 1962), cats, and dogs (Dubey et al. 1969). Kostadinova et al. (2002) suggested *A. sufrartyfex* is conspecific with *A. malayanum* (Leiper, 1911) that precedes *A. sufrartyfex* (Lane, 1915). On the other hand, *A. mehrai* Faruqui, 1930, later synonymized with *A. malayanum* by Ahluwalia (1962) and Kostadinova et al. (2002), was described in 1930 from a Hindu girl with a history of diarrhea, vomiting, anorexia, and loss of weight; worms were recovered from the vomits and from gastric contents (Raghunathan and Srinivasan 1962). Now this fluke is known to be distributed in Malaysia, Thailand, Indonesia, India, and the Philippines (Beaver et al. 1984; Maji et al. 1993; Radomyos et al. 1998; Yu and Mott 1994; Belizario et al. 2007). The first intermediate host is freshwater snails, *Indoplanorbis exustus* and *Gyraulus convexiusculus,* and the cercariae encyst in various species of snails, i.e., *Pila scutata, Lymnaea (Bullastra) cumingiana,* and *Digoniostoma pulchella* (Yu and Mott 1994). The definitive hosts of this fluke include humans, pigs, rats, cats, dogs, mice, hamsters, and house shrews (Lie 1963; Dubey et al. 1969; Premvati and Pande 1974; Yu and Mott 1994). Human infections are occasionally reported in Southeast Asia (Yu and Mott 1994; Belizario et al. 2007).

7.2.3 *Artyfechinostomum oraoni* Bandyopadhyay et al., Manna and Nandy, 1989

This species was reported from 20 human infections in a tribal community near Calcutta, India (Bandyopadhyay and Nandy 1986; Bandyopadhyay et al. 1989). The freshwater snail, *Lymnaea* sp., was experimentally proven to be a first intermediate host (Maji et al. 1995). It has been proven that *A. oraoni* provoked fatal diarrhea in naturally infected pigs (Bandyopadhyay et al. 1995).

7.2.4 Echinochasmus fujianensis Cheng et al., 1992

The first description of *E. fujianensis* was based on specimens from humans, dogs, cats, pigs, and rats in Fujian Province, China (Cheng et al. 1992a). It has a prominent head collar with 24 collar spines arranged in a single row, interrupted dorsally. *E. fujianensis* differs genetically from *Echinochasmus japonicus* as revealed by random amplified polymorphic DNA analysis (RAPD) patterns (Cheng et al. 1999). The prevalence among residents in five areas of southern Fujian Province was 3.2% (1.6–7.8%); two-third of the infected people were 3–15 years of age. In Hubei, Anhui, and Fujian Provinces, China, *E. fujianensis* is the dominant species among the three *Echinochasmus* species existing, i.e., *E. fujianensis, E. japonicus*, and *E. perfoliatus* (Cheng et al. 1999). Its first intermediate host is a species of freshwater snail, *Bellamya aeruginosa,* and the second intermediate hosts include *Pseudorasbora parva* and *Cyprinus carpio* (Yu and Mott 1994). The natural definitive hosts include dogs, cats, pigs, and rats (Cheng et al. 1992a).

7.2.5 Echinochasmus japonicus Tanabe, 1926

Tanabe (1926) first described *E. japonicus* based on specimens from experimental animals such as dogs, cats, rats, mice, and birds that have been fed the metacercariae encysted in freshwater fish in Japan (Tanabe 1926). Its characteristic morphology includes a small, plump body, the presence of a total of 24 collar spines, which are interrupted dorsally, two large and tandem testes, and a very small number (usually less than 5) of uterine eggs (Fig. 7.2) (Chai and Lee 2002). *E. japonicus* is known to exist mainly in the Far East (Chai and Lee 2002). Since an experimental human infection was successfully reported in Japan (Ujiie 1936), natural human infections have been found in various Provinces of China (Lin et al. 1985; Zhu et al. 1986), and the Republic of Korea (Seo et al. 1985a). The first intermediate host is a species of freshwater snail, *Parafossarulus manchouricus* (Lee et al. 1983; Choi et al. 2006). Eighteen species of freshwater fish have been found to be the second intermediate hosts, including *Pseudorasbora parva, Hypomesus olidus*, and *Gnathopogon strigatus* (Lee et al. 1984; Chai et al. 1985a; Choi et al. 2006). A part of its life cycle, from cercariae to adults, has been successfully completed in the laboratory (Choi et al. 2006). Natural infections in avian species such as ducks (Eom and Rim 1984) and egrets (Ryang et al. 1991), and mammalian species such as cats (Sohn and Chai 2005) have been confirmed by the recovery of adult flukes. In six counties of Fujian and Guangdong Provinces, China, the prevalence was reported to be 4.9% in humans, 39.7% among dogs, and 9.5% among cats (Yu and Mott 1994).

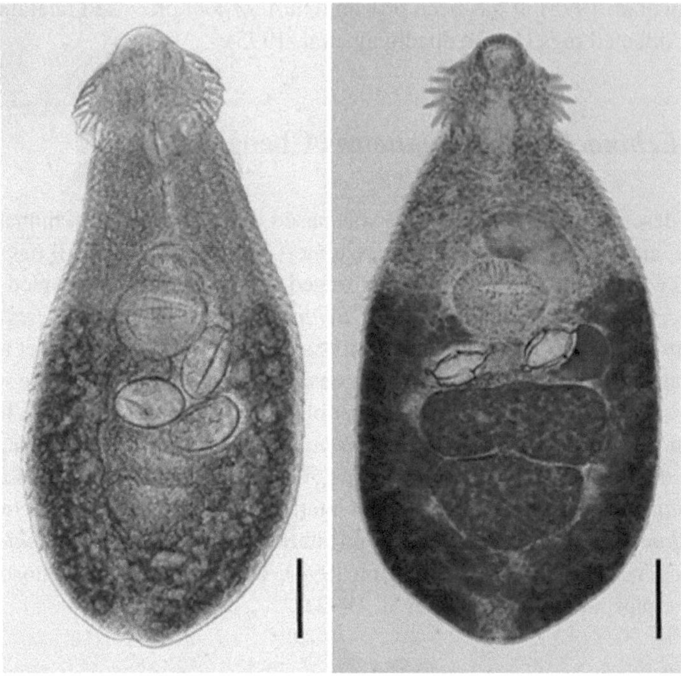

a b

Fig. 7.2 (**a**) *Echinochasmus japonicus* adult from an experimentally infected chick, containing three uterine eggs, day 10 postinfection. Acetocarmine stain. Scale bar = 0.1 mm. (**b**) *E. Japonicus* adult from an experimentally infected dog, day 12 postinfection. Acetocarmine stain. Scale bar = 0.1 mm

7.2.6 *Echinochasmus jiufoensis* Yu and Mott, 1994

In 1988, a new species of *Echinochasmus* was discovered at autopsy of a 6-month-old girl who died from pneumonia and dehydration in Guangzhou, China (Liang and Ke 1988). It was named as *Echinochasmus jiufoensis* (Liang and Ke 1988). This fluke has 24 collar spines arranged in a single row, interrupted dorsally and ventrally, and is closely allied to *Echinochasmus beleocephalus* Dietz, 1909, but distinguished from the latter in several distinct points (Liang and Ke 1988; Yu and Mott 1994). The life cycle and the route of infection in the child are unknown (Liang and Ke 1988).

7.2.7 *Echinochasmus liliputanus* (Looss, 1896) Odhner, 1910

The original description of *E. liliputanus* was based on specimens from dogs and cats in Egypt, Syria, and Palestine (Yamaguti 1958). It has a reniform head collar with 24 collar spines arranged in a single row, interrupted dorsally and ventrally, and vitellaria do not extend anteriorly up to the level of the ventral sucker (Witenberg 1932). Human infections were first discovered in Anhui Province, China, in 1991, with the prevalence rate of 13.4% among 2,426 people examined (Xiao et al. 1992). Higher infection rates were observed in age groups 3–15 years (22.7%) and 16–30 years (16.4%) than in others. In the same place, the infection rates in dogs and cats were 60 and 45%, respectively (Xiao et al. 1992). Since then, more than 2,500 human infection cases have been reported in Anhui Province, China (Xiao et al. 2005). The freshwater snail *Parafossarulus striatulus* (Yu and Mott 1994), and the freshwater fish *Pseudorasbora parva* (Yu and Mott 1994) and goldfish (Xiao et al. 2005) were the first and second intermediate hosts, respectively. The natural definitive hosts include badgers, foxes, and raccoons, in addition to dogs and cats (Yu and Mott 1994; El-Gayar 2007). It was postulated that humans could be infected with this echinostome through drinking untreated water containing the cercariae, since the infection rate was only 1.5% in the inhabitants who did not drink raw water, whereas it was 20.1% among those who used to drink raw water (Xiao et al. 1995). An experimental study was undertaken to explain the mechanism of the human oral infection with cercariae, and the phenomenon of cercarial encystment in the presence of human gastric juice was proposed as a possible mechanism (Xiao et al. 2005).

7.2.8 Echinochasmus perfoliatus (Ratz, 1908) Dietz, 1910

This echinostome is a common parasite of the small intestine of dogs and cats in Hungary, Italy, Romania, Russia, Japan, China, and Taiwan (Yu and Mott 1994; Shimalov and Shimalov 2002), and of red foxes in Denmark (Saeed et al. 2006). The body is elongated and larger than the closely related species, *E. japonicus*. It has a head collar with 24 collar spines arranged in a single row, interrupted dorsally and ventrally. An experimental human infection was reported in Japan (Tanabe 1922), and subsequently a natural human infection was discovered (Hirazawa 1928). Later, many human infections were reported from China (Guangdong, Fujian, Anhui, and Hubei Provinces), the prevalence rate being 1.8% (34/1,846), including a child who died from the infection; about 14,000 worms were found at autopsy (Yu and Mott 1994). The molluskan intermediate hosts include *Parafossarulus manchouricus, Bithynia leachi*, and *Lymnaea stagnalis* (Yamaguti 1958). Numerous species of freshwater fishes such as *Carassius* sp., *Zacco platypus, Zacco temminckii,* and *Pseudorasbora parva* harbor the metacercariae, which are encysted only on the gills (Rim 1982; Yu and Mott 1994). The natural definitive hosts are rats, cats, dogs, foxes, fowls, and wild boars (Yamashita 1964; Beaver et al. 1984; Saeed et al. 2006).

7.2.9 *Echinoparyphium recurvatum* (Von Linstow, 1873) Diez, 1909

(syn. *Echinoparyphium koidzumii* Tsuchimochi, 1924)

This echinostome is a cosmopolitan species infecting various species of birds and mammals including rats (Beaver et al. 1984; Lee et al. 1990c; Yu and Mott 1994; Betlejewska and Jorol 2002). It has a relatively small narrow body (Fig. 7.3), having a head collar with 45 collar spines arranged in two alternating rows and 4–5 end group spines on each side (Fig. 7.4), a large ventral sucker, and a short uterus

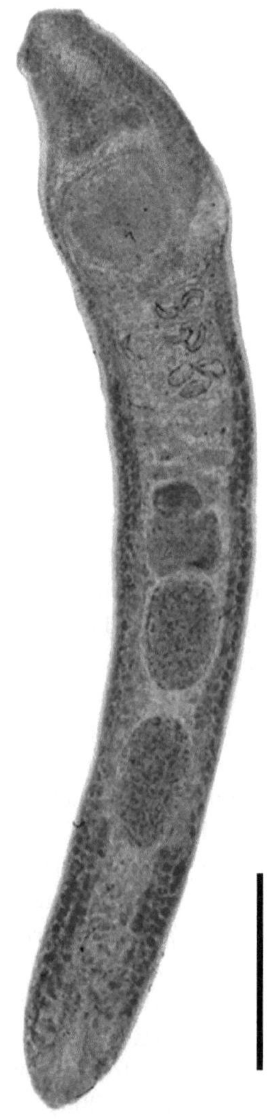

Fig. 7.3 *Echinoparyphium recurvatum* adult from a naturally infected rat caught at Yangyang-gun (County), Kangwon-do (Province), Republic of Korea. Acetocarmine stain. Scale bar = 0.6 mm

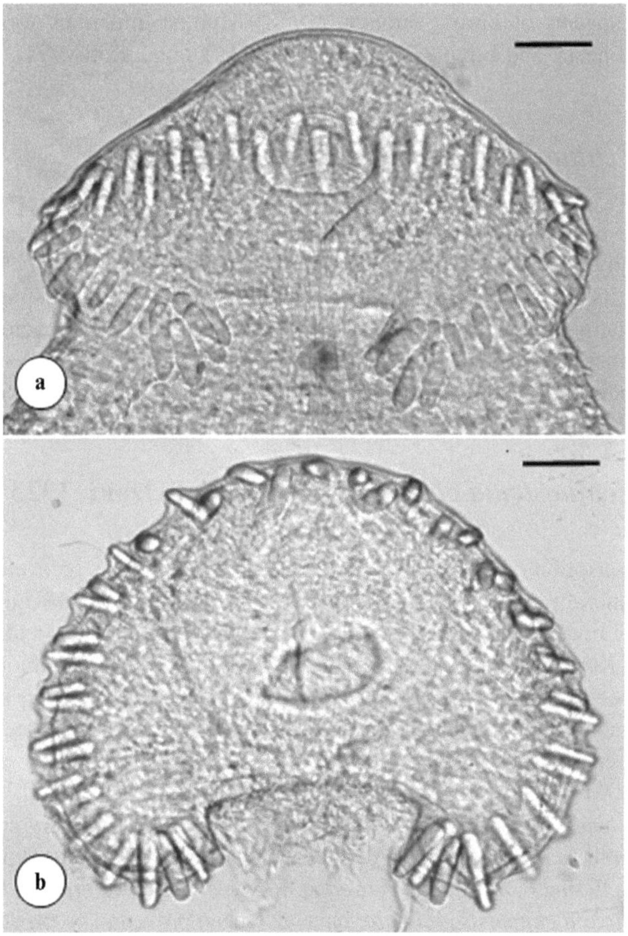

Fig. 7.4 Head collars ((**a**) Ventral view; (**b**) Head-on view) of *Echinoparyphium recurvatum* bearing total 45 collar spines. Specimens were from a naturally infected rat in Yangyang-gun (County), Kangwon-do (Province), Republic of Korea. Acetocarmine stain. Scale bar = 0.04 mm

with only a few eggs (Rim 1982; Lee et al. 1990c). It has been suggested that *E. recurvatum* complex consists of at least three species, including two other similar species, i.e., *E. pseudorecurvatum* and *E. mordwilkoi* (Fried 2001). The first intermediate host is planorbid snails, including *Physa alexandrina, Physa fontinalis, Planorbis planorbis, Lymnaea pervia, Lymnaea peregra*, *Valvata piscinalis*, and *Radix auricularia coreana* (Evans and Gordon 1983; McCarthy 1990; Sohn 1998). The metacercariae encyst in tadpoles and frogs of *Rana temporaria,* and also in snails, *P. planorbis* and *Lymnaea* sp. (Yu and Mott 1994), *R. auricularia coreana* (Sohn 1998), and *Lymnaea stagnalis* (Yurlova et al. 2006). The definitive hosts include house rats (Lee et al. 1990c), wild rats (*Arvicanthis niloticus*) (Beaver et al.

1984), and species of birds (Yamaguti 1958). Human infections were found in Taiwan, Indonesia, and Egypt (Beaver et al. 1984; Yu and Mott 1994).

7.2.10 Echinostoma angustitestis Wang, 1977

Echinostoma angustitestis was first described in 1977 based on specimens from dogs experimentally infected with metacercariae isolated from the freshwater fish (Yu and Mott 1994). Its oral sucker has a head collar with a total of 41 collar spines arranged in two alternating rows (Cheng et al. 1992b). Two human infections were reported in Fujian Province, China (Cheng et al. 1992b). The patients complained of dizziness, abdominal pain, and diarrhea (Cheng et al. 1992b). Its life history is unknown.

7.2.11 Echinostoma cinetorchis Ando and Ozaki, 1923

The first description of *E. cinetorchis* was based on specimens recovered from the small intestine of rats in Japan (Ando and Ozaki 1923). It was subsequently found in dogs in Taiwan (Sugimoto 1933; Fischthal and Kuntz 1975), and rats in the Republic of Korea (Seo et al. 1964, 1981). This fluke is also listed among the echinostome species existing in China (Wu 2004). Its characteristic morphology includes an abnormal location and/or disappearance of one or both testes (Fig. 7.5), and the presence of 36–38 collar spines (mostly 37) around the oral sucker (Chai and Lee 2002). The freshwater snails *Hippeutis cantori* (Lee et al. 1990a) and *Segmentina hemisphaerula* (Chung et al. 2001) were experimentally confirmed as the first as well as the second intermediate hosts (Lee et al. 1990a). Other freshwater snails, including *R. auricularia coreana*, *Physa acuta, Cipangopaludina* sp., and *Cipangopaludina chinensis malleata,* have also been proved to be the second intermediate hosts (Ahn et al. 1989; Chung and Jung 1999). In addition, tadpoles of *Rana nigromaculata, Rana rugosa*, and *Rana japonica* (Ando and Tsuyuki 1923), and the freshwater fish, especially the loach *Misgurnus anguillicaudatus* (Saito and Tani 1982; Seo et al. 1984), were proven to harbor the metacercarial stage. Rats (Ando and Ozaki 1923; Seo et al. 1964, 1981) and dogs (Cho et al. 1981) are the natural definitive hosts. Rats and mice are highly susceptible to experimental infection with *E. cinetorchis* (Lee et al. 1988c). Human infections were first discovered in Japan (Kawahara and Yamamoto 1933) and then in the Republic of Korea (Seo et al. 1980; Ryang et al. 1986; Lee et al. 1988a).

7.2.12 Echinostoma echinatum (Zeder, 1803) Rudolphi, 1809

(syn. *Echinostoma lindoense* Sandground and Bonne, 1940)

This echinostome species was first described based on specimens from the intestine of mammals in Germany under the name *Distoma echinatum* Zeder, 1803

Fig. 7.5 *Echinostoma cinetorchis* adult from an experimentally infected rat, day 17 postinfection. Both testes disappeared, as is a characteristic feature of this echinostome species. Acetocarmine stain. Scale bar = 0.1 mm

(Kanev 1994). It has 37 collar spines arranged in two alternating rows and closely resembles *Echinostoma revolutum*, and thus was synonymized with *E. revolutum* by Diez in 1909 (Kanev 1994). However, its taxonomic validity had been argued and is now acknowledged by various authors (Huffman and Fried 1990; Kanev 1994; Kostadinova 1995; Kostadinova and Gibson 2000). Then, *Echinostoma barbosai* Jeyarasasingam et al., 1972 and *Echinostoma lindoense* Sandground and Bonne, 1940 were synonymized with *E. echinatum* (Huffman and Fried 1990; Fried and Graczyk 2004). This fluke is known to be distributed in Europe, Asia, and

South America (Fried and Graczyk 2004). Its first intermediate is freshwater snails, including *Lymnaea, Planorbarius, Planorbis, Anisus, Gyraulus, Biomphalaria*, and *Viviparus* (Fried and Graczyk 2004). The mode of human infection is eating raw or insufficiently cooked mussels, *Corbicula lindoensis, Corbicula sucplanta,* and *Idiopoma javanica*, which contain the metacercariae (Beaver et al. 1984). In Brazil, *Biomphalaria glabrata* snails were the source of infection (Lie 1968). Rats and mice are experimental definitive hosts (Beaver et al. 1984). During 1937 and 1956, a high prevalence of 24–96% and heavy infections was reported among people in Celebes, Indonesia under the name of *E. lindoense* (Yu and Mott 1994).

7.2.13 *Echinostoma hortense* Asada, 1926

The first description of *E. hortense* was based on specimens obtained from rats in Japan (Asada 1926), and then this species was reported from rats in the Republic of Korea (Park 1938; Seo et al. 1964, 1981; Chai and Lee 2002) and China (Fan and Sun 1989). Its morphological characters include the presence of a total of 27–28 collar spines arranged in two alternative rows and a laterally located ovary (Chai and Lee 2002). Kostadinova and Gibson (2002) transferred *E. hortense* to *Isthmiophora* raising a new combination *Isthmiophora hortensis*, but here the name *E. hortense* is retained until a definite conclusion could be drawn. The first intermediate hosts are freshwater snails, *L. pervia* and *R. auricularia coreana* (Ahn and Kang 1988; Lee et al. 1991; Chai and Lee 2002). The second intermediate hosts include the loach, *M. anguillicaudatus* and *Misgurnus mizolepis*, and other freshwater fish, including *Odontobutis obscura interrupta* (Fig. 7.6), *Moroco oxycephalus, Coreoperca kawamebari*, and *Squalidus coreanus* (Chai et al. 1985b; Ryang et al. 1985; Ryang 1990; Lee et al. 1988b). Tadpoles of *Rana nigromaculata* were experimentally proven to be a second intermediate host (Fig. 7.7) (Lee et al. 1991). In China, 69.7% of the loach *M. anguillicaudatus* from a market in Liaoning Province was found infected with *E. hortense* metacercariae (Yu and Mott 1994). Rats (Park 1938; Seo et al. 1964, 1981), dogs (Cho et al. 1981), and cats (Sohn and Chai 2005) have been found to be natural definitive hosts. Mice, rats, and humans have all been determined, experimentally, susceptible to *E. hortense* infection (Seo et al. 1985b; Lee et al. 2004). Human infections have been found in Japan, Republic of Korea, and China (Seo et al. 1983; Chai and Lee 2002). In Japan, more than 20 human infections have been reported based on recovery of adult flukes (Miyamoto et al. 1983). In China, a survey was done in Liaoning Province of northeast China, and six out of ten hospitalized hepatitis patients who had eaten raw loach were

Fig. 7.7 Metacercariae of *Echinostoma hortense* encysted in an experimentally infected tadpole ((**a**) Two metacercariae in the gill; (**b**) A metacercaria isolated from the gill showing characteristic collar spines around the oral sucker, from Lee et al. (1991) with permission). Fresh preparations without stain. Scale bars = 0.12 mm (**a**) and 0.028 mm (**b**)

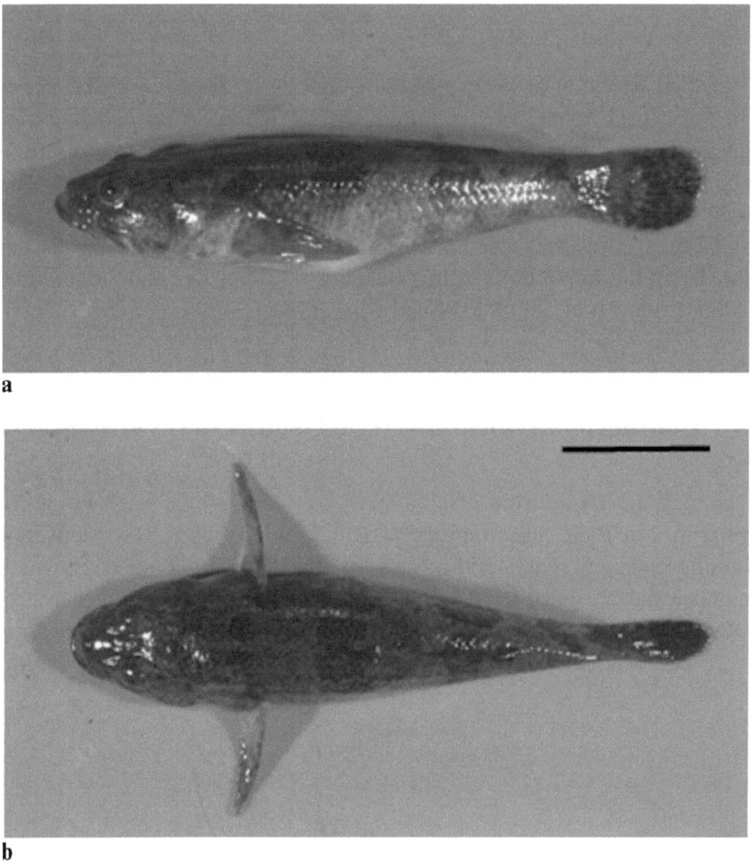

Fig. 7.6 A species of freshwater fish, *Odontobutis obcura interrupta* ((**a**) Lateral view; (**b**) Doral view), a second intermediate host for *Echinostoma hortense* in the Republic of Korea. Scale bar = 30 mm

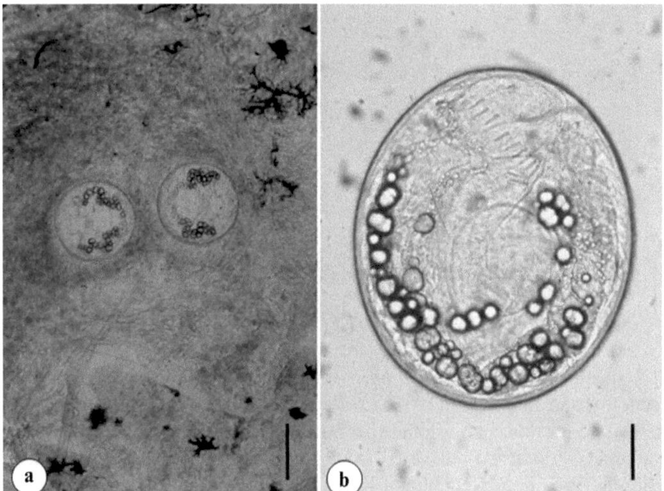

found infected (Chen et al. 1993). In the Republic of Korea, a 22.4% prevalence with heavy worm loads was reported among villagers of Cheongsong-gun, Kyongsangbuk-do (Province); three of 35 people from whom successful worm recovery was performed after anthelmintic medication and purgation revealed the recovered worm numbers of 134, 475, and 649 (Fig. 7.8), respectively (Lee et al. 1988b). Clinical cases diagnosed at gastroduodenal endoscopy have been reported occasionally in the Republic of Korea (Chai et al. 1994; Lee and Hong 2002; Cho et al. 2003; Chang et al. 2005; Park and Kim 2006).

7.2.14 Echinostoma ilocanum (Garrison, 1908) Odhner, 1911

The eggs of *E. ilocanum* were first found in the feces of five prisoners in Manila, the Philippines in 1907, and from one patient 21 adult flukes were recovered after anthelmintic treatment (Cross and Basaca-Sevilla 1986). Its characteristic morphology includes the presence of 49–51 collar spines and deeply lobed testes (Rim 1982; Radomyos et al. 1982). The first intermediate hosts are *Gyraulus* or *Hippeutis* snails (Yu and Mott 1994). The sources of human infections include large snails

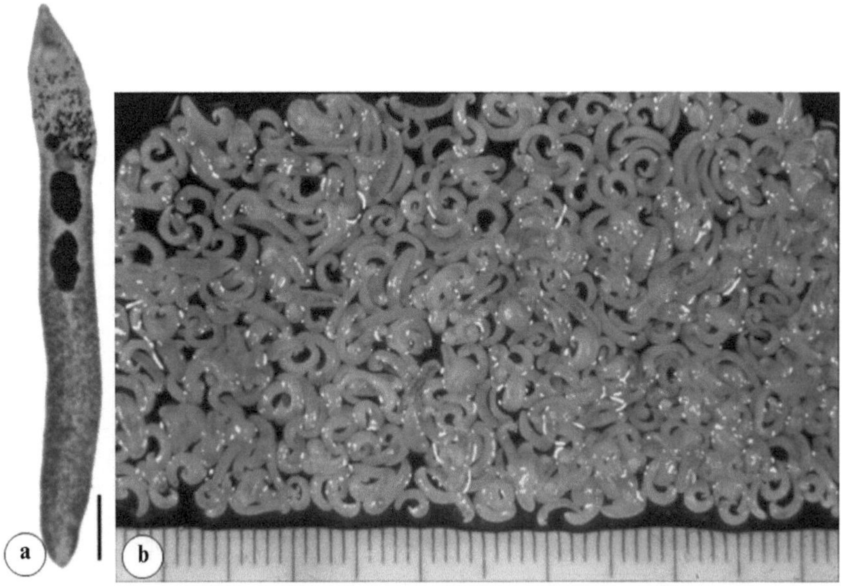

Fig. 7.8 (**a**) A specimen of *Echinostoma hortense* adult from an experimentally infected rat, day 120 postinfection (from Seo et al. (1985b) with permission). The laterally deviated ovary is a characteristic feature for this echinostome species. Acetocarmine stain. Scale bar = 0.16 mm. (**b**) *E. hortense* adults (649 worms) recovered from a heavily infected patient in Cheongsong-gun, Kyongsangnam-do (Province), Republic of Korea

Pila conica (Philippines) and *Viviparus javanicus* (Java) (Beaver et al. 1984). The Norway rat and the dog are reservoir hosts (Beaver et al. 1984). Human infections have been reported from Celebes, Java, Indonesia, China, Thailand, the Philippines, and India (Radomyos et al. 1982; Cross and Basaca-Sevilla 1981, 1986; Yu and Mott 1994; Grover et al. 1998). The prevalence among the Ilocano population in northern Luzon, Philippines, was 11% on average (7–17% by age group) (Cross and Basaca-Sevilla 1986).

7.2.15 *Echinostoma macrorchis* Ando and Ozaki, 1923

This echinostome species was first described from naturally infected rats, *Rattus rattus* and *Rattus norvegicus,* in Japan (Ando and Ozaki 1923). This has a total of 43–47, usually 45 collar spines in two alternating rows (Rim 1982). *Segmentina nitiella* and *Planorbis compressus japonicus* act as the first and second intermediate hosts (Rim 1982). Metacercarial cysts were found from freshwater snails, *Cipangopaludina malleata, Cipangopaludina japonica, S. nitiella, Viviparus malleatus*, and the frog *Rana* sp. (Yu and Mott 1994). An avian species, *Capella gallinago gallinago,* has been added to the natural definitive hosts in Japan (Yamashita 1964). Human infections were reported in Japan (Majima 1927).

7.2.16 *Echinostoma revolutum* (Froelich, 1802) Looss, 1899

(syn: *Echinoparyphium paraulum* Diez, 1909)
This is the oldest echinostome species recorded in the literature. It was originally described as *Fasciola revoluta* by Froelich in 1802 in Germany with adult specimens from the large intestine of naturally infected wild duck *Anas boschas fereae* dissected in 1798 (Kanev 1994). Now it is known to be an intestinal fluke of the duck, goose, muskrat, and humans in Asia, Europe, Africa, Australia, New Zealand, and North and South America (Kanev 1994; Yu and Mott 1994; Fried and Graczyk 2004). This species is the representative of the so-called 37-collar-spined *E. revolutum* group (Huffman and Fried 1990). The collar spines are arranged conspicuously as 5–6–15–6–5, including five corner spines and six lateral spines on each side in single rows, and 15 dorsal spines in a double row (Kostadinova et al. 2000). The snail host includes *Lymnaea* sp., *Physa* sp., *Paludina* sp., *Segmentina* sp., and *Helisoma* sp. (Beaver et al. 1984). Cercariae penetrate into the tadpoles, snails, or clam *Corbicula producta*, which are the sources of infection to definitive hosts (Beaver et al. 1984). This parasite was reported from house rats and cats in the Republic of Korea (Lee et al. 1990c; Sohn and Chai 2005). In addition, *Echinoparyphium paraulum*, a synonym of *E. revolutum* (Beaver et al. 1984), was described from dogs (Yamaguti 1958) and birds (Ashford and Crewe 2002). The first human infection with *E. revolutum* was reported from Taiwan in 1929

(Anazawa 1929), and the prevalence in Taiwan was once estimated to be between 2.8 and 6.5% (Yu and Mott 1994). This fluke was also reported from human infections in Yunnan and Guangdong Provinces of China, and also in Indonesia, Thailand, and Russia (Rim 1982; Beaver et al. 1984; Ashford and Crewe 2002).

7.2.17 Episthmium caninum (Verma, 1935) Yamaguti, 1958

Species of *Episthmium* have been reported from birds, and only *E. caninum* was described from dogs in Calcutta, India (Yamaguti 1958). *E. caninum* has a total of 24 collar spines interrupted dorsally: 12 dorsal spines are arranged in a single row and 12 lateral and ventral ones are arranged in two alternating rows (Radomyos et al. 1985). The genus *Episthmium* was suggested to be tentatively regarded as a synonym of *Echinochasmus* (Kostadinova and Gibson 2001). However, further studies seem to be needed for doing this. Human infections were reported from northeast Thailand (Radomyos et al. 1985, 1991), and the source of infection was freshwater fish (Radomyos et al. 1991).

7.2.18 Himasthla muehlensi Vogel, 1933

This fluke was described by Mühlens based on five adult specimens from a German patient who lived in Colombia and traveled to New York City where he had eaten raw clams *Venus mercenaria* (Beaver et al. 1984). It has an elongated body, and a total of 32 collar spines, including 28 spines in an uninterrupted horseshoe pattern on the dorsum and around the sides, and two on each side of the ventral angle (Beaver et al. 1984). A species of marine operculate snails, *Littorina littorea,* serves as the first intermediate host, and bivalve mollusks, *Mytilus* and *Mya* spp., as the second intermediate hosts (Beaver et al. 1984). Birds are natural definitive hosts (Yu and Mott 1994).

7.2.19 Hypoderaeum conoideum (Block, 1872) Diez, 1909

This species was discovered from birds in 1872 and is now known to be is an intestinal fluke of birds, including the duck, goose, and fowl in Europe, Japan, and Siberia (Yamaguti 1958). It has a small, inconspicuous head collar, on which a total of 47–53 (usually 49) collar spines are arranged in two alternating rows (Rim 1982). The first intermediate hosts are freshwater snails, *Planorbis corneus, Indoplanorbis exustus, L. stagnalis, Lymnaea limosa, Lymnaea ovata,* and *Lymnaea rubiginosa,* and snails and tadpoles are the second intermediate hosts (Yamaguti 1958; Harinasuta et al. 1987). *Lymnaea tumida* in Russia (Yurlova 1987) and

Lymnaea peregra and *Lymnaea corvus* in Spain (Toledo et al. 1999) were also experimentally proven to be the potential first intermediate hosts. In Thailand, *H. conoideum* was reported from humans and birds in Thailand (Yokogawa et al. 1965; Harinasuta et al. 1987). In an area of northeast Thailand, 55% of 254 residents were found infected (Yokogawa et al. 1965).

7.2.20 Isthmiophora melis (Schrank, 1788) Lühe, 1909

(syn. *Euparyphium melis* Railliet, 1919, *Euparyphium jassyense* Leon and Ciurea, 1922) *Isthmiophora melis* was described from rodents and carnivores in Europe and North America (Yamaguti 1958). *I. melis* has total 27 collar spines, 19 of which are arranged in uninterrupted double rows and four are as corner spines on each side (Rim 1982). Its taxonomic position has been unstable, either placed in *Euparyphium* (Beaver et al. 1984) or in *Echinostoma* (Harinasuta et al. 1987); however, recently it has been designated as a species of *Isthmiophora* (Fried 2001; Kostadinova and Gibson 2002). In the region of Douglas Lake, Michigan, the snail *Stagnicola emarginata angulata* is the first intermediate host, and tadpoles are the second intermediate host (Beaver et al. 1984). Domestic and wild animals were found to be infected with this fluke in Russia (Shimalov and Shimalov 2002). A human infection with *I. melis* was first confirmed by Leon in 1916 in a diarrheic patient in Romania and then in another case at autopsy of a Chinese patient (Beaver et al. 1984). Later, human infections were reported in Taiwan, and Liaoning Province, China (Yu and Mott 1994). The source of human infections in China is presumed to be the loach, *M. anguillicaudatus* (Yu and Mott 1994).

7.3 Pathology and Symptomatology of Echinostomes Infecting Humans

7.3.1 Pathogenicity of Echinostomes

The pathogenicity of echinostomes is not well understood, and complex and diverse (Yu and Mott 1994; Toledo et al. 2006; Chai 2007). The complexity is dependent upon a wide variety of parasite- and host-side factors (Toledo et al. 2006). Two major parasite-related factors, namely, mechanical irritation by the flukes and toxic effects by their metabolites, seem to be highly responsible for the pathogenesis. In this respect, it is of note that the intensity of infection, in terms of individual worm load, may be important in the pathogenicity and severity of the clinical disease (Rim 1982; Huffman and Fried 1990). This was experimentally supported by Fried et al. (1990) who compared the host pathologies between single and multiple infections with an animal echinostome, *E. caproni,* in hamsters. Certain species of echinostomes seem to be more pathogenic to humans, since a fatal human case of *A. malayanum* (under the name *A. mehrai*), the death being due to marked malnutrition and anemia, was reported in India, from whom several hundred worms were recovered in the intestine at autopsy (Reddy et al. 1964). A case of *A. malayanum* (under the name *A. mehrai*) infection with bowel perforation was also reported in an Indian child (Kaul et al. 1974). As to the host-side factors, the susceptibility of each host species, avian or mammalian, and immune responses of the host species should be important. The host immune responses against echinostome infections are discussed elsewhere in this chapter and other chapters.

7.3.2 Pathological Findings in the Intestine of Experimental Animals

The intestinal histopathology in echinostome infections was studied using laboratory rodents in a few echinostome species, namely, *E. revolutum* (Bindseil and Christensen 1984; Huffman et al. 1986), *E. hortense* (Lee et al. 1990b; Chai and Lee 2002), *E. caproni* (Kim and Fried 1989; Toledo et al. 2006), and *E. trivolvis* (Fujino et al. 1993). The worms were located in the lumen of the upper small intestine of the rodents, and the pathological changes were chiefly observed at the attachment sites of the echinostomes (Chai and Lee 2002; Toledo et al. 2006). The flukes were seen to tightly pinch and suck the mucosa with their oral suckers (Fig. 7.9), and it seemed that the mucosal layer was eventually eaten by the flukes. Villous atrophy, crypt hyperplasia, inflammation of the stroma, decreased villus/crypt ratios, and focal loss of the mucosal tissue were observed in the small intestine of experimentally infected animals (Huffman et al. 1986; Lee et al. 1990b). The damages at the mucosa were so severe that in focal areas massive destruction and detachment of villi, at times complete loss of the mucosal integrity and ulcerations, were observed (Huffman et al. 1986; Chai and Lee 2002). These pathological features were similar to those observed in severe catarrhal enteritis patients (Chai and Lee 2002). In two pigs naturally infected with *A. oraoni,* fatal diarrhea was developed in both pigs, and autopsy revealed massive parasite infections with hemorrhagic and edematous mucosa of the jejunum and duodenum, extending up to the pyloric end of the stomach (Bandyopadhyay et al. 1995).

7.3.3 Symptoms and Signs of Human Echinostomiasis

Major clinical symptoms due to echinostome infection include abdominal pain, diarrhea, easy fatigue, and loss of body weight (Rim 1982; Chai and Lee 2002; Fried et al. 2004). Considering the severer mucosal damages, including ulcerations of the mucosa, as can be seen in naturally infected patients and in experimental rats infected with *E. hortense* (Lee et al. 1990b), the symptoms in echinostomiasis seem

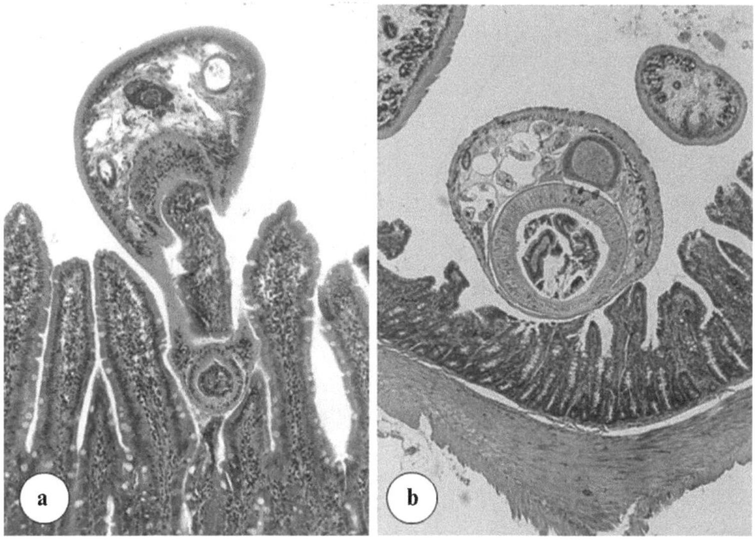

Fig. 7.9 Histopathological features of the small intestines of rats experimentally infected with *Echinostoma hortense*. (**a**) Duodenum of a rat, day 7 postinfection. The oral and ventral suckers of a worm are pinching and sucking the villus (from Lee et al. (1990b) with permission). × 100. (**b**) Duodenum of a rat, day 44 postinfection. The oral sucker of a worm is pinching the duodenal villi. × 40

to be severer than those seen in other intestinal fluke infections, for example, heterophyid fluke infections. A patient with an *E. hortense* infection complained of lower abdominal pain, diarrhea and tenesmus, easy fatigability, and urinary incontinence (Lee et al. 1986). Patients infected with *E. ilocanum* experienced intestinal colic and diarrhea (Beaver et al. 1984). Interesting to note are reports of patients with *E. hortense* infection suffering from severe epigastric discomfort and ulcerative lesions in the stomach or duodenum that were diagnosed by identification of worms through gastroduodenal endoscopy (Chai et al. 1994; Lee and Hong 2002; Cho et al. 2003; Chang et al. 2005; Park and Kim 2006). The first patient (Chai et al. 1994) was admitted to a hospital because of epigastric pain and hematemesis, and in gastroduodenoscopy an adult fluke was seen attached at the lesion, which was removed by an endoscopic clipper; after praziquantel treatment three more adult flukes were recovered from the feces. The patient had eaten raw freshwater fishes (Chai et al. 1994). The second patient (Lee and Hong 2002) experienced epigastric discomfort for a year, and acute epigastric pain and diarrhea of 1-month duration accompanied by acidic belching, dizziness, and 2-kg weight loss. The third patient complained of epigastric discomfort for several days (Cho et al. 2003), and the fourth patient suffered from upper abdominal pain and discomfort which persisted for 2 weeks (Chang et al. 2005). The fifth patient had epigastric discomfort for 1 month, and indigestion, anorexia, dizziness, headache, nausea, and vomiting occurred at times; nausea and vomiting became severer and thus the patient was hospitalized (Park and Kim 2006). The clinical signs, including laboratory findings,

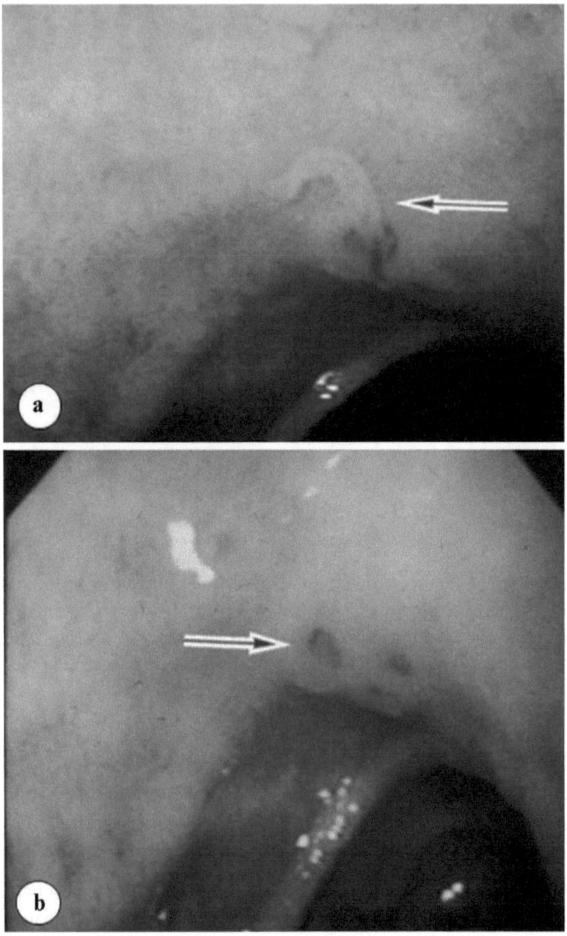

Fig. 7.10 Gastroendoscopic views of a patient infected with *Echinostoma hortense* in the Republic of Korea (from Chai et al. 1994 with permission). (**a**) A moving leaf-like worm (*arrow*) is attached to an ulcerated lesion with bleeding at the proximal duodenal wall of the patient. The worm was removed by an endoscopic forceps and morphologically identified as *E. hortense*. (**b**) Three mucosal ulcerations (*arrow*) are seen after removal of the worm

in echinostomiasis are poorly known. However, peripheral blood eosinophilia is commonly observed (Lee et al. 1988b). It has been reported that the levels of peripheral blood eosinophilia in *E. hortense* infection were dependent upon individual worm burdens; 11–24% (av. 17%) among the patients with more than 100 worms, 4–21% (av. 10%) among those with 51–100 worms, and 2–14% (av. 5%) among those with less than 50 worms (Lee et al. 1988b). The clinical symptoms in *E. cinetorchis, E. japonicus*, and *A. tyosenense* infections are not well known (Seo et al. 1980, 1985a; Chai et al. 2001), but may be similar to those of *E. hortense* infection. It is of note that among 18 American tourists to Kenya and Tanzania who were echinostome egg positive in fecal examinations (adult flukes were not recovered), ten had moderately severe abdominal cramps and loose or watery stools (Poland et al. 1985).

7.3.4 Endoscopic Findings in E. hortense *Infection*

Gastroduodenal endoscopy of a male patient revealed a moving, leaf-like *E. hortense* worm attached to an ulcerated mucosal layer of the distal part of the stomach (Fig. 7.10) (Chai et al. 1994). The lesion was accompanied by a stage IIc or stage III early gastric cancer and multiple ulcerations and bleeding in the stomach and duodenum (Chai et al. 1994). The ulcerations were strongly suggested to be caused by the flukes; mucosal bleeding may also have occurred due to worms (Chai et al. 1994). Upper endoscopy of another patient revealed two *E. hortense* worms in the stomach: one was penetrating into the mucosa on the lesser curvature of the antrum; the other was attached to the mucosal surface of the greater curvature of the antrum (Lee and Hong 2002). The worm wriggled on stimulation and became tightly stretched during retrieval with a forceps (Lee and Hong 2002). Gastroduodenal endoscopy of a third (Cho et al. 2003) and fourth (Chang et al. 2005) patients revealed two living *E. hortense* flukes (Fig. 7.11) each in the duodenal bulb area, at times accompanied by mucosal erosions (Cho et al. 2003). Still another patient revealed two living *E. hortense* flukes attached to the mucosal surface of the duodenal bulb area; the gastric mucosa revealed signs of chronic gastritis, and at the proximal part of the duodenum multiple cured ulcerative lesions were observed (Park and Kim 2006). Also in Japan, four cases of *E. hortense* infection were reported based on worm recoveries by upper endoscopy (Hamamoto et al. 1997).

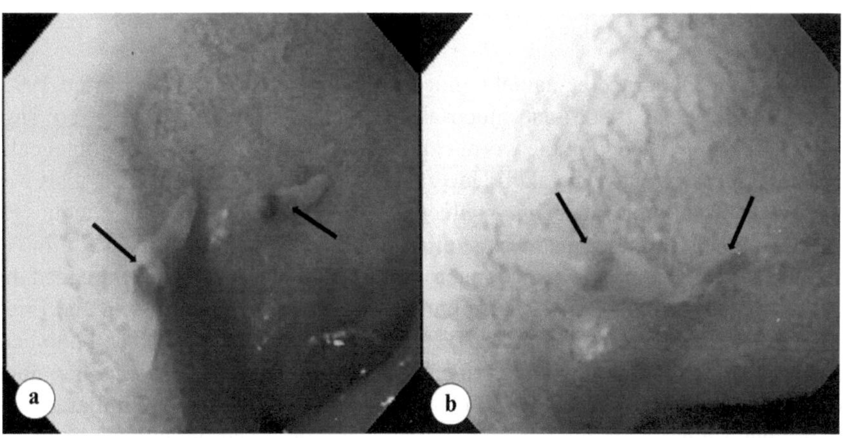

Fig. 7.11 Gastroendoscopic views (**a**,**b**) of the duodenal bulb area of another patient infected with *Echinostoma hortense* in the Republic of Korea, showing two moving worms (*arrows*) and mild mucosal erosions (from Cho et al. (2003) with permission)

7.4 Immunology of Human Echinostomiasis

7.4.1 Host Resistance to Echinostome Infection

Immunology of echinostomiasis has been studied extensively in animal-infecting species, for example, *E. caproni, E. trivolvis,* and *Echinostoma friedi* (Toledo et al. 2006). With regard to human-infecting echinostomes, *E. hortense* has been most frequently subjected to immunological studies (Chai 2007; Cho et al. 2007; Ryang et al. 2007). From previous studies, it has been known that the rodent hosts express various types of resistance to echinostome infection (Toledo et al. 2006). One type of the resistance is the spontaneous expulsion of primary infections (Toledo et al. 2006). This interesting phenomenon has been observed in various echinostome species models, i.e., *E. trivolvis, E. hortense, E. caproni,* and *E. friedi* (Fujino et al. 1993; Kim et al. 2000; Brunet et al. 2000; Toledo et al. 2006). However, the dynamics are different depending on different models of parasite species, host species, and even host strain (Fujino and Fried 1993; Toledo et al. 2006; Cho et al. 2007). For example, the elimination of *E. trivolvis* from mice occurs within 2–4 weeks postinfection (Weinstein and Fried 1991), whereas *E. revolutum* survive for long periods of time in golden hamsters (Mabus et al. 1988). *E. friedi* survive for at least 12 weeks postinfection in hamsters, but the infection is expelled at 3–4 weeks in rats (Toledo et al. 2003). *E. caproni* produce chronic infections in hamsters and mice, but the worms are expelled in 6–8 weeks in rats (Toledo et al. 2004). In the case of *E. hortense* infection, it is interesting to note that BALB/c mice rejected the worms, while C3H/HeN mice underwent a chronic infection (Lee et al. 2004).

7.4.2 Role of Goblet Cells and Mast Cells

Two types of effector cells, namely, mucosal goblet cells (Fujino and Fried 1993; Fujino et al. 1995a, 1996a) and mucosal mast cells (Kim et al. 2000), were suggested to be associated with the worm expulsion. However, the roles of goblet cells and mast cells are controversial (Toledo et al. 2006). For instance, mastocytosis was suppressed in athymic nude mice infected with *E. trivolvis*, but the kinetics of worm expulsion in these mice was similar to that of conventional mice (Fujino et al. 1993). In *E. caproni*-infected mice, chronic infection was accompanied by reduced goblet cell numbers (Bindseil and Christensen 1984; Weinstein and Fried 1991; Fujino and Fried 1993). In *E. trivolvis*-infected mice, the goblet cell numbers increased coinciding with worm expulsion, and expulsion was delayed when mice were treated with dexamethasone (Fujino et al. 1993, 1996a, 1997). Similar effects of immunosuppression by dexamethasone were observed in golden hamsters infected with *E. caproni* and *E. trivolvis* (McMaster et al. 1995). In addition, mucins increased by a primary *Nippostrongylus brasiliensis* infection in the intestine of mice were suggested to be responsible for a rapid expulsion of a challenge infection with *E. caproni* or *E. trivolvis* (Fujino et al. 1996b). However, a contradicting result was observed in the RAG-2-deficient mice, a strain of genetically lacking B- and T-lymphocytes, infected with *E. caproni*; goblet cell hyperplasia was marked in RAG-2-deficient mice but worms survived as they did in normal ICR mice (Frazer et al. 1999). This suggested that other immune mechanisms are also involved in worm expulsion (Toledo et al. 2006).

7.4.3 Role of Other Effector Cells and Antibodies

For effectors other than goblet cells and mast cells, mucosal neutrophils and mononuclear inflammatory cells in the mesentery were suggested to be essential in the course of *E. caproni* infection in mice (Muñoz-Antoli et al. 2007). In addition, factors in mouse serum were shown to induce retraction of collar spines in *E. trivolvis*, probably enhancing worm expulsion (Fujino et al. 1994). Elevated levels of antibodies, such as IgG, IgM, and IgA, have been detected in serum and the small intestine of mice infected with *E. caproni* (Agger et al. 1993; Sotillo et al. 2007). Further, it was shown that antibodies in mouse serum covered the surface of *E. revolutum* in vitro, and worms obtained from immune mice were found covered by antibodies (Simonsen and Andersen 1986). Local immune responses were also shown to be important in host resistance. Whereas the level of IgG2a in serum remained constant during the course of *E. caproni* infection, the IgG2a responses in the intestine were significantly related with worm expulsion (Sotillo et al. 2007). Since IgG2a is an important element mediating T-helper-2 (Th2) immune responses of mice (Sotillo et al. 2007; Shin et al. 2007), the expulsion of echinostomes from the host intestine is suggested to be mediated by a Th2 type immune response.

7.4.4 Cytokines and Th1/Th2 Immune Responses

In order to understand better the importance of Th2 responses in expulsion of echinostomes, Th1 and Th2 cytokines profiles were studied in rodent hosts experimentally infected with *E. caproni* (Brunet et al. 2000) or *E. hortense* (Cho et al. 2007; Ryang et al. 2007). Upon antigen stimulation, mesenteric lymph node cells from *E. caproni*-infected mice produced significantly higher levels of IFN-γ in vitro for 3 weeks; during this period they are unable to expel *E. caproni* worms (Brunet et al. 2000). However, injection of these mice with anti-IFN-γ monoclonal antibodies significantly lowered the worm burden of *E. caproni* (Brunet et al. 2000). Similarly, *E. hortense* infection in BALB/c mice, a resistant mouse strain (Cho et al. 2007), exhibited stronger mRNA expressions of IL-4 and IL-5 in the spleen, than C3H/HeN mice, a susceptible strain which undergo a chronic infection (Ryang et al. 2007). Treatment of BALB/c mice with ketotifen, an antiallergic drug, suppressed the mRNA levels of IL-4 and IL-5 and elevated the worm recoveries from this mouse strain (Ryang et al. 2007). Now it is generally accepted that echinostome worm rejection by the host is closely related to Th 2 responses mediated by IL-4 and IL-5, and chronicity of infection seems to be associated with Th 1 responses mediated by IL-12, TNF-α, and IFN-γ (Brunet et al. 2000; Toledo and Fried 2005).

7.5 Diagnosis, Treatment, and Control of Echinostomes Infecting Humans

7.5.1 Diagnosis of Human Echinostomiasis

The diagnosis of human echinostomiasis is usually based on recovery of eggs in fecal examinations (Chai 2007). The eggs are oval in most cases, yellowish, with a thin and refractory shell, and with a small, inconspicuous operculum located at the anterior end, a germ cell and yolk inside, and abopercular wrinkles or thickening at the terminal end (Fig. 7.12). The operculum, germ cell, and abopercular wrinkles may be difficult to see in Kato-Katz thick smear slides (Fig. 7.12b). The size of human-infecting echinostome eggs is in the range, 0.066–0.149 mm in length and 0.043–0.090 mm in width (Table 7.4). The eggs of *A. malayanum, A. oraoni,* and *I. melis* are larger than 0.12 mm in length, whereas those of *E. liliputanus* and *E. angustitestis* are smaller than 0.082 mm in length (Table 7.4). Upon careful microscopic observations with measurements of the eggs, specific diagnosis may be possible in known endemic areas with a single or a mixed echinostome species

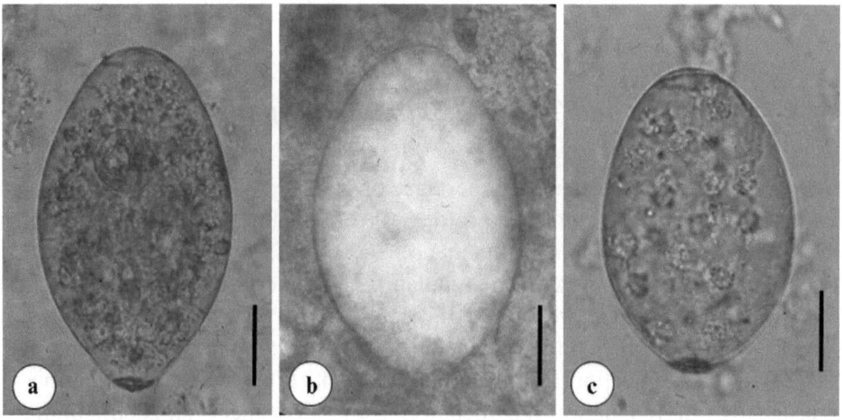

Fig. 7.12 Eggs of *Echinostoma hortense* (**a,b**) and *Echinostoma cinetorchis* (**c**) seen in the feces of an infected human volunteer (**a**: from Seo et al. (1985b) with permission) and naturally infected patients (**b,c**). They characteristically show a small, inconspicuous operculum, a germ cell and yolk inside, a thin and refractile shell, and abopercular wrinkles at the terminal end (**a,c**). These structures are difficult to see in Kato-Katz thick fecal smears (**b**). Scale bars = 0.03 mm

Table 7.4 Egg size of human-infecting echinostomes

Species	Egg size (av. length × width: mm)	References
Acanthoparyphium tyosenense	0.084–0.110 × 0.060–0.069	Chai et al. (2001)
Artyfechinostomum malayanum	0.120–0.130 × 0.080–0.090	Belizario et al. (2007)
Artyfechinostomum oraoni	0.129–0.135 × 0.060–0.075	Bandyopadhyay and Nandy (1986)
Echinochasmus fujianensis	0.100–0.114 × 0.064–0.072	Cheng et al. (1992a)
Echinochasmus japonicus	0.076–0.087 × 0.052–0.063	Seo et al. (1985a)
Echinochasmus jiufoensis	0.100–0.115 × 0.072–0.079	Liang and Ke (1988)
Echinochasmus liliputanus	0.066–0.080 × 0.043–0.046	Witenberg (1932)
Echinochasmus perfoliatus	0.090–0.135 × 0.055–0.095	Rim (1982)
Echinoparyphium recurvatum	0.108–0.110 × 0.081–0.084	Rim (1982)
Echinostoma angustitestis	0.077–0.082 × 0.052–0.055	Cheng et al. (1992b)
Echinostoma cinetorchis	0.099–0.116 × 0.065–0.076	Seo et al. (1980)
Echinostoma echinatum	0.092–0.124 × 0.065–0.076	Rim (1982) as E. lindoense
Echinostoma hortense	0.115–0.130 × 0.068–0.080	Seo et al. (1985b)
Echinostoma ilocanum	0.083–0.116 × 0.058–0.069	Rim (1982)
Echinostoma macrorchis	0.095–0.110 × 0.055–0.065	Rim (1982)
Echinostoma revolutum	0.088–0.113 × 0.061–0.074	Kanev (1994)
Episthmium caninum	0.084–0.102 × 0.056–0.069	Radomyos et al. (1985)
Himasthla muehlensi	0.114–0.149 × 0.062–0.085	Rim (1982)
Hypoderaeum conoideum	0.095–0.108 × 0.061–0.068	Rim (1982)
Isthmiophora melis	0.121–0.139 × 0.078–0.088	Kostadinova and Gibson (2002)

infection. For example, due to the apparently larger size of *E. hortense* eggs than *E. cinetorchis* eggs (Fig. 7.12), a specific diagnosis can be done in an area of mixed infection with these two echinostomes (Ryang 1990). However, recovery and identification of the adult fluke is strongly recommended, if a definite diagnosis is preferred (Chai 2007). The egg detectability in fecal examinations varies remarkably depending on different echinostome species and worm load. It is considerably high in infections with *E. hortense* (Seo et al. 1985b) and *E. cinetorchis* (Seo et al. 1984), whereas it is very low in infections with *E. japonicus* (Chai et al. 1985a) and *A. tyosenense* (Chai et al. 2001). The difference is greatly due to remarkably different numbers of intrauterine eggs and the different egg-laying capacity of each echinostome species (Chai 2007).

7.5.2 *Treatment and Control of Echinostomiasis in Humans*

Echinostome infections can be treated successfully using 10–20 mg kg^{-1} praziquantel in a single oral dose (Lee et al. 1988b; Chai et al. 1994, 2001; Chai 2007). Albendazole may also be effective (Chai 2007). Eating raw or improperly cooked freshwater fish and fresh or brackish water snails should be avoided to prevent echinostome infections (Chai 2007).

References

Agger, M. K., Simonsen, P. E., and Vennervald, B. J. 1993. The antibody response in serum, intestinal wall and intestinal lumen of NMRI mice injected with *Echinostoma caproni*. Journal of Helminthology 67:169–178.

Ahluwalia, S. S. 1962. Studies on some helminths of the domestic pig (*Sus scrofa domestica*) in western Uttar Pradesh. Journal of Helminthology 36:347–364.

Ahn, Y. K., and Kang, H. S. 1988. Development and cercarial shedding of *Echinostoma hortense* in the snail host, *Radix auricularia coreana*. Journal of Wonju Medical College 1:137–152.

Ahn, Y. K., Ryang, Y. S., Chai, J. Y., and Sohn, W. M. 1989. Cercarial shedding of *Echinostoma cinetorchis* and experimental infection of the cercariae to several kinds of snails. Korean Journal of Parasitology 27:203–211 (in Korean).

Anazawa, K. 1929. On a human case of *Echinostoma revolutum* and its infection route. Taiwan Igahkai Zasshi No. 288:221–241 (in Japanese).

Ando, R., and Ozaki, Y. 1923. On four new species of trematodes of the family Echinostomatidae. Dobutsugaku Zasshi 35:108–119 (in Japanese).

Ando, R., and Tsuyuki, H. 1923. Studies on intestinal parasites taking rats as their final host (II). On the trematode Echinostoma and its 2nd intermediate host in Nagoya City. Tokyo Iji Shinshi No. 2340:1487–1499 (in Japanese).

Asada, J. 1926. On a new echinostomatid trematode and its life history. Transactions of Japanese Society of Pathology 16:293–294 (in Japanese).

Ashford, R. W., and Crewe, W. 2002. The Parasites of *Homo sapiens*. London: Liverpool School of Tropical Medicine.

Bandyopadhyay, A. K., and Nandy, A. 1986. A preliminary observation on the prevalence of echinostomes in a tribal community near Calcutta. Annals of Tropical Medicine and Parasitology 80:373–375.

Bandyopadhyay, A. K., Manna, B., and Nandy, A. 1989. Human infection of *Artyfechinostomum oraoni* n. sp. (Paryphostominae: Echinostomatidae) in a tribal community, "Oraons' in West Bengal, India. Indian Journal of Parasitology 13:191–196.

Bandyopadhyay, A. K., Maji, A. K., Manna, B., Bera, D. K., Addy, M., and Nandy, A. 1995. Pathogenicity of *Artyfechinostomum oraoni* in naturally infected pigs. Tropical Medicine and Parasitology 46:138–139.

Beaver, P. C., Jung, R. C., and Cupp, E. W. 1984. Clinical Parasitology. Philadelphia: Lea and Febiger.

Belizario, V. Y., Geronilla, G. G., Anastacio, M. B. M., de Leon, W. U., Suba-an, A. P., Sebastian, A. C., and Bangs, M. J. 2007. *Echinostoma malayanum* infection, the Philippines. Emerging Infectious Diseases 13:1130–1131.

Betlejewska, K. M., and Jorol, E. N. 2002. Taxonomic, topical and quantitative structure of the community of intestinal flukes (Digenea) of mallards, *Anas platyrhynchos* Linnaeus, 1758 from the area of Szcecin. Wiadomosci Parazitologiczne 48:343–357.

Bindseil, E., and Christensen, N. O. 1984. Thymus-independent crypt hyperplasia and villous atrophy in the small intestine of mice infected with the trematode *Echinostoma revolutum*. Parasitology 88:431–438.

Brunet, I. R., Joseph, S., Dunne, D. W., and Fried, B. 2000. Immune responses during the acute stage of infection with the intestinal trematode *Echinostoma caproni*. Parasitology 120:565–571.

Chai, J. Y. 2007. Intestinal flukes. In Food-Borne Parasitic Zoonoses: Fish and Plant-Borne Parasites, World Class Parasites, Vol. 11, eds. K. D. Murrell and B. Fried, pp 53–115. New York: Springer.

Chai, J. Y., and Lee, S. H. 2002. Food-borne intestinal trematode infections in the Republic of Korea. Parasitology International 51:129–154.

Chai, J. Y., Hong, S. J., Son, D. W., Lee, S. H., and Seo, B. S. 1985a. Metacercariae of the *Echinochasmus japonicus* encysted in a freshwater fish, *Pseudorasbora parva*, and their development in experimental mice. Korean Journal of Parasitology 23:221–229.

Chai, J. Y., Hong, S. J., Sohn, W. M., Lee, S. H., and Seo, B. S. 1985b. Studies on intestinal trematode in Korea XVI. Infection status of loaches with the metacercariae of *Echinostoma hortense*. Korean Journal of Parasitology 23:18–23.

Chai, J. Y., Hong, S. T., Lee, S. H., Lee, G. C., and Min, Y. I. 1994. A case of echinostomiasis with ulcerative lesions in the duodenum. Korean Journal of Parasitology 32:201–204.

Chai, J. Y., Han, E. T., Park, Y. K., Guk, S. M., and Lee, S. H. 2001. *Acanthoparyphium tyosenense*: the discovery of human infection and identification of its source. Journal of Parasitology 87:794–800.

Chang, Y. D., Sohn, W. M., Ryu, J. H., Kang, S. Y., and Hong, S. J. 2005. A human infection of *Echinostoma hortense* in duodenal bulb diagnosed by endoscopy. Korean Journal of Parasitology 43:57–60.

Chen, X. Q., Feng, G. X., and Qian, Z. F. 1993. Survey on infection due to *Echinostoma hortense* in Liaoning Province. Chinese Journal of Parasitology and Parasitic Diseases 11:226 (in Chinese).

Cheng, Y. Z., Lin, J. X., Chen, B. J., Fang, Y. Y., Guo, Z. F., Xu, G. F., and Wu, Z. Y. 1992a. A new species of *Echinochasmus* parasitic in man and observation of its experimental infection. Wuyi Science Journal 9:43–48 (in Chinese).

Cheng, Y. Z., Lin, J. X., Fang, Y. Y., Lin, A. Q., Chen, B. J., and Zhuang, H. J. 1992b. Discovery of human infection with *Echinostoma angustitestis*. Chinese Journal of Zoonoses 8:7–8 (in Chinese).

Cheng, Y. Z., Zhang, Y., Lin, C., Zhan, Z., Yu, X., Lin, J. X., Cai, M. J., and Che, W. 1999. Taxonomic studies on *Echinochasmus fujianensis* and its related species by random amplified polymorphic DNA analysis and experimental infection. Chinese Journal of Parasitology and Parasitic Diseases 17:135–139 (in Chinese).

Cho, S. Y., Kang, S. Y., and Ryang, Y. S. 1981. Helminthes infections in the small intestine of stray dogs in Ejungbu City, Kyunggi Do, Korea. Korean Journal of Parasitology 19:55–59 (in Korean).

Cho, C. M., Tak, W. Y., Kweon, Y. O., Kim, S. K., Choi, Y. H., Kong, H. H., and Chung, D. I. 2003. A human case of *Echinostoma hortense* (Trematoda: Echinostomatidae) infection diagnosed by gastroduodenal endoscopy in Korea. Korean Journal of Parasitology 41:117–120.

Cho, Y. K., Ryang, Y. S., Kim, I. S., Park, S. K., Im, J. A., and Lee, K. J. 2007. Differential immune profiles following experimental *Echinostoma hortense* infection in BALB/c and C3H/HeN mice. Parasitology Research 100:1053–1061.

Choi, M. H., Kim, S. H., Chung, J. H., Jang, H. J., Eom, J. H., Chung, B. S., Sohn, W. M., Chai, J. Y., and Hong, S. T. 2006. Morphological observations of *Echinochasmus japonicus* cercariae and the in vitro maintenance of its life cycle from cercariae to adults. Journal of Parasitology 92:236–241.

Chung, P. R., and Jung, Y. 1999. *Cipangopaludina chinensis malleata* (Gastropoda: Viviparidae): a new second molluscan intermediate host of a human intestinal fluke *Echinostoma cinetorchis* (Trematoda: Echinostomatidae) in Korea. Journal of Parasitology 85:963–964.

Chung, P. R., Jung, Y., and Park, Y. K. 2001. *Segmentina hemisphaerula*: a new molluscan intermediate host for *Echinostoma cinetorchis* in Korea. Journal of Parasitology 87:1169–1171.

Cross, J. H., and Basaca-Sevilla, V. 1981. Intestinal parasitic infections in Southeast Asia. Southeast Asian Journal of Tropical Medicine and Public Health 12:262–274.

Cross, J. H., and Basaca-Sevilla, V. 1986. Studies on *Echinostoma ilocanum* in the Philippines. Southeast Asian Journal of Tropical Medicine and Public Health 17:23–26.

Dubey, J. P., Srivastava, H. O. P., and Sahasrabudhe, V. K. 1969. *Artyfechinostomum sufrartyfex* infection in cat and dog in India. Tropical and Geographical Medicine 21:210–213.

El-Gayar, A. K. 2007. Studies on some trematode parasites of stray dogs in Egypt with a key to the identification of intestinal trematodes of dogs. Veterinary Parasitology 144:360–365.

Eom, K. S., and Rim, H. J. 1984. A study on the parasitic helminths of domestic duck (*Anas platyrhynchos* var. *domestica* Linnaeus) in Korea. Korean Journal of Parasitology 22:215–221.

Evans, N. A., and Gordon, D. M. 1983. Experimental observations on the specificity of *Echinoparyphium recurvatum* toward second intermediate hosts. Zeitschrift für Parasitenkunde 69:217–222.

Fan, S. Q., and Sun, M. F. 1989. *Echinostoma hortense.* Bulletin of Harbin Medical University 23:161 (in Chinese).

Fischthal, J. H., and Kuntz, R. E. 1975. Some digenetic trematodes of mammals from Taiwan. Proceedings of the Helminthological Society of Washington 42:149–157.

Frazer, B. A., Fried, B., Fujino, T., and Sleckman, B. P. 1999. Host-parasite relationships between *Echinostoma caproni* and RAG-2-deficient mice. Parasitology Research 85:337–342.

Fried, B. 2001. Biology of echinostomes except *Echinostoma.* Advances in Parasitology 49:163–210.

Fried, B., and Graczyk, T. K. 2004. Recent advances in the biology of *Echinostoma* species in the '*revolutum*' group. Advances in Parasitology 58:139–195.

Fried, B., Huffman, J. E., and Weiss, P. M. 1990. Single and multiple worm infections of *Echinostoma caproni* (Trematoda) in the golden hamster. Journal of Helminthology 64:75–78.

Fried, B., Graczyk, T. K., and Tamang, L. 2004. Food-borne intestinal trematodiases in humans. Parasitology Research 93:159–170.

Fujino, T., and Fried, B. 1993. *Echinostoma caproni* and *E. trivolvis* alter the binging of glycoconjugates in the intestinal mucosa of C3H mice as determined by lectin histochemistry. Journal of Helminthology 67:179–188.

Fujino, T., Fried, B., and Hosier, D. W. 1994. The expulsion of *Echinostoma trivolvis* (Trematoda) from ICR mice: extension/retraction mechanisms and ultrastructure of the collar spines. Parasitology Research 80:581–587.

Fujino, T., Fried, B., Ichikawa, H., and Tada, I. 1995a. Rapid expulsion of the intestinal trematodes *Echinostoma trivolvis* and *E. caproni* from C3H mice by trapping with increased goblet cell mucins. International Journal for Parasitology 26:319–324.

Fujino, T., Takahashi, Y., and Fried, B. 1995b. A comparison of *Echinostoma trivolvis* and *E. caproni* using random amplified polymorphic DNA analysis. Journal of Helminthology 69:263–264.

Fujino, T., Ichikawa, H., Fried, B., and Fukuda, K. 1996a. The expulsion of *Echinostoma trivolvis*: suppressive effects of dexamethasone on goblet cell hyperplasia and worm rejection in C3H/HeN mice. Parasite 3: 283–289.

Fujino, T., Yamada, M., Ichikawa, H., Fried, B., Arizono, N., and Tada, I. 1996b. Rapid expulsion of the intestinal trematodes *Echinostoma trivolvis* and *E. caproni* from C3H/HeN mice after infection with *Nippostongylus brasiliensis*. Parasitology Research 82:577–579.

Fujino, T., Ichikawa, H., Fried, B., and Fukuda, K. 1997. The expulsion of *Echinostoma trivolvis*: worm kinetics and intestinal reactions in C3H/HeN mice treated with dexamethasone. Journal of Helminthology 71:257–259.

Grover, M., Dutta, R., Kumar, R., Aneja, S., and Mehta, G. 1998. *Echinostoma ilocanum* infection. Indian Pediatrics 35:549–552.

Hamamoto, T., Kawasaki, H., Maejima, J., and Hirai, K. 1997. A case of *Echinostoma hortense* infection diagnosed by the upper gastrointestinal endoscopy. Japanese Journal of Gastroenterological Society 94:487–491 (in Japanese).

Han, E. T., Kim, J. L., and Chai, J. Y. 2003. Recovery, growth and development of *Acanthoparyphium tyosenense* (Digenea: Echinostomatidae) in experimental chicks. Journal of Parasitology 89:176–179.

Harinasuta, T., Bunnag, D., and Radomyos, P. 1987. Intestinal fluke infections. In Bailliere's Clinical Tropical Medicine and Communicable Diseases, Vol. 2, No. 3: pp 695–721.

Hirazawa, I. 1928. *Echinochasmus perfoliatus* (Ratz) found in man. Tokyo Iji Shinshi No. 2577:1328–1334 (in Japanese).

Huffman, J. E., and Fried, B. 1990. *Echinostoma* and echinostomiasis. Advances in Parasitology 29:215–269.

Huffman, J. E., Michos, C., and Fried, B. 1986. Clinical and pathological effects of *Echinostoma revolutum* (Digenea: Echinostomatidae) in the golden hamster, *Mesocricetus auratus.* Parasitology 93:505–515.

Ito, J. 1957. Studies on the brackish water cercariae in Japan. III. Three new echinostome cercariae in Tokyo bay. Japanese Journal of Medical Science and Biology 10: 439–453.

Kanev, I. 1994. Life-cycle, delimitation and redescription of *Echinostoma revolutum* (Froelich, 1982) (Trematoda: Echinostomatidae). Systematic Parasitology 28:125–144.

Kanev, I., Fried, B., Dimitrov, V., and Radev, V. 1995. Redescription of *Echinostoma trivolvis* (Cort, 1914) (Trematoda: Echinostomatidae) with a discussion on its identity. Systematic Parasitology 32:61–70.

Kaul, B. K., Singhal, G. D., and Pillai, P. N. 1974. *Artyfechinostomum mehrai* infestation with bowel perforation. Journal of Indian Medical Association 63:263–265.

Kawahara, S., and Yamamoto, E. 1933. Human cases of *Echinostoma cinetorchis*. Tokyo Iji Shinshi No. 2840:1794–1796 (in Japanese).

Kim, S., and Fried, B. 1989. Pathological effects of *Echinostoma caproni* (Trematoda) in the domestic chick. Journal of Helminthology 63:227–230.

Kim, I., Im, J. A., Lee, K. J., and Ryang, Y. S. 2000. Mucosal mast cell responses in the small intestine of rats infected with *Echinostoma hortense*. Korean Journal of Parasitology 38:139–143.

Kim, Y. K., Yu, J. E., Chung, E. Y., and Chung, P. R. 2004. *Acanthoparyphium tyosenense* (Digenea: Echinostomatidae): experimental confirmation of the cercaria and its complete life history in Korea. Journal of Parasitology 90:97–102.

Kostadinova, A. 1995. *Echinostoma echinatum* (Zeder 1803) sensu Kanev (Digenea: Echinostomatidae): a note of caution. Systematic Parasitology 32:23–26.

Kostadinova, A. 1999. Cercarial chaetotaxy of *Echonostoma miyagawai* Ishii, 1932 (Digenea: Echinostomatidae), with a review of the sensory patterns in the '*revolutum*' group. Systematic Parasitology 44:201–209.

Kostadinova, A., and Gibson, D. I. 2000. The systematic of the echinostomes. In Echinostomes as Experimental Models for Biological Research, eds. B. Fried and T. K. Graczyk, pp 31–57. Dordrecht, Boston, London: Kluwer.

Kostadinova, A., and Gibson, D. I. 2001. A redescription of *Uroproctepisthmium bursicola* (Creplin, 1837) n. comb. (Digenea: Echinostomatidae), and re-evaluations of the genera *Episthmium* Lühe, 1909 and *Uroproctepisthmium* Fischthal and Kuntz, 1976. Systematic Parasitology 50:63–67.

Kostadinova, A., and Gibson, D. I. 2002. *Isthmiophora* Lühe, 1909 and *Euparyphium* Dietz, 1909 (Digenea: Echinostomatidae) re-defined, with comments on their nominal species. Systematic Parasitology 52:205–217.

Kostadinova, A., Gibson, D. I., Biserkov, V., and Chipev, N. 2000. Re-validation of *Echinostoma miyagawai* Ishii, 1932 (Digenea: Echinostomatidae) on the basis of the experimental completion of its life-cycle. Systematic Parasitology 45:81–108.

Kostadinova, A., Vaucher, C., and Gibson, D. I. 2002. Redescription of two echinostomes from birds in Paraguay, with comments on *Drepanocephalus* Diez, 1909 and *Paryphostomum* Dietz, 1909 (Digenea: Echinostomatidae). Systematic Parasitology 53:147–158.

Kostadinova, A., Herniou, E. A., Barrett, J., and Littlewood, D. T. J. 2003. Phylogenetic relationships of *Echinostoma* Rudolphi, 1809 (Digenea: Echinostomatidae) and related genera re-assessed via DNA and morphological analyses. Systematic Parasitology 54:159–176.

Lee, S. J., and Hong, S. J. 2002. Gastric echinostomiasis diagnosed by endoscopy. Gastrointestinal Endoscopy 55:440–442.

Lee, J. K., Lee, H. I., Baek, B. K., and Kim, P. G. 1983. Survey on encysted cercariae of trematodes from fresh-water fishes in Mangyeong riverside area. Korean Journal of Parasitology 21:187–192 (in Japanese).

Lee, J. K., Lee, H. I., Baek, B. K., and Kim, P. G. 1984. Survey on encysted cercariae of trematodes from fresh-water fishes in Tongjin riverside areas in Korea. Korean Journal of Parasitology 22:190–202 (in Korean).

Lee, S. K., Chung, N. S., Ko, I. H., Ko, H. I., and Chai, J. Y. 1986. Two cases of natural human infection by *Echinostoma hortense*. Korean Journal of Parasitology 24:77–81 (in Korean).

Lee, S. K., Chung, N. S., Ko, I. H., Ko, H. I., and Sohn, W. M. 1988a. A case of natural human infection by *Echinostoma cinetorchis*. Korean Journal of Parasitology 26:61–64 (in Korean).

Lee, S. K., Chung, N. S., Ko, I. H., Sohn, W. M., Hong, S. T., Chai, J. Y., and Lee, S. H. 1988b. An epidemiological survey of *Echinostoma hortense* infection in Chongsong-gun, Kyongbuk Province. Korean Journal of Parasitology 26:199–206.

Lee, S. H., Lee, J. K., Sohn, W. M., Hong, S. T., Hong, S. J., and Chai, J. Y. 1988c. Metacercariae of *Echinostoma cinetorchis* encysted in the freshwater snail, *Hippeutis* (*Helicorbis*) *cantori*, and their development in rats and mice. Korean Journal of Parasitology 26:189–197.

Lee, S. H., Chai, J. Y., Hong, S. T., and Sohn, W. M. 1990a. Experimental life history of *Echinostoma cinetorchis*. Korean Journal of Parasitology 28:39–44.

Lee, S. H., Noh, T. Y., Sohn, W. M., Kho, W. G., Hong, S. T., and Chai, J. Y. 1990b. Chronological observation of intestinal lesions of rats experimentally infected with *Echinostoma hortense*. Korean Journal of Parasitology 28:45–52 (in Korean).

Lee, S. H., Sohn, W. M., and Chai, J. Y. 1990c. *Echinostoma revolutum* and *Echinoparyphium recurvatum* recovered from house rats in Yangyang-gun, Kangwon-do. Korean Journal of Parasitology 28:235–240.

Lee, S. H., Hwang, S. W., Sohn, W. M., Kho, W. G., Hong, S. T., and Chai, J. Y. 1991. Experimental life history of *Echinostoma hortense*. Korean Journal of Parasitology 29:161–172 (in Korean).

Lee, K. J., Park, S. K., Im, J. A., Kim, S. K., Kim, G. H., Kim, G. Y., Yang, E. J., and Ryang, Y. S. 2004. Susceptibility of several strains of mice to *Echinostoma hortense* infection. Korean Journal of Parasitology 42:51–56.

Liang, C., and Ke, X. L. 1988. *Echinochasmus jiufoensis* sp. nov., a human parasite from Guangzhou (Trematoda: Echinostomatidae). Acta Zootaxonomica Sinica 13:4–8 (in Chinese).

Lie, K. J. 1963. Studies on Echinostomatidae in Malaya. III. The adult *Echinostoma malayanum* Leiper, 1911 (Trematoda) and the probable synonymy of *Artyfechinostomum sufrartyfex* Lane, 1915. Zeitschrift für Parasitenkunde 23:124–135.

Lie, K. J. 1968. Further studies on the life history of *Echinostoma lindoense* Sandground and Bonne, 1940 (Trematoda: Echinostomatidae) with a report of its occurrence in Brazil. Proceedings of the Helminthological Society of Washington 35:74–77.

Lin, J. X., Chen, Y. Z., Liang, C. Z., Yang, I. B., and Zhuang, H. J. 1985. Epidemiological investigation and experimental infection of *Echinochasmus japonicus*. Chinese Journal of Parasitology and Parasitic Diseases 3:89–91 (in Chinese).

Mabus, J., Huffman, J. E., and Fried, B. 1988. Humoral and cellular response to infection with *Echinostoma revolutum* in the golden hamster, *Mesocricetus auratus*. Journal of Helminthology 62:127–132.

Maji, A. K., Bera, D. K., Manna, B., Nandy, A., Addy, M., and Bandyopadhyay, A. K. 1993. First record of human infection with *Echinostoma malayanum* in India. Transactions of the Royal Society of Tropical Medicine and Hygiene 87:673.

Maji, A. K., Manna, B., Bandyopadhyay, A. K., Bera, D. K., Addy, M., and Nandy, A. 1995. Studies on the life cycle of *Artyfechinostomum oraoni* Bandyopadhyay, Manna and Nandy, 1989: embryogenesis and development in the intermediate host. Indian Journal of Medical Research 102:124–128.

Majima, M. 1927. On *Echinostoma macrorchis* found in man. Kumamoto Igakkai Zasshi No. 2252, 2260 (in Japanese).

Maldonado, A. Jr., Zeitone, B. K., Amado, L. A., Rosa, I. F., Machado-Silva, J. R., and Lanfredi, R. M. 2005. Biological variation between two Brazilian geographical isolates of *Echinostoma paraensei*. Journal of Helminthology 79:345–351.

McCarthy, A. M. 1990. Speciation of echinostomas: evidence for the existence of two sympatric sibling species in the complex *Echinoparyphium recurvatum* (von Linstow 1873) (Digenea: Echinostomatidae). Parasitology 101:35–42.

McMaster, R. P., Huffman, J. E., and Fried, B. 1995. The effect of dexamethasone on the course of *Echinostoma caproni* and *E. trivolvis* infection in the golden hamster (*Mesocricetus auratus*). Parasitology Research 81:518–521.

Miyamoto, K., Nakao, M., and Inaoka, T. 1983. Studies on the zoonoses in Hokkaido, Japan. 5. On the epidemiological survey of *Echinostoma hortense* Asada, 1926. Japanese Journal of Parasitology 32:261–269.

Morgan, J. A. T., and Blair, D. 1995. Nuclear rDNA ITS sequence variation in the trematode genus *Echinostoma*: an aid to establishing relationships within the 37-collar-spine group. Parasitology 111:609–615.

Muñoz-Antoli, C., Sotillo, J., Monteagudo, C., Fried, B., Marcilla, A., and Toledo, R. 2007. Development and pathology of *Echinostoma caproni* in experimentally infected mice. Journal of Parasitology 93:854–859.

Park, J. T. 1938. A rat trematode, *Echinostoma hortense* Asada, from Korea. Keijo Journal of Medicine 9:283–286.

Park, C. J., and Kim, J. 2006. A human case of *Echinostoma hortense* infection diagnosed by endoscopy in area of southwestern Korea. Korean Journal of Medicine 71:229–234 (in Korean).

Poland, G. A., Navin, T. R., and Sarosi, G. A. 1985. Outbreak of parasitic gastroenteritis among travelers returning from Africa. Archives of Internal Medicine 145: 2220–2221.

Premvati, G., and Pande, V. 1974. On *Artyfechinostomum malayanum* (Leiper, 1911) Mendheim, 1943 (Trematoda: Echinostomatidae) with synonymy of allied species and genera. Proceedings of the Helminthological Society of Washington 41:151–160.

Radomyos, P., Bunnag, D., and Harinasuta, T. 1982. *Echinostoma ilocanum* (Garrison, 1908) Odhner, 1911, infection in man in Thailand. Southeast Asian Journal of Tropical Medicine and Public Health 13:265–269.

Radomyos, P., Bunnag, D., and Harinasuta, T. 1985. Report of *Episthmium caninum* (Verma, 1935) Yamaguti, 1958 (Digenea: Echinostomatidae) in man. Southeast Asian Journal of Tropical Medicine and Public Health 16:508–511.

Radomyos, P., Charoenlarp, P., and Harinasuta, T. 1991. Human *Episthmium caninum* (Digenea: Echinostomatidae) infection: report of two more cases. Journal of Tropical Medicine and Parasitology 14:48–50.

Radomyos, B., Wongsaroj, T., Wilrairatana, P., Radomyos, P., Praevanich, R., Meesomboon, V., and Jongsuksuntikul, P. 1998. Opisthorchiasis and intestinal fluke infections in northern Thailand. Southeast Asian Journal of Tropical Medicine and Public Health 29:123–127.

Raghunathan, V. S., and Srinivasan, T. 1962. *Artyfechinostomum mehrai* infestation. Journal of Indian Medical Association 38:485–487.

Reddy, D. B., Ranganaykamma, I., and Venkataratnam, D. 1964. *Artyfechinostomum mehrai* infestation in man. Journal of Tropical Medicine and Hygiene 67:58–59.

Rim, H. J. 1982. Echinostomiasis. In CRC Handbook Series in Zoonoses, Section C: Parasitic Zoonoses, Vol. III. Trematode Zoonoses, pp 53–69. Boca Raton, Florida, USA: CRC.

Ryang, Y. S. 1990. Studies on *Echinostoma* spp. in the Chungju Reservoir and upper streams of the Namhan River. Korean Journal of Parasitology 28:221–233 (in Korean).

Ryang, Y. S., Ahn, Y. K., Lee, K. W., Kim, T. S., and Hhan, M. H. 1985. Two cases of natural human infection by *Echinostoma hortense* and its second intermediate host in Wonju area. Korean Journal of Parasitology 23:33–40 (in Korean).

Ryang, Y. S., Ahn, Y. K., Kim, W. T., Shin, K. C., Lee, K. W., and Kim, T. S. 1986. Two cases of human infection by *Echinostoma cinetorchis*. Korean Journal of Parasitology 24:71–76 (in Korean).

Ryang, Y. S., Ahn, Y. K., and Yoon, M. B. 1991. Trematode infections in the small intestine of *Egretta alba medesta* in Kangwon-do. Korean Journal of Parasitology 29:227–233 (in Korean).

Ryang, Y. S., Yang, E. J., Kim, J. L., Lee, K. J., Sung, H. J., Kim, J. B., and Kim, I. S. 2007. Immune response and inhibitory effect of ketotifen on the BALB/c and C3H/HeN mice infected with *Echinostoma hortense*. Parasitology Research 101:1103–1110.

Saeed, I., Maddox-Hyttel, C., Monrad, J., and Kapel, C. M. 2006. Helminths of red foxes (*Vulpes vulpes*) in Denmark. Veterinary Parasitology 139:168–179.

Saito, S., and Tani, S. 1982. Comparison of meracercariae of *Echinostoma hortense* Asada, 1926 and *Echinostoma cinetorchis* Ando et Ozaki, 1923 in loach, *Misgurnus anguillicaudatus*. Japanese Journal of Parasitology 31:281–287 (in Japanese).

Seo, B. S., Rim, H. J., and Lee, C. W. 1964. Studies on the parasitic helminths of Korea. I. Trematodes of rodents. Korean Journal of Parasitology 2:20–26.

Seo, B. S., Cho, S. Y., and Chai, J. Y. 1980. Studies on intestinal trematodes in Korea. I. A human case of *Echinostoma cinetorchis* infection with an epidemiological investigation. Seoul Journal of Medicine 21:21–29.

Seo, B. S., Cho, S. Y., Hong, S. T., Hong, S. J., and Lee, S. H. 1981. Studies on parasitic helminths of Korea. V. Survey on intestinal trematodes of house rats. Korean Journal of Parasitology 19:131–136.

Seo, B. S., Hong, S. T., Chai, J. Y., and Lee, S. H. 1983. Studies on intestinal trematodes in Korea. VII. A human case of *Echinostoma hortense* infection. Korean Journal of Parasitology 21:219–223.

Seo, B. S., Park, Y. H., Chai, J. Y., Hong, S. J., and Lee, S. H. 1984. Studies on intestinal trematodes in Korea. XIV. Infection status of loaches with metacercariae of *Echinostoma cinetorchis* and their development in albino rats. Korean Journal of Parasitology 22:181–189.

Seo, B. S., Lee, S. H., Chai, J. Y., and Hong, S. J. 1985a. Studies on intestinal trematodes in Korea XX. Four cases of natural human infection by *Echinochasmus japonicus*. Korean Journal of Parasitology 23:214–220.

Seo, B. S., Chun, K. S., Chai, J. Y., Hong, S. J., and Lee, S. H. 1985b. Studies on intestinal trematodes in Korea. XVII. Development and egg laying capacity of *Echinostoma hortense* in albino rats and human experimental infection. Korean Journal of Parasitology 23:24–32.

Shimalov, V. V., and Shimalov, V. T. 2002. Human parasitizing echinostomatata (Trematoda, Fasciolidae) in mammals in Belorussia forest zone. Meditsinskaia Parazitologiia i Parazitamye Bolezni, No. 1:47–50 (in Russian).

Shin, E. H., Lee, S. H., Kim, J. L., Park, Y. K., and Chai, J. Y. 2007. T-helper-1 and T-helper-2 immune responses in mice infected with the intestinal fluke *Neodiplostomum seoulense*: their possible roles in worm expulsion and host fatality. Journal of Parasitology 93:1036–1045.

Simonsen, P. E., and Andersen, B. J. 1986. *Echinostoma revolutum* in mice: dynamics of the antibody attack to the surface of an intestinal trematode. International Journal for Parasitology 16:475–482.

Sohn, W. M. 1998. Life history of *Echinoparyphium recurvatum* (Trematoda: Echinostomatidae) in Korea. Korean Journal of Parasitology 36:91–98.

Sohn, W. M., and Chai, J. Y. 2005. Infection status with helminthes in feral cats purchased from a market in Busan, Republic of Korea. Korean Journal of Parasitology 43:93–100.

Sorensen, R. E., Curtis, J., and Minchella, D. J. 1998. Intraspecific variation in the rDNA ITS loci of 37-collar-spined echinostomes from North America: implications for sequence-based diagnoses and phylogenetics. Journal of Parasitology 84:992–997.

Sotillo, J., Muñoz-Antoli, C., Macilla, A., Fried, B., Esteban, J. G., and Toledo, R. 2007. *Echinostoma caproni*: kinetics of IgM, IgA and IgG subclasses in the serum and intestine of experimentally infected rats and mice. Experimental Parasitology 116:390–398.

Sugimoto, M. 1933. On a trematode parasite (*Echinostoma cinetorchis* Ando and Ozaki, 1923) from a Formosan dog. Nippon Juigakkai Zasshi 12:231–236 (in Japanese).

Tanabe, H. 1922. Studies on trematodes taking freshwater fishes as their intermediate hosts III. On *Echinochasmus perfoliatus* (Tatz) found in man. Okayama Iggakai Zasshi No. 387:1–20 (in Japanese).

Tanabe, H. 1926. Studies on trematodes taking freshwater fishes as their intermediate hosts III. On a new species, *Echinochasmus japonicus* (n. sp.). Nippon Byori Gakkai Kaishi 16:295–296 (in Japanese).

Toledo, R., and Fried, B. 2005. Echinostomes as experimental models for interactions between adult parasites and vertebrate hosts. Trends in Parasitology 21:251–254.

Toledo, R., Muñoz-Antoli, C., Perez, M., and Esteban, J. G. 1999. Miracidial infectivity of *Hypoderaeum conoideum* (Trematoda: Echinostomatidae): differential susceptibility of two lymnaeid species. Parasitology Research 85:212–215.

Toledo, R., Muñoz-Antoli, C., and Esteban, J. G. 2000. The life-cycle of *Echinostoma friedi* n. sp. (Trematoda: Echinostomatidae) in Spain and a discussion on the relationships within the '*revolutum*' group based on cercarial chaetotaxy. Systematic Parasitology 45:199–217.
Toledo, R., Espert, A., Carpena, I., Muñoz-Antoli, C., and Esteban, J. G. 2003. An experimental study of the reproductive success of *Echinostoma friedi* (Trematoda: Echinostomatidae) in the golden hamster. Parasitology 126:433–441.
Toledo, R., Espert, A., Carpena, I., Muñoz-Antoli, C., Fried, B., and Esteban, J. G. 2004. The comparative development of *Echinostoma caproni* (Trematoda: Echinostomatidae) adults in experimentally infected hamsters and rats. Parasitology Research 93:439–444.
Toledo, R., Esteban, J. G., and Fried, B. 2006. Immunology and pathology of intestinal trematodes in their definitive hosts. Advances in Parasitology 63:285–365.
Ujiie, N. 1936. On the development, the structure of *Echinochasmus japonicus* and its parasitism in man. Taiwan Igakkai Zasshi 35:535–546 (in Japanese).
Weinstein, M. S., and Fried, B. 1991. The expulsion of *Echinostoma trivolvis* and retention of *Echinostoma caproni* in the ICR mouse: pathological effects. International Journal for Parasitology 21:255–257.
Witenberg, G. 1932. On two parasite species of the genus *Echinochasmus* (Trematoda) in dogs and cats in Palestine. Zeitschrift für Parasitenkende 5:213–216.
Wu, G. 2004. Human Parasitology. Beijing: Rin Min Hu Sheng.
Xiao, X., Wong, T. P., Lu, D. B., Gao, J. F., Xu, L. F., Wu, W. D., Mei, J. D., and Xu, F. L. 1992. The first record of human natural infection with *Echinochasmus liliputanus*. Chinese Journal of Parasitology and Parasitic Diseases 10:132–135 (in Chinese).
Xiao, X., Lu, D., Wang, T., Gao, J., Zhu, C., Zhang, B., An, J., Pong, H., Xu, M., Wu, W., Mei, J., and Xu, F. 1995. Studies on mode of human infection with *Echinochasmus liliputanus*. Chinese Journal of Parasitology and Parasitic Diseases 13:197–199 (in Chinese).
Xiao, X., Wang, T., Zheng, X., Shen, G., and Tian, Z. 2005. In vivo and in vitro encystment of *Echinochasmus liliputanus* cercariae and biological activity of the metacercariae. Journal of Parasitology 91:492–498.
Yamaguti, S. 1939. Studies on the helminth fauna of Japan. Part 25. Trematodes of birds, IV. Japanese Journal of Zoology 8:131–210.
Yamaguti, S. 1958. Systema Helminthum, Vol. I. Part I. The digenetic trematodes of vertebrates. New York: Interscience.
Yamashita, J. 1964. Echinostome. In Progress of Medical Parasitology in Japan, Vol. 1, pp 287–313. Tokyo: Meguro Parasitological Museum.
Yokogawa, M., Harinasuta, C., and Charoenlarp, P. 1965. *Hypoderaeum conoideum* (Block, 1782) Dietz, 1909, a common intestinal fluke of man in the north-east Thailand. Japanese Journal of Parasitology 14:148–153.
Yu, S. H., and Mott, K. E. 1994. Epidemiology and morbidity of food-borne intestinal trematode infections. Tropical Disseases Bulletin 91:R125–R152.
Yurlova, N. I. 1987. Morphology and biology of trematodes *Hypoderaeum cubanicum* comb. nov. and *H. conoideum* (Echinostomatidae). Parazitologiya 21:35–42 (in Russian).
Yurlova, N. I., Vodyanitskaya, S. N., Serbina, E. A., Biserkov, V. Y., Georgiev, B. B., and Chipev, N. H. 2006. Temporal variation in prevalence and abundance of metacercariae in the pulmonate snail *Lymnaea stagnalis* in Chany Lake, West Siberia, Russia: long-term patterns and environmental covariates. Journal of Parasitology 92:249–259.
Zhu, D., Lin, J., Cheng, Y., Liang, C., Wang, X., Guo, Z., and Zhuang, H. 1986. Pyquiton treatment of *Echinochasmus japonicus* in ection. Chinese Journal of Parasitology and Parasitic Diseases 4:1–4 (in Chinese).

Chapter 8
Immunology and Pathology of Echinostome Infections in the Definitive Host

Rafael Toledo

Contents

8.1 Introduction 186
8.2 Manifestations of Resistance to Infection 186
8.3 Experimental Models for the Study of the Immunobiology of Echinostome Infections 188
8.4 Pathology of the Echinostome Infections 190
8.5 Mechanisms of Resistance to Echinostome Infections 192
8.5.1 Mast Cells 192
8.5.2 Eosinophils 193
8.5.3 Goblet Cells 194
8.5.4 Antibody Responses 196
8.5.5 Cytokine Responses 198
8.6 Antigenic Characterization of Adult Echinostomes 199
8.7 Immunodiagnosis 200
8.8 Concluding Remarks 202
References 202

Abstract This chapter examines the salient literature on the immunology and pathology of members of the family Echinostomatidae in their definitive hosts, with emphasis on experimental studies that may provide useful information on factors that determine resistance to the parasites. For this purpose, several topics such as manifestations and mechanisms of resistance to infection, experimental strategies, and antigenic characterization of echinostomes are covered. Moreover, other topics such as immunodiagnosis are also analyzed. The analysis is focused on members of the genus *Echinostoma*. Although some of the nomenclature for echinostome species is disputed, the names used are those currently accepted.

R. Toledo (✉)
Departamento de Biología Celular y Parasitología, Facultad de Farmacia, Universidad de Valencia, Av. Vicent Andrés Estellés s/n, 46100 Burjassot-Valencia, Spain
e-mail: rafael.toledo@uv.es

B. Fried and R. Toledo (eds.) *The Biology of Echinostomes*,
DOI: 10.1007/978-0-387-09577-6_8, © Springer Science+Business Media, LLC 2009

8.1 Introduction

Echinostomes are important intestinal parasitic flatworms invading domestic and wildlife animals and occasionally humans. They are able to parasitize a wide range of vertebrate species as definitive hosts making them good models for the study of host–parasite interactions (Toledo and Fried 2005). Among other aspects, echinostomes have been used as models for the study of chronic and acute infections. This is based on the observation that they are rapidly expelled or develop long-lasting infections depending on the host species used. Consequently, several experimental approaches have been employed to elucidate the factors that determine the generation of host responses that cause resistance to intestinal helminth infection. Furthermore, application of recent advances in basic immunology and proteomics in echinostome infections is providing important information to the understanding of the responses that facilitate intestinal helminth resistance. This chapter considers immunological and pathological work on different species of *Echinostoma* with emphasis on experimental studies that provide useful information on the factors that determine the course of the infection.

In several papers, different names have been used for the species of *Echinostoma* due to the taxonomic confusion within the genus *Echinostoma* (see Chap. 1 of this book). The name currently most accepted for each species, regardless of what name was used in the original paper, is used in this chapter.

8.2 Manifestations of Resistance to Infection

One of the main reasons to justify interest in the study of immunological aspects of echinostome infections comes from the fact that several manifestations of resistance to these infections have been described. These manifestations have been shown to be dependent on the properties of the host immune response to the parasite. Immunity against adult stages of echinostomes is manifested by: (1) increased worm burden after immunosuppression of the host; (2) natural expulsion of primary infections; (3) generation of resistance to homologous and heterologous secondary infections; (4) reduction of fecundity of echinostomes; and (5) changes in the morphology of adult echinostomes.

(1) Increased worm burden after immunosuppression of the host

There is evidence indicating that the number of worms established in the definitive host depends on the immunological background of the host. An increase in worm recovery as a consequence of immunosuppression has often been described in *Echinostoma caproni* and *E. trivolvis*, and constitutes a major manifestation of effective immunity against these digeneans. Treatment of mice with dexamethasone resulted in an increased worm recovery of both species of echinostomes with respect to nontreated controls (McMaster et al. 1995; Fujino et al. 1996a; Fried et al. 1997). Similarly, Ryang et al. (2007) reported that treatment with ketotifen of

BALB/c and C3H/HeN mice significantly increased the worm recovery rate and the total period of worm recovery of *E. hortense*.

(2) Natural expulsion of primary infections

The pattern of worm expulsion in echinostome infections is markedly host species dependent. For each echinostome species, expulsion of primary infections occurs rapidly in a number of rodent species, while in other species chronic infections are developed (Toledo and Fried 2005). The elimination of *E. caproni* primary infections in rats occurred at 7–8 weeks postinfection (wpi) (Hansen et al. 1991; Toledo et al. 2004a), whereas worm survival in hamsters showed a persistence of at least 20 wpi (Toledo et al. 2004b), and in mice the longevity of the adult worms reached 23 wpi (Fried and Peoples 2007). The elimination of *E. trivolvis* from mice occurs within 2–3 wpi (Weinstein and Fried 1991), while in hamsters worms survived for up to 123 days postinfection (dpi) (Mabus et al. 1988). Similarly, *E. friedi* was able to survive for at least 12 wpi in hamsters, whereas the infection was expelled at 3–4 wpi in rats (Toledo et al. 2003a, 2006a; Muñoz-Antoli et al. 2004). Differences in the pattern of worm expulsion between strains of mice also have been observed. Lee et al. (2004) reported that ICR and C3H/HeN mice remained infected with *E. hortense* up to 7 and 5 wpi, respectively, whereas all the worms had been expelled at 4 wpi in BALB/c, C57BL6, and FvB strains.

There is some evidence suggesting that differences in the pattern of worm expulsion have an immunological background. There are three lines of evidence for this: (1) the survival of echinostome infections is prolonged by immunosuppression; (2) expulsion of *E. caproni* adult worms is impaired by concurrent infection with *Schistosoma mansoni* (Christensen et al. 1985); and (3) significant differences in the immunological and pathological responses between hosts in which the parasites are rapidly rejected and those in which chronic infections occur have been observed (Toledo et al. 2003b, 2004b, 2006b; Muñoz-Antoli et al. 2007).

(3) Generation of resistance to homologous and heterologous secondary infections

The generation of homologous resistance to *E. caproni* and *E. trivolvis* infections has been demonstrated. Sirag et al. (1980) showed that a 20-day-old primary *E. caproni* infection in mice was sufficient to generate a complete resistance to reinfection from 8 days after elimination of the primary infection by antihelminthic treatment. A marked resistance to *E. caproni* secondary infection occurred in NMRI and SVS mice for up to at least 12–14 and 6 weeks, respectively, following the expulsion of *E. caproni* primary infections (Christensen et al. 1986; Odaibo et al. 1989). Hosier et al. (1988) showed that ICR mice can develop resistance against homologous challenge infection with either *E. caproni* or *E. trivolvis*. After homologous challenge, established worm burdens of *E. caproni* and *E. trivolvis* were 42 and 87.5%, respectively, with respect to those observed in primary infections. Heterologous resistance was also demonstrated since there was a 17% reduction of the worm burden in an *E. caproni* challenged infection upon a primary *E. trivolvis* infection and a 55% reduction of *E. trivolvis* challenge infection upon a primary infection with *E. caproni* (Hosier et al. 1988). Furthermore, crossresistance against *E. caproni* induced by *S. mansoni* infection in mice has been described. Sirag et al. (1980) reported that 79- and 99-day-old infection with *S. mansoni* conferred complete resistance to infection with *E. caproni*.

(4) Reduction of fecundity of echinostomes

Reduced fecundity may be a result of immunity against the parasite. There is evidence suggesting a role of the host immune response in the regulation of the fecundity of echinostomes. Differences in egg counts in relation to the host species have been reported for *E. caproni* (Toledo et al. 2004a), *E. trivolvis* (Franco et al. 1986), and *E. friedi* (Muñoz-Antoli et al. 2004; Toledo et al. 2006a). In some cases, the differences observed in the egg output could be attributed to the different number of worms established in each host species in relation to the host–parasite compatibility. However, Muñoz-Antoli et al. (2004) found that the number of eggs of *E. friedi* released weekly was significantly lower in rats than in hamsters, though the worm burden was similar in both host species (see Chap. 4 of this book for further information on this topic). Lee et al. (2004) infected five strains of mice (ICR, C3H/HeN, BALB/c, C57BL6, and FvB) with *E. hortense*. Although the five strains became infected, eggs only were detected in the feces of ICR and C3H/HeN mice. All these facts suggest that host immunological background may be an important determinant of echinostome fecundity.

(5) Changes in the morphology of adult echinostomes

Changes in the morphology of echinostome adult worms can be another manifestation of resistance to infection. In fact, increases in the body area have been reported as a consequence of immunosuppression of the host. Fujino et al. (1996a) observed that the body area of *E. trivolvis* adult worms from dexamethasone-treated C3H/HeN mice increased markedly on 12 wpi and were more than 2.5 times greater than those worms collected from nontreated mice. These results were confirmed by Fried et al. (1997). In the case of *E. caproni*, the adult worms collected from RAG-2-deficient mice, a strain genetically altered mice lacking B- and T-lymphocytes, showed greater body areas than those from conventional ICR mice (Frazer et al. 1999).

8.3 Experimental Models for the Study of the Immunobiology of Echinostome Infections

Experimental studies on the immunology and pathology of echinostome infections have been performed both in rodents and domestic chicks. However, most of the recent knowledge on these topics comes from studies using different rodent models. The *Echinostoma*-rodent model has proved to be an excellent one for studying the host-intestinal trematode relationships (Toledo and Fried 2005). In general, echinostomes are able to infect a wide range of rodent hosts, although its compatibility differs considerably among rodent species. In highly compatible rodent hosts, echinostome infections become chronic, whereas the worms are rapidly expelled from rodent hosts with low compatibility (see Chap. 4 of this book). This enables comparative studies of one species of echinostomatid within different experimental rodent hosts in which the course of the infection differs markedly. This experimental approach may facilitate the establishment of the host-dependent factors determining the rejection of the worms or, by contrast, the establishment of chronic infections. This type of study has been performed using different strategies: (1) immunosuppressed mice; (2) different strains of mice; and (3) different rodent species.

(1) Immunosuppressed mice

Much of the information on echinostome immunobiology is derived from comparative studies on the infection on immunosuppressed and immunocompetent mice. Immunosuppression can be accomplished by the use of genetically altered mice or immunosuppressant treatment. This approach has been used for *E. caproni, E. trivolvis*, and *E. hortense*. Mice are hosts of high compatibility with *E. caproni* in which chronic infections occur (Hosier and Fried 1991; Fried and Peoples 2007; Muñoz-Antoli et al. 2007). This limits the use of immunodeficient mice to study the factors determining worm rejection. However, Frazer et al. (1999) compared the characteristics of *E. caproni* infections in RAG-2-deficient and ICR mice. Although significant goblet cell hyperplasia was detected in the infected vs. noninfected RAG mice, this could not be related to the elimination of the parasite. *E. trivolvis* is rapidly rejected from conventional mice (Weinstein and Fried 1991), and the use of immunosuppressed mice has been useful. Fujino et al. (1993) studied the *E. trivolvis* infection in conventional and congenitally athymic BALB/c mice. There were no marked differences in the infectivity of *E. trivolvis* in both mice strains. The drastic expulsion of the worms occurred between 10 and 13 dpi in both groups. Apart from this fact, marked increases in goblet cells and mast cells were observed in conventional but not in athymic mice in response to the infection. Treatment of mice with immunosuppressants, such as dexamethasone or FK506, (tacrolimus) has also been used to study *E. trivolvis* infections (McMaster et al. 1995; Fujino et al. 1996a, 1998a,1998b). The results obtained suggest that the expulsion of *E. trivolvis* is related to the increases in the goblet cell numbers in the mouse small intestine. Ryang et al. (2007) studied the effect of ketotifen on *E. hortense* infection in BALB/c and C3H/HeN. The worm recovery was higher and lasted longer in the treated mice in comparison to the untreated mice. In addition, ketotifen inhibited the increase of eosinophils and goblet cells and the expression of several cytokines in various degrees depending on the host strain. These results may be of interest in studies on factors determining worm rejection.

(2) Different strains of mice

Various strains of mice, such as ICR, C3H, Swiss Webster, CBA, SCID, RAG, C3H/HeN or athymic, have been used as experimental hosts for *E. caproni* and *E. trivolvis* (Bindseil and Christensen 1984; Hosier and Fried 1986; Odaibo et al. 1988; Fujino and Fried 1993a,b; Baek et al. 1996; Frazer et al. 1999; Muñoz-Antoli et al. 2007). However, the results obtained in relation to the worm rejection on the basis of the mice strain are not conclusive and further studies are required. Lee et al. (2004) studied the susceptibility of five strains of mice to *E. hortense* infection. This study showed that *E. hortense* is highly infectious to ICR and C3H/HeN mice, but not to the other strains. Thereafter, the host immune response against *E. hortense* in a high (C3H/HeN) vs a low (BALB/c) compatible strain of mice was compared, suggesting that the immune sensitivity depends on the host strain. The BALB/c mice were more resistant probably in relation to the higher induction of protective Th2 immunity (Cho et al. 2007; Ryang et al. 2007).

(3) Different host species

As mentioned earlier, the course and the characteristics of echinostome infections are largely dependent on the host species. Thus, analysis of the immune response against a single parasite species in different host species may be helpful to identify the host-related factors responsible for either worm rejection or worm maintenance. This experimental approach has been extensively used for *E. caproni* (Toledo and Fried 2005). Apart from the biological features, several immunological and pathological parameters are also dependent on the host species in *E. caproni* infections. Chronic infections in mice and hamsters have been associated with high levels of systemic IgG responses and strong local infiltration of inflammatory mononuclear cells. In contrast, infections in rats are characterized by a rapid parasite expulsion together with low levels of IgG in serum and weaker infiltration of inflammatory cells (Toledo et al. 2003b, 2004b, 2006a,b; Muñoz-Antoli et al. 2007; Sotillo et al. 2007). Further studies on these topics may be useful for the understanding of the factors responsible for the differences in the host responses and their consequences in the parasite and the course of the infection.

8.4 Pathology of the Echinostome Infections

The pathology of echinostome infections represents a complex set of reactions. The complexity is dependent on a wide variety of factors including the parasite and the host species (Toledo et al. 2006c). Clinical disease is modulated by the intensity of the infection, and the main signs in heavy infections are weakness, diarrhea, weight loss, and unthriftiness (Huffman and Fried 1990; Toledo et al. 2006c).

Most studies on the intestinal pathology of echinostome infections use rodents as experimental models. *E. caproni* and *E. trivolvis* induce similar lesions in golden hamsters. Damage has been observed at the attachment sites of adult worms. Such areas showed marked dilation and lymphocytic infiltration (Huffman et al. 1988; Mabus et al. 1988; Fried et al. 1990; Weinstein and Fried 1991; Fujino and Fried 1993a; Toledo et al. 2006b). Destruction and erosion of the villi also are common in these infections in hamsters. This feature has been observed both in hamsters and rats infected with *E. caproni* (Toledo et al. 2006b). However, the villous damage is almost completely regenerated in hamsters at 30 dpi, whereas in rats the villous damage is greater at 30 dpi than earlier (Toledo et al. 2006b). Probably, differences in the cellular responses observed between both host species may be involved in the earlier villous regeneration in hamsters. Increases in goblet cell numbers were observed in *E. caproni*-infected rats, whereas the goblet cell hyperplasia was significantly lower in hamsters. Eosinophilic infiltration also was higher in rats than in hamsters infected with *E. caproni* (Toledo et al. 2006b). Greater differences between *E. caproni*-infected rats and hamsters have been observed in relation to the presence of mucosal neutrophils and mesenteric inflammatory cells. A marked inflammatory cell infiltration together with increased numbers of mucosal neutrophils was observed in hamsters, but not in rats. This suggests that greater local inflammatory responses may be associated with *E. caproni* chronic infections (Toledo et al. 2006b).

Several studies have shown that the mouse model is very suitable for research on the pathology of echinostome infections. *E. caproni* induces chronic infections in mice, whereas *E. trivolvis* is expelled from mice at 4 wpi (Hosier and Fried 1991; Fried and Peoples 2007). Thus, comparative studies on the pathology induced by each parasite species in mice may be useful. The histopathological effects of *E. caproni* in mice occur mainly in the surrounding area of the attachment sites of the worms. The intestine of mice infected with *E. caproni* showed a marked dilation in which the majority of the worms were found (Muñoz-Antoli et al. 2007). Marked villous atrophy with fused and eroded villi and crypt hyperplasia may be seen (Bindseil and Christensen 1984; Weinstein and Fried 1991; Fujino and Fried 1993a). The total mucosal thickness increased simultaneously with a hypertrophy of the tunica muscularis (Bindseil and Christensen 1984). In the hyperplastic crypts, the mitotic rate was increased over that seen in normal crypts (Bindseil and Christensen 1984; Fujino and Fried 1993a). An increase of collagen fibers and fibroblasts was also observed in the epithelium of the atrophic villi (Bindseil and Christensen 1984; Fujino and Fried 1993a). Cellular infiltration of lymphocytes, eosinophils, and plasma cells was observed in lamina propria and submucosa (Weinstein and Fried 1991; Muñoz-Antoli et al. 2007). Mast cell counts showed an increase at 15 and 30 dpi similar to those observed in rats and hamsters (Toledo et al. 2006b; Muñoz-Antoli et al. 2007). The number of eosinophils in *E. caproni*-infected mice rapidly increased, reaching a maximum at 15 dpi (Bindseil and Christensen 1984; Muñoz-Antoli et al. 2007). The number of eosinophils was higher than that observed in hamsters which may be correlated with the worm burden observed in each host species (Toledo et al. 2006b; Muñoz-Antoli et al. 2007). The number of goblet cells in *E. caproni*-infected mice was also higher than in *E. caproni*-infected hamsters supporting the idea that these cells may be important in regulatory mechanisms of worm expulsion (Toledo et al. 2006b,2006c; Muñoz-Antoli et al. 2007). Interestingly, the numbers of mucosal neutrophils and mononuclear inflammatory cells in the mesentery were lower than in hamsters and higher than in rats. This fact, together with the worm recovery observed in each host species, suggests that the increases in these populations of cells may be associated with a higher survival of *E. caproni* adult worms.

The pathology of the related species *E. trivolvis* in mice shows marked differences which may be of interest considering the different survival of each echinostome in this host species. No significant dilation of the gut or villous atrophy has been associated with *E. trivolvis* infections in mice (Weinstein and Fried 1991; Fujino and Fried

1993a; Fujino et al. 1993). Differences in the enzymatic activity in the intestine of infected mice were also observed. The alkaline phosphatase activity was less reduced in *E. trivolvis*-infected than in noninfected mice (Fujino et al. 1993). In *E. trivolvis*-infected mice there was hyperplasia of Paneth cells, mast cells, and goblet cells (Weinstein and Fried 1991; Fujino and Fried 1993a; Fujino et al. 1993).

One of the factors determining these differences may be that effector mechanisms mediating the histopathological changes in *E. caproni* and *E. trivolvis* infections in mice are different. Bindseil and Christensen (1984) showed that pathological changes in *E. caproni*-infected mice were thymus independent. In contrast, Fujino et al. (1993) showed that increases in mast cell number did not occur in athymic mice infected with *E. trivolvis*, and also the number of goblet cells was lower in athymic and in conventional mice. These facts suggest that thymus-dependent mechanisms operate in *E. trivolvis* infections in mice.

8.5 Mechanisms of Resistance to Echinostome Infections

As mentioned in previous sections of this chapter, resistance to echinostome infections can be manifested in several ways. This indicates that an immune response may be generated after a threshold level of stimulus to the host has occurred. This stimulus can be antigenic, though stimulation may also involve mechanical or enzymatic damage to the host tissue. To date, mechanisms responsible for the manifestations of resistance are not fully understood. However, some features have been observed in relation to the resistance to echinostome infections. This section will provide a description of the typical effector immune mechanisms associated with echinostome infections and a review of the studies describing the possible role of these mechanisms in parasite rejection.

8.5.1 Mast Cells

Mast cells reside and mature in tissues and consist of a heterogeneous population, though mucosal mast cells have been the most extensively studied in relation to intestinal helminth infections. The mucosal mast cells can reside in the lamina propria or the epithelium. Intraepithelial mucosal mast cells are considered to be end-stage effector, ideally located to discharge their products into the lumen and epithelium where they can affect the microenvironment of the parasite with minimal effect on the underlying tissue layers. Increase of mast cells in parasitized animals is dependent on cytokines produced by activated T cells, particularly IL-3, IL-9, and IL-18 (Lantz et al. 1998; Yoshimoto and Nakanishi 2006; Bischoff and Krämer 2007). Mast cells may release an array of biologically active mediators and the effect produced includes: smooth muscle contraction, increased vascular permeability, increased mucus secretion, and degradation of blood vessels and epithelial cell basement membranes (Balic et al. 2000; Sayed and Brown 2007).

Mast cell hyperplasia in the intestinal tissues of rodents infected with *Echinostoma* spp. is well recognized. However, there are conflicting data in relation to the role of mast cells in the course of the infection. Although several studies have shown that maximal mastocytosis coincides with worm expulsion, other effector responses appear to be needed to explain the rejection of adult echinostomes. In *E. trivolvis*-infected mice, the number of mast cells rapidly increased to reach a maximum at 2–3 wpi, coinciding with worm rejection (Tani and Yoshimura 1988; Fujino et al. 1993, 1996b, 1998a). However, in athymic mice and mice immunosuppressed with FK506, the mastocytosis was suppressed, but the kinetics of worm expulsion did not change with respect to nontreated mice (Fujino et al. 1993, 1998b).

In the *E. caproni*-rodent model, increases in the mast cell counts have been observed during the first weeks of the infection in all the host species studied. However, this fact does not appear to be a determinant for worm expulsion. Infected mice and hamsters develop similar mast cell kinetics than rats, though the worm survival differs greatly in these host species (Toledo et al. 2006b; Muñoz-Antoli et al. 2007). A similar situation occurs in the *E. hortense*-rodent model. Significant mast cell hyperplasia has been observed in rats and mice experimentally infected, and was greatest in the duodenum, where most of the worms were found. However, this response does not seem effective since *E. hortense* induces long-lasting infections in rats ,whereas the parasite is rejected earlier in mice (Kim et al. 2000; Park et al. 2005).

8.5.2 Eosinophils

Eosinophils consist of a uniform population of cells with mature eosinophils being recruited at a low level from the bone narrow into the blood and from there to the tissues where they may survive from several days to weeks. Although eosinophils are present in a low number in peripheral blood, they are present in a large number in the tissues, particularly in the gastrointestinal tract. The number of eosinophils may increase dramatically after helminth infections both in blood and tissues and can show a directional migration toward a parasite target (Balic et al. 2000, 2006).

The key cytokine for the T-cell dependent increase in eosinophils after helminth infection is IL-5. Activated eosinophils can produce oxygen-derived metabolites and synthesize lipid mediators such as prostaglandins and leucotrienes (Weller 1997). Moreover, they can also produce a range of cytokines and chemokines. Eosinophil desgranulation is initiated by crosslinking of surfaces since they have receptors for complement factors, IgG, IgA, and IgE. Moreover, they also express receptors for the secretory component of IgA, which may be of importance in gastrointestinal infections (Lamkhioued et al. 1995). Their major action during helminth infections may be a direct cytotoxic action (Balic et al. 2000).

Blood eosinophilia and eosinophil infiltration in the intestinal tract have been often associated with echinostome infections suggesting their involvement in worm expulsion mechanisms. Increases in the number of blood eosinophils have been observed in rats infected with *E. caproni* (Hansen et al. 1991), and mice (BALB/c and C3H/HeN strains) infected with *E. hortense* (Ryang et al. 2007). The degree of blood eosinophilia was closely related with the worm burden in each case. Eosinophilic infiltration of the intestinal tissue also is a characteristic of rodents infected with *E. caproni* (Bindseil and Christensen 1984; Toledo et al. 2006b; Muñoz-Antoli et al. 2007), *E. trivolvis* (Fujino et al. 1996a), and *E. hortense* (Ryang et al. 2007). Intestinal eosinophilia rapidly increases after infection to reach a maximum at 1–2 wpi and thereafter decreases. The use of the *E. caproni* model has allowed for comparison to the eosinophilic infiltration in rodent hosts displaying different compatibility with the parasite. The worm burden appears to be related with the levels of eosinophilia in each host species. The highest levels were observed in rats in which the parasite is rapidly expelled. In mice, the levels of eosinophilia were higher than in hamsters concomitantly with a lower worm burden (Toledo et al. 2006b; Muñoz-Antoli et al. 2007).

The aforementioned studies suggest that eosinophils may play a role in the regulatory mechanisms against adult echinostomes. However, other mechanisms are probably involved based on: (1) increased eosinophilia is common in mice infected with *E. caproni* or *E. trivolvis* (Fujino et al. 1996a,b) though the life span of each echinostome species in mice differs markedly (2–4 wpi in *E. trivolvis* vs up to 29 wpi in *E. caproni*); (2) immunosuppressant treatment inhibits the eosinophilia, but only slight changes in the worm recovery were observed (Fujino et al. 1996a; Ryang et al. 2007); and (3) eosinophilia usually peaks at 1–2 wpi though worm expulsion occurs later in the most cases.

8.5.3 Goblet Cells

Goblet cells are specialized mucus-producing glands located within the epithelium of the intestinal tract among other tissues. They secrete mucus to form a protective gel overlaying the surface of epithelial cells (Specian and Oliver 1991). Apart from its role in lubrication and protection of mucosal surfaces, the mucus layer also provides a microenvironment for the interaction of cellular and molecular components with pathogenic organisms. It has been proposed that mucin proteins mediate intestinal helminth expulsion by enveloping the parasites and/or inhibit adhesion (Nawa et al. 1994).

Goblet cell hyperplasia often occurs simultaneously with mastocytosis, although both processes are differentially regulated (Balic et al. 2000). Goblet cell hyperplasia is associated with Th2 cytokine responses (Else and Finkelman 1998) and appears to be IL-4-independent (Lawrence 2003). It has been shown that IL-3 plays a crucial role in goblet cell hyperplasia and mucus production which is correlated with worm expulsion of *Nippostrongylus braziliensis* independent of IgE and mast cells (McKenzie et al. 1998).

Several reports have shown that increased mucus production in association with goblet cell hyperplasia is involved in echinostome adult worm rejection. In mice and hamsters, *E. caproni* induces chronic infections concomitantly with reduced goblet cell numbers (Bindseil and Christensen 1984; Weinstein and Fried 1991; Fujino and Fried 1993a; Toledo et al. 2006b; Muñoz-Antoli et al. 2007). In contrast, in *E. trivolvis*-infected hosts, the goblet cell counts increased coinciding with worm expulsion (Fujino et al. 1993). Treatment with dexamethasone inhibited the goblet cell hyperplasia and delayed worm expulsion. Furthermore, worm expulsion coincided with the cessation of dexamethasone treatment and increases in the goblet cell numbers (Fujino et al. 1996a, 1997). As in normal mice, dexamethasone administration to immunodeficient SCID mice delayed the expulsion of *E. trivolvis* (Fujino et al. 1998). Fujino et al. (1996c) demonstrated that primary infection with *N. braziliensis* in C3H/HeN mice induced a rapid expulsion of either *E. trivolvis* or *E. caproni*. Increased secretion of mucus by hyperplasic goblet cells associated with *N. braziliensis* primary infection was suggested to be responsible for earlier expulsion.

In *E. hortense*-infected mice, worm expulsion has been associated with goblet cell hyperplasia (Park et al. 2005; Ryang et al. 2007). In both C3H/HeN and BALB/c mice, goblet cell infiltration in the intestine increased after infection. Ketotifen treatment reduced the goblet cell hyperplasia resulting in a slight increase in worm survival (Ryang et al. 2007).

Although the involvement of goblet cells in the expulsion of echinostome infections seems evident, Frazer et al. (1999) showed that the situation is more complex. These authors observed a marked goblet cell response in RAG-2 mice, though *E. caproni* adult worms survived as they did in conventional ICR mice. Moreover, Park et al. (2005) did not find a relationship between the differences in the kinetics of goblet cell numbers and the *E. hortense* worm expulsion in C3H/HeN and BALB/c mice. These facts suggest that worm expulsion of echinostome infections cannot be explained exclusively on the basis of goblet cell numbers. In this context, intestinal worm rejection was noted to be regulated by the alteration of goblet cell function through modification of the mucin's terminal sugar, specially the expression of *N*-acetyl-D galactosamine (Ishikawa et al. 1993, 1994). Fujino and Fried (1993b) studied the lectin-labeling patterns in the small intestine of C3H mice infected with either *E. trivolvis* or *E. caproni*. They noted marked differences in the distribution of glycoconjugates in infected hosts. In *E. trivolvis*-infected mice the amount of mucins, including *N*-acetyl-d galactosamine, sialic acid, and *N*-acetyl-D glucosamine, was strongly expressed together with the increase in goblet cell numbers. In *E. caproni*-infected mice, the binding of most of the lectins was reduced in association with a low number of goblet cells. Interestingly, *E. trivolvis*-infected hamsters did not show goblet cell hyperplasia or increase in the glycoconjugates and chronic infections were developed (Fujino and Fried 1996). Similarly, immunostaining of the intestine of *E. hortense*-infected mice showed a significant increase of lectin-binding goblet cells suggesting that these cells may regulate worm expulsion by altering the mucin terminal sugar (Park et al 2005). A similar situation was described for *N. braziliensis*. Ishikawa et al. (1994) showed that the expulsion of this nematode is associated with T-cell-dependent goblet cell hyperplasia together with a T-cell-independent alteration of goblet cell mucins.

Accordingly, the goblet cell number and the alteration of sugar residues of these cells seem to play a critical role in the expulsion of adult echinostome worms.

8.5.4 Antibody Responses

Several antigen preparations have been used to study the antibody responses in primary echinostome infections, i.e., adult crude worms (Simonsen et al. 1991; Agger et al. 1993; Toledo et al. 2003a, 2004c), excretory–secretory products (Toledo et al. 2003a,b, 2004b,c, 2005; Cho et al. 2007; Carpena et al. 2007; Sotillo et al. 2007), or glycocalix membrane from adult worms (Graczyk and Fried 1994, 1995). The results showed that the generation of systemic antibody responses depends on the host–parasite combination. *E. caproni* induced rapid and strong IgG responses in mice (Agger et al. 1993; Graczyk and Fried 1994; Toledo et al. 2005; Sotillo et al. 2007). Significant levels of IgG were detected from 1 to 2 wpi and the values progressively increased over the course of the infection. In hamsters, the responses were greater but slower than in mice, and positive levels of IgG were observed from 7 wpi (Simonsen et al. 1991; Toledo et al. 2004b). In contrast, rats developed a weak IgG response against *E. caproni* and only low levels of IgG were detected from 7 wpi (Toledo et al. 2003b, 2004c; Sotillo et al. 2007). Similarly, *E trivolvis* and *E. hortense* elicited strong serum IgG responses in mice (Graczyk and Fried 1994; Cho et al. 2007). *E. friedi* induced greater responses in hamsters than in rats (Toledo et al. 2003a; Carpena et al. 2007).

Cho et al. (2007) studied the IgG1, IgG2a, IgE, and IgA responses in the sera of C3H/HeN and BALB/c mice infected with *E. hortense*. The levels of IgG1 were higher than those of IgG2a, IgE, and IgA. The titers peaked at 3–4 wpi and were higher in BALB/c mice at every week postinfection. Increased IgE and IgA were also observed, but there were no differences in both strains of mice.

Sotillo et al. (2007) compared the kinetics of IgM, IgA, and IgG subclasses in two host species of *E. caproni* displaying a different degree of compatibility with the parasite, i.e., mice in which the parasite induces chronic infections and rats, in which the parasite is rejected earlier. The early response was of the IgM class in both host species showing the maximum level at 1–2 wpi. The intense response of IgM suggested the presence of thymus-independent antigens on *E. caproni* that interact early with the host immune system. This was supported by the intense IgG3 response observed in mice. Responses mediated by IgG3 are related to carbohydrate thymus-independent antigens that can induce an IgG3 class switch (Snapper et al. 1992). The kinetics of serum total IgG response were markedly greater in mice than in rats. Furthermore, the kinetics of serum IgG1 and IgG2a were also different in both host species. The results suggested a markedly biased response toward a Th2 phenotype in mice, characterized by a strong production of IgG1 from 3 wpi and beyond. In rats, an initial dominance of IgG2a response occurred, though a slight increase of IgG1 levels was detected from 7 wpi and beyond, suggesting that worm expulsion might be associated with balanced Th1/Th2 systemic responses and the development of long-lasting infections associated with a dominance of systemic Th2 responses with high levels of serum IgG1. The kinetics of circulating IgA were also different in mice and rats. In the serum of mice, a progressive increase was observed over the course of the infection, whereas in rats IgA peaked at 1–2 wpi and returned thereafter to negative values. This appears to confirm the dominance of systemic Th2 systemic responses in relation to chronic infections, since expression of IgA is mediated by Th2 cytokines such as IL-5, IL-6, or IL-10 (Ramsay 1995).

The target antigens of circulating antibodies in *E. caproni* infections were studied by Sotillo et al. (2008) using an immune proteomic approach. A total of four proteins (enolase, actin, HSP-70, and aldolase) were recognized by serum IgM, IgA, IgG, and or IgG1 of mice, though the recognition profile was specific for each isotype. However, this issue will be discussed in further detail in Sect. 8.6 of this chapter.

The influence of systemic antibody responses is not known. In *E. caproni* infections, stronger responses have been observed in host species in which long-lasting infections developed (Toledo et al. 2006c). This suggests that circulating antibody responses constitute a collateral consequence of the infection. Toledo et al. (2004b) suggested that differences in the systemic antibody responses were related to differences in the local inflammatory responses. Juvenile and adult worms secrete antigens that can cross the intestinal mucosa, to reach the circulatory system and induce serum responses by B-cell stimulation. It is known that the passage of antigens through the intestinal mucosa is mediated by local inflammation (Yu and Perdue 2001). Thus, differences in mucosal inflammatory responses may result in differences in systemic antibody responses. In this sense, high levels of *E. caproni* seroantigens concomitantly with strong serum antibody responses have been observed in mice and hamsters, whereas low levels of antigens and antibodies were observed in the serum of rats (Toledo et al. 2004b, 2005). Toledo et al. (2006b) detected marked differences in local inflammatory responses in *E. caproni*-infected hamsters and rats. A strong inflammatory cell infiltration together with increased numbers of mucosal neutrophils was observed in hamsters, but not in rats. Under these conditions, the maintenance of epithelial barriers may be disrupted resulting in an increased passage of worm antigens through the intestinal mucosa. Furthermore, Toledo et al. (2006b) performed an immunohistochemical study using polyclonal anti-excretory–secretory products of *E. caproni* antibodies. The results showed greater antibody binding in the intestine of *E. caproni*-infected hamsters than in *E. caproni*-infected rats confirming the greater antigen passage. This was suggested as the responsible factor for the higher systemic antibody responses.

In this context, the local antibody response appears to be of greater importance. However, this topic has been poorly studied. Agger et al. (1993) analyzed the antibody response in the intestinal wall and lumen of mice infected with *E. caproni*. Significant increases of IgG, IgA, and IgM were detected in the intestinal tissue, whereas only IgA was detected in the lumen. Sotillo et al. (2007) studied the antibody response in the intestine of *E. caproni*-infected mice and rats. A response of IgM was observed by 8 wpi in mice, but not in rats. This late rise in local IgM suggests that this response may be involved in the reduced worm burden observed from

8 wpi and beyond in mice (Muñoz-Antoli et al. 2007). The IgM class is a major complement-fixing antibody, and antibody has been suggested as one of the potential effector immune mechanisms against *E. caproni* (Simonsen and Andersen 1986). However, there are two reasons that prevent us to consider IgM as a major immune mechanism involved in the rejection of *E. caproni* adult worms: (1) The effective role of complement in the intestinal lumen appears to be limited; and (2) the lack of IgM response in rats together with the earlier expulsion of *E. caproni* suggest that the role of IgM in parasite clearance can only be secondary.

The IgG1 and IgG2a local responses in mice and rats were slower and less intense than in serum. In mice, increases of both subclasses were observed probably as reflected in a balanced Th1/Th2 response. In contrast, rats only developed an IgG2a response which suggests that local cellular mechanisms can be of great importance in the parasite expulsion. Interestingly, worm expulsion coincides with a maximum level of IgG2a at 7–8 wpi (Sotillo et al. 2007).

Local IgA was only observed in mice. Positive values were observed by 2 wpi with a maximum at 8 wpi. This indicates that IgA secretion is not sufficient for *E. caproni* rejection (Sotillo et al. 2007). The induction of IgA might be the consequence of local production of Th2 cytokines, which suggests that the development of local Th2 responses may be associated with chronic *E. caproni* infections.

Little is known about the role of antibodies against echinostomes. Simonsen and Andersen (1986) showed that antibodies in the serum of mice infected with *E. caproni* bound to surface antigens of the parasite. However, they were rapidly expelled due to the turnover rate of surface antigens which was suggested as a strategy to avoid antibody attack (Simonsen and Andersen 1986; Andresen et al. 1989; Simonsen et al. 1990).

8.5.5 Cytokine Responses

The identification and manipulation of cytokine responses has had a major impact on the understanding of immune mechanisms involved in helminth infections. However, only few studies have examined the cytokine responses in echinostome infections, mainly using ELISA or RT-PCR technology to detect cytokine specific mRNA expression.

Brunet et al. (2000) examined the preferential development of either Th1 or Th2 responses during early stages of *E. caproni* infection in mice by studying cytokine production in spleen and mesenteric lymph node cells using an ELISA technique. Whereas spleen cells failed to respond to antigen stimulation, mesenteric lymph node cells produced IFN-γ and to a lesser extent IL-4. Furthermore, IL-5 levels were elevated during the period of study (3 wpi), suggesting a balanced Th1/Th2 phenotype at local level. Moreover, Brunet et al. (2000) treated mice with a single injection of anti-IFN-γ monoclonal antibodies at 2 wpi to assess the role of this cytokine in protective immunity. A significant reduction in worm burden was found which suggested that IFN-γ may be important in the establishment of *E. caproni* chronic infections. However, Noland et al. (2008) did not detect significant increases in the levels of IFN-γ in *E. caproni*-infected mice as determined by ELISA using the supernatant of splenocyte cultures.

The production of Th1/Th2 cytokines in the splenocytes of *E. hortense*-infected C3H/HeN and BALB/c mice was studied using semiquantitative RT-PCR by Cho et al. (2007). Regardless the mouse strain, the mRNA expression of the Th1 cytokines IFN-γ and IL-12 were weak. In contrast the expression of IL-4 and IL-5 mRNA was increased in both strains. The mRNA expression of IL-4 peaked at 6 h after antigenic stimulation. The expression of IL-5 lasted longer than did that of IL-4. The secretion of cytokines from splenocytes was studied by ELISA, and the results were similar to those observed by RT-PCR (Cho et al. 2007). Ryang et al. (2007) also studied the levels of TNF-α and IL-1β in by RT-PCR in the spleen of the same strains of *E. hortense*-infected mice, though only weak increases in the levels of mRNA expression of these Th1 cytokines were observed, confirming the dominance of Th2 responses in *E. hortense*-infected mice.

8.6 Antigenic Characterization of Adult Echinostomes

The antigenic composition of adult echinostomes has been little studied compared to other intestinal helminths. Andresen et al. (1989) suggested that the surface antigens were mainly responsible for the antibody responses against *E. caproni* in mice. SDS–PAGE and western-blot analysis of in vitro shed detergent solubilized antigens indicated that the four major antigens released from adult worms had molecular weights (MW) ranging from 26 to 88 kDa. Toledo et al. (2004c) also defined by western-blot analysis, using sera from *E. caproni*-infected rats, 11 and 7 major antigenic polypeptides in the somatic and excretory–secretory products, respectively. Moreover, Toledo et al. (2004c) demonstrated that some of these polypeptides were specifically expressed in the early phases of the infection suggesting a role in parasite establishment.

Recently, Sotillo et al. (2008) identified several antigenic proteins from *E. caproni* excretory–secretory products recognized by IgM, IgA, and IgG subclasses from the sera of *E. caproni*-infected mice using an immunoproteomic approach. A total of 11 protein spots from those recognized by these immunoglobulin classes after two-dimensional SDS–PAGE and western-blot analysis of excretory–secretory products of *E. caproni* were accurately identified. These spots corresponded to four different proteins (enolase, 70 kDa heat-shock protein, actin, and aldolase). Enolase was recognized in eight different spots, of which seven were detected in the expected MW (46 kDa) and were recognized by IgA, IgG, or IgG and IgG1. The remaining spot identified as enolase was detected at 72 kDa and only was recognized by IgM. Digestion with *N*-glycosidase F of the 72-kDa band rendered a polypeptide with an apparent MW similar to that expected for the enolase of *E. caproni* and recognized by antienolase antibodies. The authors suggested that glycosylated forms of *E. caproni* enolase may be involved in the early thymus-independent responses against the parasite. The early IgM responses were also generated by actin and the 70-kDa heat-shock protein which suggested that these proteins were exposed early to the host and may be of importance in parasite establishment. The IgA responses also appeared to be mediated by the 70 kDa heat-shock protein and aldolase which could be related to close contact of these proteins with the host mucosal surface (Sotillo et al. 2008).

Regarding other species of *Echinostoma*, Carpena et al. (2007) compared the antigenic properties of the somatic and excretory–secretory products of *E. friedi* in two host species (hamsters and rats) displaying different degrees of compatibility with the parasite. Analysis by SDS–PAGE and western blot using polyclonal antibodies raised against excretory–secretory products of *E. friedi* showed some common and a number of specific bands for each type of antigen. Furthermore, a total of 11 polypeptides were exclusively detected in materials from hamster and seven polypeptides appeared to be specifically detected in worms infecting rats. This differential profile appears to be induced by the host species.

8.7 Immunodiagnosis

Echinostomiasis in vertebrate hosts has been usually diagnosed by the presence of eggs in stool samples. However, this procedure is tedious and eggs are not always present in feces because worms may be preovigerous or mature worms may not be voiding eggs. Moreover, parasite maturation and egg production patterns may vary depending on different factors such as the crowding effect or host species (Huffman and Fried 1990; Toledo et al. 2004a; Muñoz-Antoli et al. 2004). In this context, immunological methods may be useful for diagnosis and monitoring of echinostome infections (Toledo et al. 2006c). Several techniques have been developed based on: (1) antibody detection; (2) seroantigen detection; and (3) coproantigen detection. Although the use of these methods has provided interesting results, the information is limited to experimental infections since the potential crossreactivity with other helminths has not been evaluated.

(1) Antibody detection

Immunodiagnostic test based on indirect ELISA has been developed to detect anti-*E. caproni*, anti-*E. trivolvis*, anti-*E. hortense*, and anti-*E.friedi* antibodies (Simonsen et al. 1991; Agger et al. 1993; Graczik and Fried 1994, 1995; Toledo et al. 2003a, 2004c, 2005; Cho et al. 2007). Indirect ELISA provides an early immunodiagnosis of *E. caproni* in mice. Using glycocalix membrane crude antigen, Graczyk and Fried (1994) demonstrated that IgG anti-*E. caproni* can be detected by 8 dpi in mice. Toledo et al. (2005) detected positive levels of IgG anti-*E. caproni* in mice by 2 wpi, using excretory–secretory antigens. Graczyk and Fried (1995) showed that mice present positive levels of

IgG anti-*E. trivolvis* beginning at 10 dpi using glycocalix membrane as antigen. However, a high degree of crossreactivity with *E. caproni* was found. Because of these facts, the application of serologic diagnosis has major limitations. For example, antibody titers persist after worm loss and there is also a dependence on the host species (Toledo et al. 2003b, 2004b,c; 2005). Moreover, animal handling to obtain sufficient amounts of serum is often difficult which raises the need for alternative tools.

(2) Seroantigen detection

Toledo et al. (2004b) demonstrated that detection of antigens of *E. caproni* in sera of experimentally infected rodents is feasible by capture ELISA (Toledo et al. 2004b, 2005). However, this technique does not seem to be adequate for diagnostic purposes since the presence of circulating antigens is host dependent. High levels of circulating antigens were detected in hamsters from 3 wpi and beyond. In contrast, low levels of seroantigens were detected in rats (Toledo et al. 2004b). In *E. caproni*-infected mice the levels were positive from 1–2 to 9 wpi (Toledo et al. 2005). These results appear to be associated with the dynamics of worm antigen absorption in relation to different inflammatory local responses induced by *E. caproni* in each host species as described in Sect. 8.5.4 of this chapter. Thus, the study of circulating antigens appears to be more adequate for the analysis of host–parasite relationships than for diagnostic purposes.

(3) Coproantigen detection

Toledo et al. (2003b) developed an antibody capture ELISA for detection of *E. caproni* antigens in the feces of infected rats using polyclonal anti-*E. caproni* excretory–secretory products antibodies. Coproantigens were detected from 1 dpi and, thereafter, the values were correlated with those of egg output and remained positive until 49–56 dpi, coinciding with the disappearance of eggs in the stool samples. Toledo et al. (2004b) compared the kinetics of *E. caproni* coproantigen excretion in hamsters and rats. The levels of antigens in feces were indicative of the course of the infection in each host species. Positive levels were detected from 1–2 dpi in both hosts and remained positive in hamsters until the end of the experiment at 20 wpi, reflecting the course of the infection in this host. In rats, the values reverted to negative coinciding with the loss of the infection. These facts show that coproantigen detection provides a rapid and sensitive method for the diagnosis and the monitoring of *E. caproni* infections. The main advantages of this procedure over other diagnostic methods are: (1) coproantigens are detected earlier that eggs, antibodies, and seroantigens; (2) coproantigen detection does not depend on host species; (3) the presence of antigens in feces indicates active and not previous infections; (4) the assay requires fecal samples avoiding animal handling and possible damage from serum collection; and (5) a large number of animals can be screened without time-consuming parasitological examination procedures.

8.8 Concluding Remarks

The present chapter introduces the reader to the pathological and immunological aspects of echinostome infection in the definitive host, with emphasis on experimental rodent models. Although several topics remain to be studied further, most of the information presented indicates that echinostomes may be useful to analyze immune mechanisms involved in the rejection of intestinal helminths. Under laboratory conditions, studies show that immunopathological events in echinostome infections depend on several factors related to both the parasite and the host species. Echinostomes may be particularly useful for the analysis of host-related factors determining the course of the infection. Comparative studies of an echinostome species in hosts with low and high compatibility could enable the establishment of the host-related variables, and their consequences to the parasite, determining the rejection of the intestinal worms or, in contrast, the establishment of chronic infections. In this context pathological and immunological, together with proteomic studies, may be of great use.

References

Agger, M.K., Simonsen, P.E., and Vennervald, B.J. 1993. The antibody response in serum, intestinal wall and intestinal lumen of NMRI mice infected with *Echinostoma caproni*. Journal of Helminthology 67: 169–178.

Andresen, K., Simonsen, P.E., Andersen, B.J., and Birch-Andersen, A. 1989. *Echinostoma caproni* in mice, shedding of antigens from the surface of an intestinal trematode. International Journal for Parasitology 19: 111–118.

Baek, J., Simonsen, P.E., Friis, H., and Christensen, N. 1996. Zinc deficiency and host response to helminth infection: *Echinostoma caproni* infections in CBA mice. Journal of Helminthology 70: 7–12.

Balic, A., Bowles, V.M., and Meeusen, E.N. 2000. The immunobiology of gastrointestinal nematode infections in ruminants. Advances in Parasitology 45: 181–241.

Balic, A., Cunningham. C.P., and Meeusen, E.N. 2006. Eosinophil interactions with *Haemonchus contortus* larvae in the ovine gastrointestinal tract. Parasite Immunology 28: 107–15.

Bindseil, E., and Christensen, N.Ø. (1984). Thymus-independent crypt hyperplasia and villous atrophy in the small intestine of mice infected with the trematode *Echinostoma revolutum.* Parasitology 88: 431–438.

Bischoff, S.C., and Krämer, S. 2007. Human mast cells, bacteria, and intestinal immunity. Immunological Reviews 217: 329–337.

Brunet, L.R., Joseph, S., Dunne, D.W., and Fried, B. 2000. Immune responses during the acute stages of infection with the intestinal trematode *Echinostoma caproni.* Parasitology 120: 565–571.

Carpena, I., Espert A., Muñoz-Antoli, C., Marcilla, A., Esteban, J.G., and Toledo, R. 2007. Antigenic characterization of *Echinostoma friedi* (Trematoda) in hamsters and rats experimentally infected. Revista Ibérica de Parasitología 67: 127–124.

Cho, Y.K., Ryang, Y.S., Kim, I.S., Park, S.K., Im, J.A., and Lee, K.J. 2007. Differential immune profiles following experimental *Echinostoma hortense* infection in BALB/c and C3H/HeN mice. Parasitology Research 100: 1053–1061.

Christensen, N.Ø., Knudsen, J., Fagbemi, B.O., and Nansen, P. 1985. Impairment of primary expulsion of *Echinostoma revolutum* in mice concurrently infected with *Schistosoma mansoni*. Journal of Helminthology 59: 333–335.

Christensen, N.Ø., Knudsen, J., and Andreassen, J. 1986. *Echinostoma revolutum*: resistance to secondary and superimposed infections in mice. Experimental Parasitology 61: 311–318.

Else, K.J., and Finkelman, F.D. 1998. Intestinal nematode parasites, cytokines and effector mechanisms. International Journal for Parasitology 28: 1145–1158.

Franco, J., Huffman, J.E., and Fried, B. 1986. Infectivity, growth and development of *Echinostoma revolutum* (Digenea: Echinostomatidae) in the golden hamster, *Mesocricetus auratus*. Journal of Parasitology 72: 142–147.

Frazer, B.A., Fried, B., Fujino, T., and Sleckman, B.P. 1999. Host-parasite relationships between *Echinostoma caproni* and RAG-2-deficient mice. Parasitology Research 85: 337–342.

Fried, B., and Peoples, R.C. 2007. Longevity of *Echinostoma caproni* in Balb/c mice. Journal of Helminthology 82: 95–96.

Fried, B., Huffman, J.E, and Weiss, P.M. 1990. Single and multiple worm infections of *Echinostoma caproni* (Trematoda) in the golden hamster. Journal of Helminthology 64: 75–78.

Fried, B., Nanni, T.J., Reddy, A., and Fujino, T. 1997. Maintenance of the life cycle of *Echinostoma trivolvis* (Trematoda) in dexamethasone-treated ICR mice and laboratory-raised helisoma trivolvis (Gastropoda). Parasitology Research 83: 16–19.

Fujino, T., and Fried, B. 1993a. Expulsion of *Echinostoma trivolvis* (Cort, 1914) Kanev, 1985 and retention of *E. caproni* Richard, 1964 (Trematoda, Echinostomatidae) in C3H mice: pathological, ultrastructural, and cytochemical effects on the host intestine. Parasitology Research 79: 286–292.

Fujino, T., and Fried, B. 1993b. *Echinostoma caproni* and *E. trivolvis* alter the binding of glycoconjugates in the intestinal mucosa of C3H mice as determined by lectin histochemistry. Journal of Helminthology 67: 179–188.

Fujino, T., and Fried, B. 1996. The expulsion of *Echinostoma trivolvis* from C3H mice: differences in glycoconjugates in mouse versus hamster small intestinal mucosa during infection. Journal of Helminthology 70: 115–121.

Fujino, T., Fried, B., and Tada, I. 1993. The expulsion of *Echinostoma trivolvis*: worm kinetics and intestinal cytopathology in conventional and congenitally athymic BALB/c mice. Parasitology 106: 297–304.

Fujino, T., Ichikawa, H., Fried, B., and Fukuda, K. 1996a. The expulsion of *Echinostoma trivolvis*: suppressive effects of dexamethasone on goblet cell hyperplasia and worm rejection in C3H/HeN mice. Parasite 3: 283–289.

Fujino, T., Fried, B., Ichikawa, H., and Tada, I. 1996b. Rapid expulsion of the intestinal trematodes *Echinostoma trivolvis* and *E. caproni* from C3H mice by trapping with increased goblet cell mucins. International Journal for Parasitology 26: 319–324.

Fujino, T., Yamada,M., Ichikawa, H., Fried, B., Arizono, N., and Tada, I. 1996c. Rapid expulsion of the intestinal trematodes *Echinostoma trivolvis* and *E. caproni* from C3H/HeN mice after infection with *Nippostrongylus brasiliensis*. Parasitology Research 82: 577–579.

Fujino, T., Ichikawa, H., Fried, B., and Fukuda, K. 1997. The expulsion of *Echinostoma trivolvis*: worm kinetics and intestinal reactions in C3H/HeN mice treated with dexamethasone. Journal of Helminthology 71: 257–259.

Fujino T., Ichikawa H., and Fried, B. 1998a. The immunosuppressive compound FK506 does not affect expulsion of *Echinostoma trivolvis* in C3H mice. Parasitology Research 84: 519–521.

Fujino T., Ichikawa H, Fukuda, K., and Fried, B. 1998b. The expulsion of *Echinostoma trivolvis* caused by goblet cell hyperplasia in severe combined immunodeficient (SCID) mice Parasite 5: 219–222.

Graczyk, T.K., and Fried, B. 1994. ELISA method for detecting anti-*Echinostoma caproni* (Trematoda: Echinostomatidae) immunoglobulins in experimentally infected ICR mice. Journal of Parasitology 80: 544–549.

Graczyk, T.K., and Fried, B. 1995. An enzyme-linked immunosorbent assay for detecting anti-*Echinostoma trivolvis* (Trematoda) IgG in experimentally infected ICR mice. Cross-reactivity with *E. caproni*. Parasitology Research 81: 710–712.

Hansen, K., Nielsen, J.W., Hindsbo, O., and Christensen, N.Ø. 1991. *Echinostoma caproni* in rats: worm population dynamics and host blood eosinophilia during primary infections with 6, 25 and 50 metacercariae and resistance to secondary and superimposed infections. Parasitology Research 77: 686–690.

Hosier, D.W., and Fried, B. 1986. Infectivity, growth, and distribution of *Echinostoma revolutum* (Trematoda) in Swiss webster and ICR mouse. Proceedings of the Helminthological Society of Washington. 53: 173–176.

Hosier, D.W., and Fried, B. 1991. Infectivity, growth, and distribution of *Echinostoma caproni* (Trematoda) in the ICR mouse. Journal of Parasitology 77: 640–642.

Hosier, D.W., Fried, B., and Szewczak, J.P. (1988). Homologous and heterologous resistance of *Echinostoma revolutum* and *E. liei* in ICR mice. Journal of Parasitology 74: 89–92.

Huffman, J.E., and Fried, B. 1990. *Echinostoma* and echinostomiasis. Advances in Parasitology 29: 215–269.

Huffman, J.E., Iglesias, D., and Fried, B. 1988. *Echinostoma revolutum*: pathology of extraintestinal infection in the golden hamster. International Journal for Parasitology 18: 873–874.

Ishikawa, N., Horii, Y., and Nawa, Y. 1993. Immune-mediated alteration of the terminal sugars of goblet cell mucins in the small intestine of *Nippostrongylus brasiliensis*-infected rats. Immunology 78: 303–307.

Ishikawa, N., Horii, Y., Oinuma, T., Suganuma, T., and Nawa Y. 1994. Goblet cell mucins as the selective barrier for the intestinal helminths: T-cell-independent alteration of goblet cell mucins by immunologically 'damaged' *Nippostrongylus brasiliensis* worms and its significance on the challenge infection with homologous and heterologous parasites. Immunology 81: 480–486.

Kim, I., Im, J.A., Lee, K.J., and Ryang, Y.S. 2000. Mucosal mast cell responses in the small intestine of rats infected with *Echinostoma hortense*. The Korean Journal of Parasitology 38: 139–143.

Lamkhioued, B., Gounni, A.S., Gruart, V., Pierce, A., Capron, A., and Capron M. 1995. Human eosinophils express a receptor for secretory component. Role in secretory IgA-dependent activation. European Journal of Immunology 25: 117–125.

Lantz, C.S., Boesiger, J., Song, C.H., Mach, N., Kobayashi, T., Mulligan, R.C., Nawa, Y., Dranoff, G., and Galli, S.J. 1998. Role for interleukin-3 in mast-cell and basophil development and in immunity to parasites. Nature 392: 90–93.

Lawrence, C.E. 2003. Is there a common mechanism of gastrointestinal nematode expulsion? Parasite Immunology 25: 271–281.

Lee, K.J., Park, S.K., Im, J.A., Kim, S.K., Kim, G.H., Kim, G.Y., Yang, E.J., and Ryang, Y.S. 2004. Susceptibility of several strains of mice to *Echinostoma hortense* infection. The Korean Journal of Parasitology 42: 51–56.

Mabus, J., Huffman, J.E., and Fried, B. 1988. Humoral and cellular response to infection with *Echinostoma revolutum* in the golden hamster, *Mesocricetus auratus*. Journal of Helminthology. 62: 127–132.

McKenzie, G.J., Bancroft, A., Grencis, R.K., and McKenzie. A.N. 1998. A distinct role for interleukin-13 in Th2-cell-mediated immune responses. Current Biology 8: 339–342.

McMaster, R.P., Huffman, J.E., Fried. B. 1995. The effect of dexamethasone on the course of *Echinostoma caproni* and *E. trivolvis* infections in the golden hamster (*Mesocricetus auratus*). Parasitology Research 81: 518–521.

Muñoz-Antoli, C., Carpena, I., Espert, A., Esteban, J.G., and Toledo R. 2004. The effect of host species on the development of *Echinostoma friedi* (Trematoda: Echinostomatidae) adult worms. Revista Ibérica de Parasitología 64: 81–87.

Muñoz-Antoli, C., Sotillo, J., Monteagudo, C., Fried, B., Marcilla, A., and Toledo, R. 2007. Development and pathology of *Echinostoma caproni* in experimentally infected mice. Journal of Parasitology 93: 854–859.

Nawa, Y., Ishikawa, N., Tsuchiya, K., Horii, Y., Abe, T., Khan, A.I., Bing-Shi, Itoh, H., Ide, H., and Uchiyama, F. 1994. Selective effector mechanisms for the expulsion of intestinal helminths. Parasite Immunology 16: 333–338.

Noland, G.S., Urban, J.F., Fried B., Kumar, N. 2008. Counter-regulatory anti parasite cytokine responses during concurrent *Plasmodium yoelii* and intestinal helminth infections in mice. Experimental Parasitology 119: 272–278.

Odaibo, A.B., Christensen, N.Ø, and Ukoli, F.M.A. 1988. Establishment, survival and fecundity in *Echinostoma caproni* (Trematoda) infections in NMRI mice. Proceedings of the Helminthological Society of Washington 55: 265–269.

Odaibo, A.B., Christensen, N. Ø, and Ukoli, F.M.A. 1989. Further studies on the population regulation in *Echinostoma caproni* infections in NMRI mice. Proceedings of the Helminthological Society of Washington 56: 192–198.

Park, K.Y., Lee, K.J., Kim, I.S., Yang, E.J., Lim, S.J., Lim, B.H., and Ryang, Y.S. 2005. Reaction of mast cells and goblet cells in the small intestine of C57BL/6 and C3H/HeN mice infected with *Echinostoma hortense*. The Korean Journal of Biomedical Laboratory Sciences 11: 259–266.

Ramsay, A.J. 1995. Genetic approaches to the study of cytokine regulation of mucosal immunity. Immunology and Cell Biology 73: 484–488.

Ryang, Y.S., Yang, E.J., Kim, J.L., Lee, K.J., Sung, H.J., Kim, J.B., and Kim, I.S. 2007. Immune response and inhibitory effect of ketotifen on the BALB/c and C3H/HeN mice infected with *Echinostoma hortense*. Parasitology Research 101: 1103–1110.

Sayed, B.A., and Brown, M.A. 2007. Mast cells as modulators of T-cell responses. Immunological Reviews 217: 53–64.

Simonsen, P.E., and Andersen, B.J. 1986. *Echinostoma revolutum* in mice; dynamics of the antibody attack to the surface of an intestinal trematode. International Journal for Parasitology 16: 475–482.

Simonsen, P.E., Vennervald, B.J., and Birch-Andersen, A. 1990. *Echinostoma caproni* in mice: ultrastructural studies on the formation of immune complexes on the surface of an intestinal trematode. International Journal for Parasitology 20: 935–491.

Simonsen, P.E., Estambale, B.B., and Agger, M. 1991. Antibodies in the serum of golden hamsters experimentally infected with the intestinal trematode *Echinostoma caproni*. Journal of Helminthology 65: 239–247.

Sirag, S.B., Christensen, N. Ø, Frandsen, F., Monrad. J., and Nansen, P. 1980. Homologous and heterologous resistance in *Echinostoma revolutum* infections in mice. Parasitology 80: 479–486.

Snapper, C.M., McIntyre, T.M., Mandler, R., Pecanha, L.M.T., Finkelman, F.D., and Mond, J.J. 1992. Induction of IgG3 secretion by interferon-γ: a model for T cell-independent class switching in response to T cell-independent Type 2 antigens. Journal of Experimental Medicine 175: 1367–1371.

Sotillo, J., Munoz-Antoli, C., Marcilla, A., Fried, B., Esteban, J.G. and Toledo, R. 2007. *Echinostoma caproni*: Kinetics of IgM, IgA and IgG subclasses in the serum and intestine of experimentally infected rats and mice. Experimental Parasitology 116: 390–398.

Sotillo, J., Valero, L., Sánchez del Pino, M.M., Fried, B., Esteban, J.G., Marcilla, A., and Toledo, R. 2008. Identification of antigenic proteins from *Echinostoma caproni* (Trematoda: Echinostomatidae) recognized by mouse immunoglobulins M, A and G using an immunoproteomic approach. Parasite Immunology 30: 271–279.

Specian, R.D., and Oliver, M.G. 1991. Functional biology of intestinal goblet cells. American Journal of Physiology 260: C183–C193.

Tani, S., and Yoshimura K. 1988. Spontaneous expulsion of *Echinostoma hortense* Asada, 1926 (Trematoda: Echinostomatidae) in mice. Parasitology Research 74: 495–497.

Toledo, R., and Fried, B. 2005. Echinostomes as experimental models in adult parasite-vertebrate host interactions. Trends in Parasitology 21: 251–254.

Toledo, R., Espert, A., Carpena, I., Muñoz-Antoli, C., and Esteban, J.G. 2003a. An experimental study of the reproductive success of *Echinostoma friedi* (Trematoda: Echinostomatidae) in the golden hamster. Parasitology 126: 433–441.

Toledo, R., Espert, A., Muñoz-Antoli, C., Marcilla, A., Fried, B., and Esteban, J.G. 2003b. Development of an antibody-based capture enzyme-linked immunosorbent assay for detecting *Echinostoma caproni* (Trematoda) in experimentally infected rats: kinetics of coproantigen excretion. Journal of Parasitology 89: 1227–1231.

Toledo, R., Espert, A., Carpena, I., Muñoz-Antoli, C., Fried, B and Esteban, J.G. 2004a. The comparative development of *Echinostoma caproni* (Trematoda: Echinostomatidae) adults in experimentally infected hamsters and rats. Parasitology Research 93: 439–444.

Toledo, R., Espert, A., Muñoz-Antoli, C., Marcilla, A., Fried, B., and Esteban, J.G.. 2004b. Kinetics of *Echinostoma caproni* (Trematoda: Echinostomatidae) antigens in feces and serum of experimentally infected hamsters and rats. Journal of Parasitology 90: 752–758.

Toledo, R., Espert, A., Muñoz-Antoli, C., Fried, B., and Esteban, J.G. 2004c. Immunological characterization of somatic and excretory/secretory antigens of *Echinostoma caproni* (Trematoda: Echinostomatidae) in experimentally infected rats. Comparative Parasitology 71: 42–48.

Toledo, R., Espert, A., Muñoz-Antoli, C., Marcilla, A., Fried, B., and Esteban, J.G. 2005. Kinetics of antibodies and antigens in serum of mice experimentally infected with *Echinostoma caproni* (Trematoda: Echinostomatidae). Journal of Parasitology 91: 978–980.

Toledo, R., Carpena, I., Espert, A., Sotillo, J., Muñoz-Antoli, C., and Esteban JG. 2006a. A quantitative approach to the experimental transmission success of *Echinostoma friedi* (Trematoda: Echinostomatidae) in rats. Journal of Parasitology 92: 16–20.

Toledo, R., Monteagudo, C., Espert, A., Fried, B., Esteban J.G., and Marcilla, A. 2006b. *Echinostoma caproni*: intestinal pathology in the golden hamster, a highly compatible host, and the Wistar rat, a less compatible host. Experimental Parasitology 112: 164–171.

Toledo, R., Esteban, J.G., and Fried, B. 2006c. Immunology and pathology of intestinal trematodes in their definitive hosts. Advances in Parasitology 63: 285–365.

Weinstein, M.S., and Fried, B. 1991. The expulsion of *Echinostoma trivolvis* and retention of *Echinostoma caproni* in the ICR mouse: pathological effects. International Journal for Parasitology 21: 255–257.

Weller, P.F. 1997. Human eosinophils. Journal of Allergy and Clinical Immunology 100: 283–287.

Yoshimoto, T., and Nakanishi, K. 2006. Roles of IL-18 in basophils and mast cells. Allergology International 55: 105–13.

Yu, L.C., and Perdue, M.H. 2001. Role of mast cells in intestinal mucosal function: studies in models of hypersensitivity and stress. Immunological Reviews 179: 61–73.

Chapter 9
Echinostomes: Genomics and Proteomics

Antonio Marcilla

Contents

9.1 Introduction ... 208
9.2 Echinostome Genomics ... 209
9.2.1 DNA and RNA Studies: Current Status of Genomic Studies on Echinostomes ... 210
9.2.2 Prospects and Limitations of Genomics in Echinostomes ... 213
9.3 Echinostome Proteomics ... 214
9.3.1 Identification of Echinostome Proteins ... 214
9.3.2 Analytical Procedures Used in Echinostome Proteomics ... 217
9.3.3 Advances in Echinostome Proteomics ... 220
9.3.4 Prospects and Limitations in Proteomics of Echinostomes ... 222
9.4 Concluding Remarks ... 224
References ... 224

Abstract Echinostomes are excellent models to study host-parasite interactions in intestinal helminthiases. In this chapter, data available from genomic and proteomic studies are analyzed in detail to indicate advantages and pitfalls in the research using these organisms. The lack of a genome sequence project and the low number of sequences deposited in the databases not only affect the genomic studies but also debilitate proteomic findings since in most cases, although good spectrometric data can be obtained, the available database often fail to identify the protein from the peptides obtained.

A. Marcilla (✉)
Departamento de Biología Celular y Parasitología, Facultad de Farmacia.
Universidad de Valencia, Av. Vicent Andrés Estellés s/n, 46100 Burjassot-Valencia, Spain
e-mail: antonio.marcilla@uv.es

B. Fried and R. Toledo (eds.) *The Biology of Echinostomes*,
DOI: 10.1007/978-0-387-09577-6_9, © Springer Science+Business Media, LLC 2009

9.1 Introduction

Complete sequencing of genomes has attracted the attention of many researchers because the information obtained was expected to reveal the molecular code for organisms, and provide the tools to understand physiological and pathological processes.

The term "genomics" was originally used in 1920 by Winkler to describe the complete set of chromosomes and their associated genes (McKusick 1997). Biron et al. (2006), have recently described three periods with regard to the understanding of the DNA molecule and its functionality, terming them "pregenomic," "genomic," and "postgenomic" periods. Since the beginning of the postgenomic era, the focus of molecular biology gradually moved from genes and genomes to proteins and proteomes and their functionality. Now that complete genome sequences of several organisms have been determined, the biggest task in the postgenomic era will be to identify the functions, reactions, interactions, and the location of the gene products in tissues and/or the cells of living organisms.

Until now, several complete genomes are available, but only two of them correspond to helminths: the model organism *Caenorabditis elegans* (The *C. elegans* Consortium et al. 1998) and the filarial parasite *Brugia malayi* (Ghedin et al. 2007). Other sequencing projects for parasitic helminths are under way, including those from species of the genus *Schistosoma* (Haas et al. 2007; http://www.tigr.org/; http://www.ebl.ac.uk/genomes; ftp://ftp.sanger.ac.uk/pub/pathogens/Schistosoma). Helminth genomes are an integral part of the five-year plan at the Wellcome Trust Sanger Institute (WTSI) (http://www.sanger.ac.uk/Projects/Helminths/), which was started in 2006, with *Schistosoma haematobium*. Availability of different genome sequences would provide considerable benefits, especially in the sphere of gene identification and evolution.

There is no current full genome sequencing project in the Echinostomatidae family, and very little information is available in the data banks. In fact, there are only 474 DNA sequences and 62 protein sequences deposited (all of them deduced from the DNA sequences). Most of those sequences correspond to ribosomal or mitochondrial molecules used in taxonomic and/or phylogenetic studies (see later). From the deposited sequences, 434 (91.5%) of DNA and 45 (72.5%) of proteins correspond to *Echinostoma* species. The available expressed sequence tags (EST), which could be considered as transcriptome data (Wilson et al. 2007), are represented by 358 entries from *Echinostoma paraensei* (Nowak and Loker 2005). No microarrays or reverse genetics have been reported for echinostomes (see later).

With respect to proteomics, most of the studies with echinostomes deal with the relationship between echinostomes and the host immune system. It is important to note that some echinostomes may share intermediate hosts with *Schistosoma mansoni*, and in contrast to *S. mansoni*, echinostomes interfere with the snail immune system in an immunosuppressive way (Lie and Heyneman 1977, 1979; Lie et al. 1982). In this context, various studies have shown that hemocytes exposed to echinostome excretory/secretory products (ESP) lose several defense functions such as adhesion, phagocytosis, and encapsulation, as well as cytotoxic activity (Loker et al. 1986, 1992; Noda and Loker 1989; Sapp and Loker 2000; Humbert and Coustau 2001). Moreover, in coinfection experiments, the immune suppression activity of echinostomes permits the survival of schistosomes in *Schistosoma*-resistant host strains (Lie and Heyneman 1977). The evasion strategies of trematodes are related to their ability to produce ESP. These products contain immunoregulatory molecules that can interfere with host innate immunity (Damian 1997; Salzet et al. 2000).

Recent studies in echinostomes have used proteomic technology, mainly by two-dimensional gel electrophoresis and mass spectrometry (MS), to characterize the ES proteome of parasites from distinct definitive hosts (Bernal et al. 2006). This technology has also been used to gain insight into the molecular basis of the sporocysts immune evasive strategies during the first hour of interaction, as well as to understand the molecular crosstalk between *S. mansoni* or *E. caproni* sporocysts and the snail host *Biomphalaria glabrata* (Guillou et al. 2007).

9.2 Echinostome Genomics

As mentioned before, there is no full genome sequencing project for echinostomes in progress and only 474 DNA sequences are available in the databases (Table 9.1), 358 of them corresponding to the only available EST project in echinostomes, performed with *Echinostoma paraensei* sporocysts RNA (Nowak and Loker 2005). The rest of the deposited sequences mainly represent molecules used in taxonomic and/or phylogenetic studies, like ribosomal (rDNA) or mitochondrial molecules (mtDNA) (Table 9.1). Few exceptions include the enolase gene, originally identified in studies characterizing spliced leader RNAs in flatworms (Davies 1997). The full-length molecule has been recently cloned from mRNA obtained either from adults or sporocysts (Marcilla et al. 2007; Guillou et al. 2007).

Most of the ribosomal molecules sequenced correspond to the internal transcribed spacers (ITS-1 and ITS-2), with 13 species sequenced; 18S and 5.8S genes (ten species), as well as to partial sequences of the large 28S gene (11 species) (http://www.ncbi.nlm.nih.gov/Taxonomy). With respect to mitochondrial markers, they include the sequences of the NADH dehydrogenase subunit 1 (ND1) from 14 species and the sequences of the cytochrome oxidase 1 (CO-1) from seven species (Table 9.1).

Both ribosomal and mitochondrial markers have been widely used for taxonomic and phylogenetic studies on echinostomes. Since their original description in 1899, the family Echinostomatidae has shown great taxonomic diversity which is associated with its broad range of final hosts and a wide geographical distribution. Echinostomes have been widely used as experimental models for studying host-parasite relationships and population regulation of intestinal flukes (Toledo and Fried 2005), but still present a confused taxonomic status due to both the morphological similarity between biologically distinct taxa and historical nomenclature problems (reviewed in Kostadinova and Gibson 2000). So it is not surprising that most of the DNA data available have been generated to help solve these taxonomic problems.

Table 9.1 Distribution of echinostome molecular data on GenBank, January 2008

Species	DNA		Protein
	Molecules	No. sequences	No. sequences
Artyfechinostomum			
A. sufrartyfex	3	18S, ITS, CO-1	1
Drepanocephalus			
D. spathans	1	18S	
Echinoparyphium			
E. aconiatum	3	ND1	3
E. cinctum	3	28S[a]	
E. recurvatum	8	ITS, ND1	7
E. sp.	2	28S	
Echinostoma			
E. paraensei	365 (358 ESTs)	ITS, ND1, CO-1	3
E. revolutum	17	18S, ITS, 28S, ND1, CO-1	9
E. caproni	16	18S, ITS, 28S, ND1, CO-1, others	9
E. trivolvis	11	ITS, ND1, others	7
E. hortense	5	ITS, 28S, ND1, CO-1	3
E. friedi	3	ITS, 28S, ND1	1
E. cf. friedi	1	ND1	1
E. liei	1	ITS	
E. species (*E. sp*)	15	ITS, ND1, CO-1	12
Euparyphium			
E. albuferensis	3	ITS, 28S, ND1	1
E. melis	2	18S, 28S	
Himasthla			
H. rhigedana	1	nk-1	1
Hypoderaeum			
H. conoideum	4	ITS, 28S, ND1	2
Isthmiophora			
I. hortense	3	18S, ITS, 28S, ND1, CO-1	1
I. melis	2	ITS, 28S, ND1	1
Mesorchis			
M. denticulatus	1	18S	
Microparyphium			
M. sp.	1	18S	
Paryphostomum			
P. radiatum	1	18S	
Petasiger			
P. phalacrocoracis	1	18S	
Protechinostoma			
P. sp.	1	ITS	
Total	474 (358 EST)		62

[a]Partial sequences

9.2.1 DNA and RNA Studies: Current Status of Genomic Studies on Echinostomes

As mentioned before, molecular analyses of some echinostomes species, like the "*revolutum*" group, are very preliminary with studies testing its taxonomic framework, as suggested by Kanev (1994) and Kanev et al. (1995a,b). Morgan and Blair

(1995) used ribosomal ITS sequence data obtained from laboratory-maintained strains, and distinguished five species of *Echinostoma*, observing rather low sequence divergence among rDNA sequences within the "*revolutum*" group, preventing the resolution of the position of the Euro-Asian species, *E. revolutum* Frölich, 1802. Later on, these authors (Morgan and Blair 1998a) found that the same *Echinostoma* species exhibited higher sequence divergence across their mtDNA genes and concluded that the ND1 gene would be a more suitable marker for species and strain detection within the group (Morgan and Blair 1998a).

We have used the ITS-2 marker to compare the echinostome sequence data available in GenBank, and we have found intraspecific differences in the available *Echinostoma revolutum* (Sorensen et al. 1998) entries (data not shown). This had already been noted by Kostadinova et al. (2003), who indicated that this fact could account for uncertainty with respect to the identification of the adult worms used in the molecular studies of Morgan and Blair (1995, 1998a,b). Moreover, Kostadinova et al. (2003) used the sequences of both nuclear ITS and mitochondrial ND1 genes, to evaluate the relationship of *E. revolutum* (*sensu* Kosupko 1971a,b, 1972) to a wide number of isolates from Europe, North America, and Australia identified as *E. revolutum* by Morgan and Blair (1995, 1998a,b) and Sorensen et al. (1998). This study, therefore, extended Morgan and Blair's approach to the echinostome relationships inferred from the ND1 gene, with additional sequence data (Kostadinova et al. 2003). Furthermore, Kostadinova and coworkers carried out an integrated analysis: (1) by attempting to identify the voucher material available for four of the isolates utilized by Morgan and Blair (1998b) and by examining the experimental material of *E. hortense* used in their studies using traditional comparative morphology methods; and (2) by adding additional ND1 sequence data from 17 isolates of the family Echinostomatidae collected from naturally infected mollusks from Europe (Kostadinova et al. 2003). These materials were identified on the basis of a morphological study prior to sequencing and represented six species of the commonly occurring genera *Echinostoma, Echinoparyphium* Dietz, 1909, *Hypoderaeum* Dietz, 1909 and *Isthmiophora* Lühe, 1909, all with a cosmopolitan distribution (Kostadinova et al. 2003).

We have also examined the utility of ITS-2 and ND1, along with partial 28S sequences, to add some insight into the taxonomic status of Echinostomatidae. The results obtained with ITS-2 sequences confirm the confusion on the taxonomic status of *Echinostoma hortense*, which was synonymized to *Isthmiophora hortensis* by Kostadinova et al. (2003), based on the very few nucleotide differences between the deposited sequences of both specimens (Trelis 2004). Our results using partial sequences of the 28S gene suggested that it may be good marker for taxonomic studies at a supraspecific level within the Echinostomatidae family (Trelis 2004). Our results confirmed the validity of both *Echinostoma caproni* and *Echinostoma friedi* species, further confirmed by an isoenzymatic analysis with two diagnostic loci found, Mannose phosphate isomerase (MPI) and Malic enzyme (ME) (Trelis 2004). It is important to note that our DNA sequence analysis showed an elevated degree of variation, suggesting higher diversity, within the *revolutum* group (Trelis 2004), confirming previous observations by Kostadinova et al. (2003).

More recently, Nowak and Loker (2005) have generated a subtractive library of *E. paraensei* (Nowak and Loker 2005). The authors used suppression subtractive hybridization (SSH) to isolate genes transcribed at different rates between two different samples, simultaneously normalizing the representation of rare and abundant differentially expressed transcripts. As mentioned before, the expressed sequence tags (EST) generated from the *E. paraensei* SSH library (SSH-ESTs) constitute the only EST data available for echinostomes (Nowak and Loker 2005). Sequencing of the subtracted sporocyst library produced 358 sequence entries corresponding to 145 ESTs, with an average nucleotide length of the unique sequences of 422 bp (GenBank accession numbers CK329176 to CK329320). Cluster analysis revealed 69 unique sequences, with 41 of the unique sequences being recovered only once (Nowak and Loker 2005).

In summary, Nowak and Loker (2005) first demonstrated in vivo expression of putative sporocyst genes and compared the relative levels of each transcript using in vitro and in vivo conditions. These authors used quantitative PCR to measure the differences between gene expression in natural and cultured sporocysts. Three of the six tested transcripts demonstrated statistically significant reductions of the mean expression level in the cultured sporocysts compared to the dissected sporocysts (Nowak and Loker 2005). This underscores the ability of in vitro environments to alter the normal transcription patterns of larval trematodes. Such conditions should be considered when interpreting expression studies based on cultured parasites. Three other transcripts had similarity to dynein light chain, which has also been identified to be a protein inhibitor of neuronal nitric oxide synthase (Chang et al. 2000; Jaffrey and Snyder 1996). Additionally, two other sequences demonstrated matches to ribosomal proteins.

Another transcript was putatively identified as glyceraldehyde 3-phosphate dehydrogenase (GAPDH) and presented variation in expression between distinct life cycle stages (Nowak and Loker 2005). In addition to its well-known role converting glyceraldehyde 3-phosphate into 1,3-bisphosphoglycerate during glycolysis, GAPDH may function as a redox buffer, absorbing oxidative potential and subsequently protecting against oxidative cytotoxicity (Ravichandran et al. 1994; Seres et al. 1996). We have recently identified this molecule in the excretory–secretory products (ESP) of *Echinostoma friedi* following a proteomic approach (Bernal et al. 2006). Previous studies localized GAPDH on the surface of *S. mansoni* (Goudot-Crozel et al. 1989). Our studies did not detect variation in the amount of GAPDH among parasites obtained from low- or high-compatible hosts (Bernal et al. 2006).

The remaining identifiable sequences reported by Loker and coworkers (2005) included genes whose expression was upregulated in the sporocyst stage of *E. paraensei*, one similar to the 26S protease regulatory subunit 4 and two other SSH-ESTs with nonoverlapping matches to actin. We have also detected actin in *E. friedi* ESP, presenting differences in secretion depending on the host (Bernal et al. 2006).

Finally, the vast majority of the SSH ESTs failed to produce informative matches (Nowak and Loker 2005). This fact underscores the novelty of the molecular processes that are active during the early intramolluskan stage of this parasite (Nowak and Loker 2005).

More recent studies by Coustau and coworkers have identified genes in the intermediate host *Biomphalaria glabrata* involved in host-parasite interactions by means of similar SSH experiments (Bouchut et al. 2007). Moreover, these researchers have identified gene transcripts which accumulate in the intermediate host in response to the parasite (Guillou et al. 2007). Those results are discussed in detail in chapter two of this book.

In contrast to what happens with echinostomes, EST sequencing efforts have provided vast coverage of the transcriptomes of the medically important trematodes *Schistosoma japonicum* and *S. mansoni* (Hu et al. 2004; Oliveira and Johnston 2001, Verjovski-Almeida et al. 2004; ftp://www.sanger.ac.uk/Projects/S_mansoni). Currently, GenBank contains information for over 650 Schistosoma species with at least 1,000 ESTs (reviewed by Oliveira 2007). These studies, however, do not include data generated from the mother or primary sporocyst stage, leaving the characterization of trematode gene transcription in intramolluskan stage relatively incomplete (Oliveira 2007). Furthermore, half of the genes deduced from the schistosome transcriptome analysis do not have a known function (Verjovski-Almeida et al. 2003; Wilson et al. 2007).

9.2.2 Prospects and Limitations of Genomics in Echinostomes

From the data presented, we can conclude that several limitations affect the genomic studies of echinostomes. They are summarized in Table 9.2, where a comparison between echinostome and schistosome resources is presented. Apart from

Table 9.2 "Echinostomics" versus "Schistosomics"

	Echinostomes	Schistosomes
Genome		
Size	NA	300 Mbp
DNA sequence entries	454	11,932
Sequencing Project	NA	Yes
Transcriptome		
EST	358	>1,82,000
Projects	NA	Microarrays
DNA tools		
cDNA libraries	Few	Several
Transfection protocols	NA	Yes
RNAi protocols	NA	Yes
Proteome		
Publications	<8	Several
Protein sequence entries	62	11,535
Proteins identified	6%	38%
Specific antibodies available	Few	Various
Glycome	NA	Yes
Immunome	Preliminary	Various studies

NA not available

the lack of microarrays or reverse genetics for echinostomes: (1) there are very few available DNA sequence entries in data banks with only one EST project available; (2) the use of the only EST project available shows that it may not contain sequences homologous to the ones we are interested in; (3) wc cannot underestimate the artifacts the DNA-searching tools may originate when looking for homologous sequences in the data banks; (4) along with an improvement in the bioinformatic tools available, we also need the development of genetic tools for echinostomes, such as efficient transformation techniques for the different developmental stages of these trematodes along with sensitive analyses of mutants.

In relation to the prospects of echinostome genomics, new cDNA and genomic libraries are required (in addition to the existing for *E. paraensei* and *E. caproni*) and some are now under construction, which could allow the annotation of new EST sequences. In addition, comparison between distinct species would be required in order to study differences that could account for important physiological differences observed in those species. As mentioned earlier, most of the DNA data available for echinostomes have been generated for phylogenetic and/or taxonomic studies.

9.3 Echinostome Proteomics

Several definitions have been provided for proteome, being one of the most accepted that which defines it as the total protein complement of an organism, tissue, cell, or organelle (Ashton et al. 2001). The recently developed suite of techniques collectively referred to as proteomics provides a way to link any protein to its encoding DNA. It separates and identifies individual proteins from an initial protein mixture (reviewed by Morphew et al. 2006).

The characterization of echinostome proteins is difficult due mainly to two facts: (1) the difficulty in obtaining sufficient material for biochemical methods and, (2) echinostomes are not suitable for culturing *in vitro*. These inconveniences are shared among most helminthes, including schistosomes (van Hellemond et al. 2007). In this chapter, I will discuss the data obtained in echinostomes using different proteomic approaches, starting with a reference to the early studies (pioneering studies), which first described echinostome proteins, continuing with a brief summary of technical improvements, which have made the use of proteomics to address echinostome problems suitable, and finally a review of the recent data will be presented.

9.3.1 Identification of Echinostome Proteins

The first reports on the study of echinostome proteins may be the studies by Taft and Fried (1968), which analyzed the oxygen consumption by *Echinostoma revolutum* adults and showed indirect evidence for the presence of cytochrome oxidase. The investigations on echinostome antigens by Heyneman et al. (1971)

using *Echinostoma lindoense* larval antigens obtained from its intermediate host *Biomphalaria glabrata* (Heyneman et al. 1971) should also be considered pioneering studies. The studies by Haque and Siddiqui (1982), on the histochemical and isoenzymatic profiles of phosphatases in various trematodes including *Echinostoma malayum* constitute the first application of isoenzyme studies on echinostomes (Haque and Siddiqui 1982). Later on, Voltz and coworkers applied the isoenzymatic technology (using up to 23 different enzymatic systems) to the study of taxonomic status of different species of the genus *Echinostoma* (Voltz et al. 1986, 1987, 1988). Other researchers have extended the use of isoelectrofocusing to compare *Echinostoma* species (Kristensen and Fried 1991; Sloss et al. 1995).

The use of heterologous antibodies to identify and immunolocalize echinostome proteins was first described by Thorndyke and Whitfield (1987), who demonstrated the presence of a vasoactive intestinal polypeptide-like molecule in the tegument of *Echinostoma liei* (Thorndyke and Whitfield 1987). Using various heterologous antibodies Richard et al. (1989) identified P granules in *Echinostoma caproni* (Richard et al. 1989). More recent studies include the localization of proteins tubulin and actin in *E. caproni* (Iomini and Justine 1997; Iomini et al. 1998; Sebelová et al. 2004); the identification of tubulin in *E. paraensei* (Pinheiro et al. 2004); the identification of Sm14-like molecules in *E. paraensei* (Thaumaturgo et al. 2002); as well as the identification of different proteins in *E. friedi* (Bernal et al. 2006) or in *E. caproni* (Marcilla et al. 2007; Guillou et al. 2007).

The purification of the mitochondrial enzyme cytochrome c oxidase and its localization studies in different trematode species including *Echinostoma trivolvis* constitute a work of reference for protein characterization (Fujino et al. 1995). These studies revealed distinct mitochondrial patterns depending on the type of trematode. Whereas intestinal species like *E. trivolvis* had some variations in mitochondrial enzyme activity in the tegumental and parenchymal cells, suggesting that they possess both aerobic and anaerobic metabolic systems, the opposite happens with *Fasciola gigantica*, which showed weak or no activity for cytochrome c oxidase, suggesting that this worm is well adapted to an anaerobic environment in the host bile duct (Fujino et al. 1995).

In relation to the identification of ligand-binding molecules in echinostomes, the first studies of lectin-binding proteins were reported by Uchikawa and Loker (1991) in *Echinostoma paraensei*. These authors characterized the pattern of lectin binding of living sporocysts (*in vitro* transformed) of *E. paraensei* and compared it with that observed with *Schistosoma mansoni* sporocysts (Uchikawa and Loker 1991). They showed that *S. mansoni* specifically bound 8 lectins, while *E. paraensei* bound only six out of the 11 lectins evaluated, with a distinct response in both species parasitizing the same intermediate host (Uchikawa and Loker 1991). More recently, we have reported the identification of enolase as a plasminogen-binding molecule present in the ESP of *E. caproni*, suggesting its role as an interacting molecule with the host extracellular matrix (Marcilla et al. 2007).

The first studies describing the use of sera to identify echinostome immunogenic proteins were published in the early 1990s. Simonsen and coworkers reported a distinct pattern of reactivity between different hosts parasitized with *E. caproni* (Simonsen et al. 1991). They used different techniques, such as ELISA, immunofluorescence, and western immunoblotting, to demonstrate these differences (Simonsen et al. 1991). Later on, Graczyk and Fried developed ELISA methods to detect circulating anti-*echinostoma* immunoglobulins in sera from experimentally infected ICR mice using *E. caproni* or *E. trivolvis* glycocalyx antigens (Graczyk and Fried 1994, 1995).

In most of the aforementioned studies the materials analyzed have been the echinostome ESP, because these materials constitute the initial contact with the environment, i.e., the host, and that is why they are so important. Pioneering studies on ESP characterization were the studies by Loker, Cimino and Hertel (1992), which used pulse-chase experiments with radioactively labeled amino acids to characterize polypeptides of *E. paraensei* ESP, as well as to study the effect of these materials on the intermediate host *Biomphalaria glabrata* (Loker et al. 1992). Some ESP fractions obtained from juvenile parasites depressed the ability of snail hemocytes to phagocytose sheep red blood cells, but had little effect on hemocyte structure or behavior (Loker et al. 1992). ESP released by sporocysts in older cultures strongly inhibited hemocyte uptake of these red cells, and also, hemocytes pretreated with ESP changed their shape and failed to spread normally. SDS–polyacrylamide gel electrophoresis (SDS–PAGE) analysis identified the active components present in the ESP as molecules greater than 100 kDa in their native configuration, heat- and trypsin-labile, and capable of binding anti-ESP antibodies (Loker et al. 1992).

Recent studies by Trouvé and Coustau (1998) have identified two polypeptides (with molecular masses of 85 and 119 kDa, respectively), which exhibit distinct expression patterns depending on the origin of *E. caproni* isolates, suggesting that intraspecific ESP differences could be involved in evolutionary processes. Furthermore, Coustau et al. (2003) observed different response patterns in snails when exposed to either *S. mansoni* or *E. caproni* ESPs (Coustau et al. 2003).

In the last few years, our group has focused on the study and characterization of ESP from *E. caproni* and *E. friedi* using different approaches. By western immunoblot from monodimensional electrophoresis (1DE) using sera from experimentally infected rats, Toledo et al. (2004) initiated the characterization of antigens present in ESP from *E. caproni* (Toledo et al. 2004). These studies have been recently completed by the use of proteomics mainly by two-dimensional electrophoresis (2DE), which allowed the identification of various proteins, including structural molecules and enzymes. Some of those molecules are also recognized as antigens by sera from experimentally infected mice (Marcilla et al. 2007; Sotillo et al. 2008; Table 9.3). In relation to *E. friedi* ESP, following a proteomic approach, we have identified structural proteins like actin, tropomyosin, and paramyosin; glycolytic enzymes like enolase, glyceraldehyde 3P dehydrogenase (GAPDH), and aldolase; detoxifying enzymes like GSTs; and the stress-related protein Hsp70 (Bernal et al. 2006, Table 9.3). Our studies also showed that actin and Hsp70 exhibited differential expression patterns depending on the definitive host (either low- or high-compatible hosts), suggesting that these proteins may play a role in the survival within the host (Bernal et al. 2006).

Moreover, we have detected posttranslational modifications in the protein pattern of echinostome proteins including tyrosine phosphorylation in response to different stimulus (Marcilla et al. 2004), or glycosylation processes (unpublished data). Very recent studies on schistosomes have shown that tyrosine phosphorylation of parasite proteins may be correlated with the adaptation to intermediate hosts (Walker and Rollinson 2007). Future studies will address whether these phenomena also occur in echinostomes.

9.3.2 Analytical Procedures Used in Echinostome Proteomics

The analysis of the proteome of echinostomes follows a general protocol already described for other trematodes (Morphew et al. 2006; Van Hellemond et al. 2007). In Fig. 9.1, a classical scheme is presented showing the steps required for such analysis. The first step separates a complex mixture into its constituent proteins. The main separative techniques include 2DE of intact proteins and liquid chromatography (LC) of peptide digests (Fig. 9.1). The second step involves MS analysis of the isolated proteins or peptides and database searching to obtain their identities (Fig. 9.1). This technology is less useful when the only target database available consists in a small collection of ESTs (see earlier), or when we can only compare our data with scarce individual protein entries or proteins from related organisms (Morphew et al. 2006), and that is exactly the case of echinostomes.

9.3.2.1 Sample Preparation

The proteomic protocols start with sample preparation, which is very often problematic. As already mentioned, the amount of starting material is limited, and samples also contain proteases which interfere with protein purification (Morphew et al. 2006).

I have mentioned before that most of the studies on echinostome proteins have been carried out using ESP, and in this case it is critical to obtain those materials in conditions where proteolysis is minimal. For such purposes, parasite cultures to obtain ESP are carried out in the presence of protease inhibitor cocktails (there are several commercial preparations containing PMSF, aprotinin, leupeptin, chelants, etc.).

Other important issues to consider before separating the sample are the concentration and purity (especially for electrophoresis), which are usually achieved either by a simple precipitation with trichloroacetic acid, or with commercially available systems. This step allows the elimination of contaminants and the enrichment of the sample.

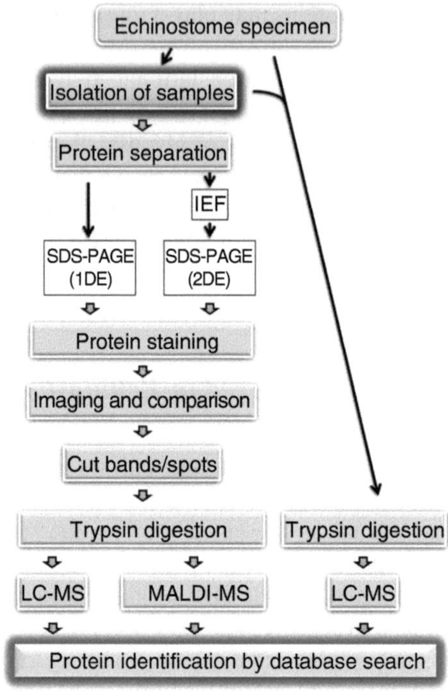

Fig. 9.1 Scheme of proteomic technology applied for the identification of echinostome proteins. In classical proteome experiments separation of the protein mixture is achieved either by conventional SDS–PAGE (1DE) or bidimensional electrophoresis (2DE), which requires the previous separation of proteins by isoelectrofocusing (IEF). Isolated bands (1DE) are excised from the gels and digested with trypsin, and peptides obtained are separated by liquid chromatography (LC) and subjected to mass spectrometry (MS). Spots of interest (2DE) are excised from the gels, digested with trypsin, and subjected to mass spectrometry (usually by matrix-assisted laser desorption/ionization) (MALDI). Obtained spectra are compared with data present in databases or either peptides identified by sequencing in MS/MS analysis. Recent approaches do not require protein separation, and either purified fractions or whole specimens are subjected to tryptic digestion and LC-MS (by multidimensional protein identification or by the use of isotopes to label proteins in different conditions). Crucial steps (initial sample preparation and data banks search) are shadowed

9.3.2.2 Protein Separation and Staining

The second step in proteomics deals with the separation of individual proteins from a mixture. Most protocols use 2DE analysis, which is carried out in two phases (Fig. 9.1). Firstly, the mixture is separated by charge in immobilized pH gradient (IPG) strips by isolectricfocusing (IEF). The second dimension is carried out after IEF by regular SDS–PAGE, which separates proteins according to their molecular weight (Fig. 9.1). Once individual proteins are separated by 2DE, they are stained with commercial reagents (like Coomassie brilliant blue and silver) compatible with spectrometric techniques. Silver staining allows for the detection of 0.5–1.2 ng, whereas Coomassie blue detects 10 ng per spot. Successful identification of echinostome proteins has been achieved using both stains (Bernal et al. 2006; Guillou et al. 2007). More sensitive staining methods for proteins include fluorescent dyes like Sypro Ruby, which has a sensitivity of 1–2 ng per spot, being also compatible with MS. We have recently reported its applicability to echinostome proteins (Sotillo et al. 2008).

9.3.2.3 Imaging and Comparison

After staining proteins, images of different stained gels must be analyzed in order to check reproducibility and to identify particular interesting spots. This analysis should confirm that spots from identical proteins are in the same place and also allows for the comparison of staining intensities (Morphew et al. 2006). Different commercial systems are available, including software packages with distinct capabilities, but, as noted by Morphew et al. (2006), even the most expensive commercial software has difficulties resolving overlapping spots (Morphew et al. 2006).

9.3.2.4 Protein Identification by Mass Spectrometry

The last step in proteomics is the identification of the protein of interest. Selected spots are excised from the reproducible stained gels and subjected to trypsin digestion (Fig. 9.1). The peptides originated by this proteolysis are now subjected to mass spectrometry analysis using different systems (Matrix-assisted Laser Desorption/Ionization Time-of-flight, MALDI-ToF; or electrospray ionization mass spectrometry, ESI MS). Sometimes tandem mass spectrometry (MS/MS) is used to determine the sequence of a particular peptide originated by trypsin digestion. MALDI-ToF via peptide mass fingerprinting is the routine analysis for genome-verified organisms. Therefore, it has little application for echinostomes. In this case, protein identification can be enhanced by sequencing the peptides by tandem MS/MS analysis (Bernal et al. 2006). In any case, the bottleneck in these investigations is the absence of enough echinostome entries in the data banks to obtain significant homology to allow for a proper identification of echinostome proteins.

Alternatives to 2DE protocols (proteomics without 2DE) include multidimensional protein identification or the use of isotopes to label proteins in different conditions. In both cases, these techniques are based on LC (Fig. 9.1). In the first case, LC is applied to all peptides originated by trypsin digestion of the original protein mixture. This approach has been successfully used for other trematodes (Skelly and Wilson 2006) and is currently being applied to *E. caproni* (Toledo, R., personal communication). In the second case, the use of distinct isotope labeling differentially tags proteins depending, for example, on the parasite stage. Once the proteins are digested, they are analyzed by LC allowing relative quantifications

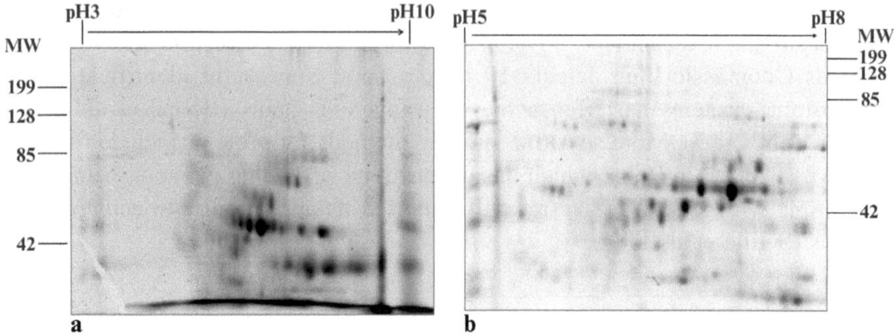

Fig. 9.2 Analysis of *Echinostoma caproni* excretory/secretory products by two-dimensional electrophoresis (2DE) at different pH gradients and stained with Sypro Ruby. (**a**) IPG strips of ph 3–10; (**b**) IPG strips of ph 5–8

between proteins from various sample stages. This technology, known as isotope-coded affinity tagging (ICAT), is now commercially available.

An example of 2DE analysis of *E. caproni* proteins is shown in Fig. 9.2. Samples of 300 μg of ESP obtained from cultured adults were analyzed by 2DE and stained with Sypro Ruby. As shown in the Fig. 9.2, initial experiments using IPG strips in the range of pH 3–10 demonstrated that most of the protein spots were present at neutral pH (Fig. 9.2a). This observation was confirmed when identical samples were subjected to IEF on IPG strips pH 5–8 (Fig. 9.2b).

9.3.3 Advances in Echinostome Proteomics

The application of proteomics to solve biological questions related to echinostome proteins is very recent, and there are still very few publications. These reports include the identification of *E. trivolvis* hemozoin by laser desorption mass spectrometry (Pisciotta et al. 2005) and the characterization of ESP proteins from adults of *E. friedi* (Bernal et al. 2006), and *E. caproni* (Marcilla et al. 2007; Higón 2007), as well as from *E. caproni* primary sporocysts (Guillou et al. 2007) (Table 9.3). As mentioned before, major groups include glycolytic and antioxidant enzymes, as well as proteins involved in structure and motility. Other identified proteins in echinostomes include chaperones, proteases, signaling molecules, and calcium-binding proteins (see Table 9.3).

In addition to these studies, our group has recently made use of immunoproteomics to identify the major antigens of *E. caproni* ESP recognized by distinct immunoglobulins (Sotillo et al. 2008), constituting the first report in echinostomes of a combination of proteomics and serology, which has been denominated

Table 9.3 Echinostome proteins identified by proteomics

Function	Protein	Species	Stage	Identified	References
Structural molecules	Actin	*E. friedi*	A	MS-WB	Bernal et al. (2006)
		E. caproni	A,S	MS-WB	Marcilla et al. (2007), Guillou et al. (2007)
	Paramyosin	*E. friedi*	A	MS/MS-WB	Bernal et al. (2006)
	Tropomyosin	*E. friedi*	A	MS	Bernal et al. (2006)
Glycolytic enzymes	Enolase	*E. friedi*	A	MS-WB	Bernal et al. (2006)
		E. caproni	A,S	MS-WB	Marcilla et al. (2007), Guillou et al. (2007)
	Glyceraldehyde 3P dehydrogenase	*E. friedi*	A	MS/MS-WB	Bernal et al. (2006)
	Aldolase	*E. friedi*	A	WB	Bernal et al. (2006)
		E. caproni	S	MS	Guillou et al. (2007)
	Triose phosphate isomerase	*E. caproni*	S	MS	Guillou et al. (2007)
	NADH dehydrogenase chain 5	*E. caproni*	S	MS	Guillou et al. (2007)
Antioxidant enzymes	Superoxide dismutase	*E. caproni*	S	MS	Guillou et al. (2007)
	Aldo-keto reductase	*E. caproni*	S	MS	Guillou et al. (2007)
	Glutathione *S*-transferase	*E. friedi*	A	MS/MS-WB	Bernal et al. (2006)
		E. caproni	A	MS	Marcilla et al. (2007)
	Thiol antioxidant protein	*E. friedi*	A	MS	Bernal et al. (2006)
Calcium binding	Ca-binding protein	*E. friedi*	A	MS	Bernal et al. (2006)
Signaling molecule	MICAL-like	*E. caproni*	S	MS	Guillou et al. (2007)
Chaperones	Heat shock protein (Hsp)-70	*E. friedi*	A	MS/MS-WB	Bernal et al. (2006)
		E. caproni	A	MS-WB	Higón (2007)
	Ubiquinone cyt C chaperone	*E. caproni*	S	MS	Guillou et al. (2007)
Proteolytic enzyme	Procathepsin L3	*E. friedi*	A	MS	Bernal et al. (2006)
Others	Hemozoin	*E. trivolvis*	A	LD-MS	Pisciotta et al. (2005)
	Acetyl coA transferase	*E. caproni*	S	MS	Guillou et al. (2007)
	Histone H1	*E. caproni*	S	MS	Guillou et al. (2007)

Abbreviations: *A* adult; *S* sporocyst; *MS* mass spectrometry; *WB* western immunoblot; *LD* laser desorption "immunome" (Wilson et al. 2007). In addition to the immunome, the application of mass spectrometry to the study of glycans, which has been called "glycomics," constitutes another potential field of investigations in echinostomes. Our group has obtained preliminary data about the glycosylation state of specific proteins like enolase (Sotillo et al. 2008). The combination of 1DE or 2DE with lectin-binding and mass spectrometry could help to identify and characterize most of the echinostome proteins capable of binding lectins, similar to that already described for *Schistosoma bovis* (Ramajo-Hernández et al. 2007a).

As mentioned before, we have also used proteomics to identify *E. caproni* proteins capable of interacting with host matrix proteins like plasminogen. We identified the enzyme enolase as being one of the most abundant and reactive proteins present (Marcilla et al. 2007). These results confirm previous observations in other trematodes like *Fasciola hepatica* (Bernal et al. 2004) or more recently in *Schistosoma bovis* (Ramajo-Hernández et al. 2007b).

We are now conducting assays to compare the surface proteome of adult *E. caproni* from low- or high-compatible hosts using multidimensional liquid chromatography mass spectrometry (LC-MS), and preliminary results show clear differences in the pattern of proteins released in both samples (unpublished data). The architecture of *Schistosoma* species surface proteins has been investigated using different proteomic approaches, including sophisticated protocols to purify tegumentary fractions of adults generically called "molecular shaving" (Braschi et al. 2006a,b; Skelly and Wilson 2006). These studies have yielded a wide array of information including the identification of major antigens and accurate localization of several proteins in the different stratus of the tegument (revised by Wilson et al. 2007). This type of analysis is therefore expected to provide similar details of information in echinostomes, allowing researchers to specifically design tools for control and diagnosis of these trematode infections.

9.3.4 Prospects and Limitations in Proteomics of Echinostomes

In the case of proteomic studies, the major limitation is without doubt the absence of a full genome project, which should provide enough annotated DNA sequence data to produce deduced amino acid sequences with which data obtained from mass spectrometry could be compared with. In the case of schistosomes, with two genome projects almost finished and another under way (Haas et al. 2007; http://www.tigr.org/; http://www.ebl.ac.uk/genomes; ftp://ftp.sanger.ac.uk/pub/pathogens/Schistosoma), and a great number of DNA entries deposited in the gene banks, the chances of a correct identification of the peptides are high. It is interesting to note that most of identifications of echinostome proteins have been possible because they were homologous to schistosome proteins. However, we cannot underestimate the number of proteins which may not present in sufficient homology (estimated as > 75%) to allow for the proper identification in echinostomes. As mentioned before, in the case of schistosomes, half of the proteins identified from their transcriptomes do not have a known function, or are even species specific (Verjovski-Almeida et al. 2003; Wilson et al. 2007), thus increasing the difficulties when working with other trematodes like echinostomes. Furthermore, not all the available data for other species like schistosomes are useful in the searches using routine bioinformatic tools, thus further modification of such data sets is needed to facilitate these searches.

These problems are augmented in the case of echinostomes by the lack of transcriptome projects available which could help the correct identification of specific proteins (see Table 9.2). As already mentioned in the first part of this chapter, there is only one EST collection (from *E. paraensei*) available with 358 sequences (Nowak and Loker 2005). Again, if we compare this situation to that of other trematodes such as *Schistosoma haematobium* or even with *Fasciola hepatica* with more of 15,000 EST each (http://www.sanger.ac.uk/Projects/Helminths/), we definitively consider abandoning our investigations. In our experience, even after obtaining good spectra, no good identification can be achieved in more than 90% of the cases (unpublished data), which means that only a small percentage of the molecules currently analyzed are properly identified using the proteomic technology available.

We cannot forget that proteomic studies are limited by mass spectrometry sensitivity, and it depends ultimately on the amount and quality of material available, and we are aware that parasite material is limited, with no possibility of echinostome cultures *in vitro*.

Some help can be obtained in the identification by combining mass spectrometry data with western blotting using heterologous (crossreactive) antibodies. In this case, the blot is at least two orders of magnitude more sensitive than gel staining (i.e., Sypro Ruby), and sometimes strong reactions have no visible counterparts on the comparable gel. In addition, this technology is feasible only for a few conserved proteins; moreover, it is expensive (commercially available antibodies are not cheap reagents) and can also be time consuming (requiring noncommercial antibodies). However, in echinostomes it is sometimes the only way to get information about the proteome, and an example of the applicability of this technology has been reported for *E. friedi*, where we used commercial antibodies against GST, actin, aldolase, Hsp-70, and enolase, as well as noncommercial antibodies for GAPDH, which successfully confirmed the identity of echinostome proteins present in the ESP (Bernal et al. 2006).

In relation to prospects of echinostome proteomics, we again have to compare our studies to what is being carried out with schistosomes. Firstly, the attempts to define the host-parasite interface could provide targets for vaccination, which in turn could serve as model for other intestinal helminthiases. Secondly, pull-down assays have been suggested to be useful for identifying molecular crosstalk between parasite and hosts. These assays require considerable amounts of protein material, so they are suitable only for adult parasites or purified recombinant proteins. Finally, new isotope technology should help calculating relative and absolute protein present in different materials and biological situations, helping to identify potentially important molecules for the parasite.

9.4 Concluding Remarks

Echinostome genomics and proteomics, which could be called "echinostomics," are in their infancy, with no genome sequencing projects under way and very scarce data on the identification of either DNA or protein molecules. In addition, no current protocols are available to transfect echinostomes (larval or adults), nor microarray assays are in progress. Major advances in these fields rely on the use of the few cDNA libraries accessible and the obtainment of ESTs to be used for the identification of echinostome proteins. It is expected that advances in the knowledge of other trematode molecules (i.e., completion of *Schistosoma* spp. genome sequencing), as well as in bioinformatic technologies could contribute to the progress in "echinostomics" studies.

Acknowledgments Dr. Lynne Yenush is thanked for critically reading the manuscript. Financial support was obtained from the Spanish Ministry of Education and Science (Madrid, Spain), grant CGL2005–02321/BOS; Conselleria d'Educació, Generalitat Valenciana (Valencia, Spain), grants GV07/006 and ACOMP07/071; and Universitat de València (Valencia, Spain) grant Accions Especials 20050201.

References

Ashton, P.D., Curwen, R.S., and Wilson, R.A. 2001. Linking proteome and genome: how to identify parasite proteins. Trends in Parasitology 17: 198–202.

Bernal, D., De la Rubia, J.E., Carrasco-Abad, A.M., Toledo, R., Mas-Coma, S., and Marcilla, A. 2004. Identification of enolase as a plasminogen-binding protein in excretory–secretory products of *Fasciola hepatica*. FEBS Letters 563: 203–206.

Bernal D., Carpena I., Espert A.M., De la Rubia, J.E., Esteban, J.G., Toledo, R., and Marcilla, A. 2006. Identification of proteins in excretory/secretory extracts of *Echinostoma friedi* (Trematoda) from chronic and acute infections. Proteomics 6: 2835–2843.

Biron, D.G., Brun, C., Lefevre, T., Lebarbenchon, C., Loxdale, H.D., Chevenet, F., Brizard, J.P., and Thomas, F. 2006. The pitfalls of proteomics experiments without the correct use of bioinformatics tools. Proteomics 6: 5577–5596.

Bouchut, A., Coustau, C., Gourbal, B., and Mitta, G. 2007. Compatibility in the *Biomphalaria glabrata/Echinostoma caproni* model: new candidate genes evidenced by a suppressive subtractive hybridization approach. Parasitology 134: 575–88.

Braschi, S., Borges, W.C., and Wilson, R.A. 2006a. Proteomic analysis of the schistosome tegument and its surface membranes. Memorias do Instituto Oswaldo Cruz 101: S205–S212.

Braschi, S., Curwen, R.S., Ashton, P.D., Verjovski-Almeida, S., and Wilson, R.A. 2006b. The tegument surface membranes of the human blood parasite *Schistosoma mansoni*: a proteomic analysis after differential extraction. Proteomics 6: 1471–1482.

Chang, Y.-W.E., Jakobi, R., McGinty, A., Foschi, M., Dunn, M.J., and Sorokin, A. 2000. Cyclooxygenase 2 promotes cell survival by stimulation of dynein light chain expression and inhibition of neuronal nitric oxide synthase activity. Molecular and Cell Biology 20: 8571–8579.

Coustau, C., Mitta, G., Dissous, C., Guillou, F., Galinier, R., Allienne, J.F., and Modat, S. 2003. *Schistosoma mansoni* and *Echinostoma caproni* excretory–secretory products differentially affect gene expression in *Biomphalaria glabrata* embryonic cells. Parasitology 127: 533–542.

Damian, R.T. 1997. Parasite immune evasion and exploitation: reflections and projections. Parasitology 115: S169–S175.

Davies, R.E. 1997. Surprising diversity and distribution of spliced leader RNAs in flatworms. Molecular and Biochemical Parasitology 87: 29–48.

Fujino, T., Fried, B., and Takamiya, S. 1995. Cytochemical localization of cytochrome c oxidase activity in mitochondria in the tegument and tegumental and parenchymal cells of the trematodes *Echinostoma trivolvis, Zygocotyle lunata, Schistosoma mansoni, Fasciola gigantica* and *Paragonimus ohirai.* Journal of Helminthology 69: 195–201.

Ghedin, E., Wang, S., Spiro, D., Caler, E., Zhao, Q., Crabtree, J., Allen, J.E., Delcher, A.L., Guiliano, D.B., Miranda-Saavedra, D., Angiuoli, S.V., Creasy, T., Amedeo, P., Hass, B., El-Sayed, N.M., Wortman, J.R., Feldblyum, T., Tallon, L., Shumnway, M., Koo, H., Salzberg, S.l., Schobel, S., Pertea, M., Pop, M., White, O., Barton, G.J., Carlow, C.K., Crawford, M.J., Daub, J., Dimmic, M.W., Estes, C.F., Foster, J.M., Ganatra, M., Gregory, W.F., Johnson, N.M., Jin, J., Komuniecki, R., Korf, I., Kumar, S., Laney, S., Li, W., Lindblom, T.H., Lustigman, S., Ma, D., Maina, C.V., Martin, D.M., McCarter, J.P., McReynolds, L., Mitreva, M., Nutman, T.B., Parkinson, J., Peregrín-Alvarez, J.M., Poole, C., Ren, Q., Saunders, L., Sluder, A.E., Smith, K., Stanke, M., Unnasch, T.R., Ware, J., Wei, A.D., Weil, G., Williams, D.J., Zhang, Y., Williams, S.A., Fraser-Liggett, C., Slatko, B., Blaxter, M.L., and Scott, A.L. 2007. Draft genome of the filarial nematode parasite *Brugia malayi.* Science 317: 1756–1760.

Goudot-Crozel, V., Caillol, D., Djabali, M., and Dessein, A.J. 1989. The major parasite surface antigen associated with human resistance to schistosomiasis is a 37-kDa glyceraldehydes-3-phosphate dehydrogenase. Journal of Experimental Medicine 170: 2065–2080.

Graczyk, T.K., and Fried, B. 1994. ELISA method for detecting anti-*Echinostoma caproni* (Trematoda: Echinostomatidae) immunoglobulins in experimentally infected ICR mice. Journal of Parasitology 80: 544–549.

Graczyk, T.K., and Fried, B. 1995. An enzyme-linked immunosorbent assay for detecting anti-*Echinostoma trivolvis* (Trematoda) IgG in experimentally infected ICR mice. Cross-reactivity with *E. caproni.* Parasitology Research 81: 710–712.

Guillou, F., Roger, E., Moné, Y., Rognon, A., Grunau, C., Therón, A., Mitta, G., Coustau, C., and Gourbal, B.E. 2007. Excretory–secretory proteome of larval *Schistosoma mansoni* and *Echinostoma caproni,* two parasites of *Biomphalaria glabrata.* Molecular and Biochemical Parasitology 155: 45–56.

Haas, B.J., Berriman, M., Hirai, H., Cerqueira, G.G., LoVerde,P.T., and El-Sayed, N.M. 2007. *Schistosoma mansoni* genome: closing in on a final gene set. Experimental Parasitology 117: 225–228.

Haque, M., and Siddiqui, A.H. 1982. Histochemical and electrophoretic studies on phosphatases of some Indian trematodes. Journal of Helminthology 56: 111–116.

Heyneman, D., Faulk, W.P., and Fudenberg, H.H. 1971. *Echinostoma lindoense*: larval antigens from the snail intermediate host, *Biomphalaria glabrata.* Experimental Parasitology 29: 480–492.

Higón, M. (2007). Cloning and molecular characterization of *Echinostoma caproni* Hsp70. Master Thesis, Universidad de Valencia, 74 pp (In Spanish).

Hu, W., Brindley, P.J., McManus, D.P., Feng, Z., and Han, Z.G. 2004. Schistosome transcriptomes: new insights into the parasite and schistosomiasis. Trends in Molecular Medicine 10: 217–225.

Humbert E., and Coustau C. 2001. Refractoriness of host haemocytes to parasite immunosuppressive factors as a putative resistance mechanism in the *Biomphalaria glabrata–Echinostoma caproni* system. Parasitology 122: 651–660.

Iomini, C., and Justine, J.L. 1997. Spermiogenesis and spermatozoon of *Echinostoma caproni* (Platyhelminthes, Digenea): transmission and scanning electron microscopy, and tubulin immunocytochemistry. Tissue Cell 29: 107–118.

Iomini, C., Bré, M.H., Levilliers, N., and Justine, J.L. 1998. Tubulin polyglycylation in Platyhelminthes: diversity among stable microtubule networks and very late occurrence during spermiogenesis. Cell Motility and the Cytoskeleton 39: 318–330.

Jaffrey, S.R., and Snyder, S.H. 1996. PIN: an associated protein inhibitor of neuronal nitric oxide synthase. Science 274: 774–777.

Kanev, I. 1994. Life-cycle, delimitation and redescription of *Echinostoma revolutum* (Frölich, 1802) (Trematoda: Echinostomatidae). Systematic Parasitology 28: 125–144.

Kanev, I., Dimitrov, V., Radev, V., and Fried, B. 1995a. Redescription of *Echinostoma trivolvis* (Cort, 1914) with a discussion of its identity. Systematic Parasitology 32: 61–70.

Kanev, I., Fried, B. Dimitrov, V., and Radev, V. 1995b. Redescription of *Echinostoma jurini* (Skvortzov, 1924) with a discussion of its identity and characteristics. Annalen des Naturhistorischen Museums in Wien Serie B Botanik und Zoologie 97B: 37–53.

Kostadinova, A., and Gibson, D.I. 2000. The systematics of the echinostomes. In: Echinostomes as experimental models for biological research, eds. B. Fried and T.K. Graczyk, pp. 31–57, Dordrecht: Kluwer Academic Publishers.

Kostadinova, A., Herniou, E.A., Barrett, J., and Littlewood, D.T. 2003. Phylogenetic relationships of *Echinostoma* Rudolphi, 1809 (Digenea: Echinostomatidae) and related genera re-assessed via DNA and morphological analyses. Systematic Parasitology 54: 159–176.

Kosupko, G.A. 1971a. New data in the bioecology and morphology of *Echinostoma revolutum* and *E. miyagawai* (Trematoda: Echinostomatidae). Byulleten' Vsesoyuznogo Instituta Gel'mintologii im. K.I. Skrjabina 5: 43–49. (In Russian).

Kosupko, G.A. 1971b. Criteria of the species *Echinostoma revolutum*, demonstrated on experimental material. In Sbornik rabot po gel'mintologii posvyashchen 90-letiyu so dnya rozhdeniya akademika, ed. K.I. Skrjabin, pp. 167–175. Moscow: 'Kolos' (In Russian).

Kosupko, G.A. 1972. Morphology and biology of *Echinostoma revolutum* Frölich, 1802 and *Echinostoma miyagawai* Ishii, 1932 (Trematoda: Echinostomatidae) studied on experimental material. Ph.D. Thesis, Moscow, 258 pp. (In Russian).

Kristensen, A.R., and Fried, B. 1991. A comparison of *Echinostoma caproni* and *Echinostoma trivolvis* (Trematoda: Echinostomatidae) adults using isoelectrofocusing. Journal of Parasitology 77: 496–498.

Lie K.J., and Heyneman, D. 1977. Studies on resistance in snails: interference by nonirradiated echinostome larvae with natural resistance to *Schistosoma mansoni* in *Biomphalaria glabrata*. Journal of Invertebrate Pathology 29: 118–125.

Lie K.J., and Heyneman, D. 1979. Acquired resistance to echinostomes in four *Biomphalaria glabrata* strains. International Journal for Parasitology 9: 533–537.

Lie K.J., Jeong K.H., and Heyneman, D. 1982. Further characterization of acquired resistance in *Biomphalaria glabrata*. Journal of Parasitology 68: 529–531.

Loker E.S., Bayne C.J., and Yui M.A. 1986. *Echinostoma paraensei*: hemocytes of *Biomphalaria glabrata* as targets of echinostome mediated interference with host snail resistance to *Schistosoma mansoni*. Experimental Parasitology 62: 149–154.

Loker E.S., Cimino D.F., and Hertel L.A. 1992. Excretory–secretory products of *Echinostoma paraensei* sporocysts mediate interference with *Biomphalaria glabrata* hemocyte functions. Journal of Parasitology 78: 104–115.

Marcilla, A., De la Rubia, J.E., Espert, A., Carpena, I., Esteban, J.G., and Toledo, R. 2004. Specific tyrosine phosphorylation in response to bile in *Fasciola hepatica* and *Echinostoma friedi*. Experimental Parasitology 106: 56–58.

Marcilla, A., Pérez-García, A., Espert, A., Bernal, D., Muñoz-Antolí, C., Esteban, J.G., and Toledo, R. 2007. *Echinostoma caproni*: identification of enolase in excretory/secretory products, molecular cloning, and functional expression. Experimental Parasitology 117: 57–64.

McKusick, V.A. (1997). Genomics: structural and functional studies of genomes. Genomics 45: 244–249.

Morgan, J.A.T. and Blair, D. 1995. Nuclear rDNA ITS sequence variation in the trematode genus *Echinostoma*: an aid to establishing relationships within the 37 collar-spine group. *Parasitology* 111: 609–615.

Morgan, J.A.T., and Blair, D. 1998a. Relative merits of nuclear ribosomal internal transcribed spacers and mitochondrial CO1 and ND1 genes for distinguishing among *Echinostoma* species (Trematoda). Parasitology 116: 289–297.

Morgan, J.A.T., and Blair, D. 1998b. Mitochondrial ND1 gene sequences used to identify echinostome isolates from Australia and New Zealand. International Journal for Parasitology 28: 493–502.
Morphew, R.M., Barrett, J., and Brophy, P.M. 2006. Flatworm parasite proteomics. In Parasitic flatworms. Molecular biology, biochemistry, immunology and physiology, eds. A.G. Maule and N.J. Marks, pp. 327–347. Wallingford: Commonwealth Agricultural Bureau International Publishing.
Noda S., and Loker E.S. 1989. Effects of infection with *Echinostoma paraensei* on the circulating haemocyte population of the host snail *Biomphalaria glabrata*. Parasitology 98: 35–41.
Nowak, T.S., and Loker, E.S. 2005. *Echinostoma paraensei*: differential gene transcription in the sporocyst stage. Experimental Parasitology 109: 94–105.
Oliveira, G. (2007). The *Schistosoma mansoni* transcriptome: an update. Experimental Parasitology 117: 229–235.
Oliveira, G., and Johnston, D.A. 2001. Mining the schistosome DNA sequence database. Trends in Parasitology 17:501–503.
Pinheiro, J., Maldonado, A., and Lanfredi, R.M. 2004. Light and scanning electron microscopy of the miracidium of *Echinostoma paraensei* (Trematoda, Echinostomatidae). Veterinary Parasitology 121: 265–275.
Pisciotta, J.M., Ponder, E.L., Fried, B., and Sullivan, D. 2005. Hemozoin formation in *Echinostoma trivolvis* rediae. International Journal for Parasitology 35: 1037–1042.
Ramajo-Hernández, A., Oleaga, A., Ramajo-Martín, V. and Pérez-Sánchez, R. 2007a. Carbohydrate profiling and protein identification of tegumental and excreted/secreted glycoproteins of adult Schistosoma bovis worms. Veterinary Parasitology 144: 45–60.
Ramajo-Hernández, A., Pérez-Sánchez, R., Ramajo-Martín, V., and Oleaga, A. 2007b. *Schistosoma bovis*: plasminogen binding in adults and the identification of plasminogen-binding proteins from the worm tegument. Experimental Parasitology 115: 83–91.
Ravichandran, V., Seres, T., Moriguchi, T., Thomas, J.A., and Johnston, R.B. Jr. 1994. S-thiolation of glyceraldehydes-3-phosphate dehydrogenase induced by the phagocytosis-related respiratory burst in blood monocytes. Journal of Biological Chemistry 269: 25010–25015.
Richard, J., Klein, M.J., and Stoeckel, M.E. 1989. Neural and glandular localisation of substance P in *Echinostoma caproni* (Trematoda: Digenea). Parasitology Research 75: 641–648.
Salzet M., Capron A., and Stefano G.B. 2000. Molecular crosstalk in host–parasite relationships: schistosome– and leech–host interactions. Parasitology Today 16: 536–540.
Sapp K.K., and Loker E.S. 2000. Mechanisms underlying digenean-snail specificity: role of miracidial attachment and host plasma factors. Journal of Parasitology 86: 1012–1019.
Sebelová, S., Stewart, M.T., Mousley, A., Fried, B., Marks, N.J., and Halton, W. 2004. The musculature and associated innervation of adult and intramolluscan stages of *Echinostoma caproni* (Trematoda) visualised by confocal microscopy. Parasitology Research 93: 196–206.
Seres, T., Ravichandran, V., Moriguchi, T., Rokutan, K., Thomas, J.A., and Johnston, R.B. Jr. 1996. Protein S-thiolation and dethiolation during the respiratory burst in human monocytes: a reversible post-translational modification with potential for buffering the effects of oxidant stress. Journal of Immunology 156: 1973–1980.
Skelly, P.J., and Wilson, R.A. 2006. Making sense of the schistosome surface. Advances in Parasitology 63: 186–285.
Simonsen, P.E., Estambale, B.B., and Agger, M. 1991. Antibodies in the serum of golden hamsters experimentally infected with the intestinal trematode *Echinostoma caproni*. Journal of Helminthology 65: 239–247.
Sloss, B., Meece, J., Romano, M., and Nollen, P. 1995. The genetic relationships between *Echinostoma caproni*, *E. paraensei*, and *E. trivolvis* as determined by electrophoresis. Journal of Helminthology 69: 243–246.
Sorensen, R.E., Curtis, J., and Minchella, D.J. 1998. Intraspecific variation in the rDNA ITS loci of 37-collar-spined echinostomes from North America: implications for sequence-based diagnoses and phylogenetics. Journal of Parasitology 85: 992–997.

Sotillo, J., Valero, L., Sánchez del Pino, M.M., Fried, B., Esteban, J.G., Marcilla, A., and Toledo, R. 2008. Identification of antigenic proteins from *Echinostoma caproni* (Trematoda: Echinostomatidae) recognized by mouse immunoglobulins M, A and G using an immunoproteomic approach. Parasite Immunology 30: 271–279.

Taft, J.L., and Fried, B. 1968. Oxygen consumption in adult *Echinostoma revolutum*. Experimental Parasitology, 23: 183–186.

Thaumaturgo, N., Vilar, M.M., Edelenyi, R., and Tendler, M. 2002. Characterization of Sm14 related components in different helminths by sodium dodecyl sulphate-polyacrylamide gel electrophoresis and Western blotting analysis. Memorias do Instituto Oswaldo Cruz 97: S115–S116.

The *C. elegans* Consortium 1998. Genome sequence of the nematode *C. elegans*: a platform for investigating biology. Science 282: 2012–2018.

Thorndyke, M.C., and Whitfield, P.J. 1987. Vasoactive intestinal polypeptide-like immunoreactive tegumental cells in the digenean helminth *Echinostoma liei*: possible role in host–parasite interactions. General and Comparative Endocrinology 68: 202–207.

Toledo, R., and Fried, B. 2005. Echinostomes as experimental models for interactions between adult parasites and vertebrate hosts. Trends in Parasitology 21: 251–254.

Toledo, R., Espert, A., Muñoz-Antoli, C., Fried, B., and Esteban, J.G. 2004. Immunological characterization of somatic and excretory/secretory antigens of *Echinostoma caproni* (Trematoda: Echinostomatidae) in experimentally infected rats. Comparative Parasitology 71: 42–48.

Trelis, M. 2004. (*Echinostoma friedi* Toledo et al., 2000 (Trematoda: Echnostomatidae): experimental biologic study of the binomial miracidium-gasteropoda and adult molecular characterization. Ph.D. Thesis, Universidad de Valencia, 269 pp) (In Spanish).

Trouvé, S., and Coustau, C. 1998. Differences in adult excretory–secretory products between geographical isolates of *Echinostoma caproni*. Journal of Parasitology 84: 1062–1065.

Uchikawa, R., and Loker, E.S. 1991. Lectin-binding properties of the surfaces of in vitro-transformed *Schistosoma mansoni* and *Echinostoma paraensei* sporocysts. Journal of Parasitology 77: 742–748.

Van Hellemond, J.J., van Balkom, B.W.M., and Tielens, A.G.M. 2007. Schistosome biology and proteomics: progress and challenges. Experimental Parasitology 117: 267–274.

Verjovski-Almeida, S., DeMarco, R., Martins, E.A.L., Guimarães, P.E.M., Ojopi, E.P.B., Paquola, A.C.M., Piazza, J.P., Nishiyama Jr., M.Y., Kitajima, J.P., Adamson, R.E., Ashton, P.D., Bonaldo, M.F., Coulson, P.S., Dillon, G.P., Farias, L.P., Gregorio, S.P., Ho, P.L., Leite, R.A., Malaquias, L.C.C., Marques, R.C.P., Miyasato, P.A., Nascimento, A.L.T.O., Ohlweiler, F.P., Reis, E.M., Ribeiro, M.A., Sá, R.G., Stukart, G.C., Soares, M.B., Gargioni, C., Kawano, T., Rodrigues, V., Madeira, A.M.B.N., Wilson, R.A., Menck, C.F.M., Setubal, J.C., Leite, L.C.C., Dias-Neto, E., 2003. Transcriptome analysis of the acoelomate human parasite *Schistosoma mansoni*. Nature Genetics 35: 148–157.

Verjovski-Almeida, S., Leite, L.C., Dias-Neto, E., Menck, C.F., and Wilson, R.A. 2004. Schistosome transcriptome: insights and perspectives for functional genomics. Trends in Parasitology 20: 304–308.

Voltz, A., Richard, J., Pesson, B., and Jourdane, J. 1986. Chemicotaxonomic study of the genus *Echinostoma*: comparison of a strain isolated in the Cameroon (E. sp.) and 2 African species (*E. caproni* and *E. togoensis*). Annales de Parasitologie Humaine et Comparée 61: 617–623. (In French).

Voltz, A., Richard, J., and Pesson, B. 1987. A genetic comparison between natural and laboratory strains of *Echinostoma* (Trematoda) by isoenzymatic analysis. Parasitology 95: 471–477.

Voltz, A., Richard, J., Pesson, B., and Jourdane, J. 1988. Isoenzyme analysis of *Echinostoma liei*: comparison and hybridization with other African species. Experimental Parasitology 66: 13–17.

Walker, A.J., and Rollinson, D. 2007. Specific tyrosine phosphorylation induced in *Schistosoma mansoni* miracidia by haemolymph from schistosome susceptible, but not resistant, *Biomphalaria glabrata*. Parasitology 6: 1–9.

Wilson, R.A., Ashton, P.D., Braschi, S., Dillon, G.P., Berriman, M., and Ivens, A. 2007. "Oming" in on schistosomes: prospects and limitations for post-genomics. Trends in Parasitology, 23: 14–20.

Chapter 10
Studies of Echinostomes Using Chromatography and Atomic Spectrometry

Joseph Sherma and Bernard Fried

Contents

10.1 Introduction 230
10.2 Thin Layer Chromatography Principles and Methods 230
10.2.1 Principles 230
10.2.2 Sample Preparation 231
10.2.3 Layers and Mobile Phases 232
10.2.4 Standard and Sample Solution Application 233
10.2.5 Plate Development with the Mobile Phase 233
10.2.6 Methods for Detection of Separated Zones 234
10.2.7 Documentation of Chromatograms 234
10.2.8 Identification of Zones 234
10.2.9 Quantification of Chromatograms 234
10.2.10 Statistical Analysis of Data 235
10.3 Thin Layer Chromatography Applications 235
10.3.1 Introduction 235
10.3.2 Lipids in Echinostomes 235
10.3.3 Amino Acids in Echinostomes 239
10.3.4 Carbohydrates in Echinostomes 240
10.3.5 Lipophilic Pigments in Echinostomes 240
10.4 Gas Chromatography 241
10.4.1 Principles 241
10.4.2 Methods and Applications 241
10.5 Column Liquid Chromatography 243
10.5.1 Principles 243
10.5.2 Methods and Applications 243
10.6 Inductively Coupled Plasma-Optical Emission Spectrometry and Atomic Absorption Spectrometry 244
10.6.1 Principles 244
10.6.2 Methods and Applications 244
10.7 Concluding Remarks 245
References 246

J. Sherma (✉)
Department of Chemistry, Lafayette College, 326 Hugel Science Center, Easton, PA 18042, USA
e-mail: shermaj@lafayette.edu

B. Fried
Department of Biology, 111 Kunkel Hall, Lafayette College Easton, PA 18042, USA

B. Fried and R. Toledo (eds.) *The Biology of Echinostomes*,
DOI: 10.1007/978-0-387-09577-6_10,

Abstract This chapter describes chromatographic and atomic spectrometric methods used for the analysis of echinostomes. Most of the analyses are concerned with larval and adult echinostomes, but some consider host tissues infected with these stages. Coverage includes the chromatographic methods used for determination of different classes of organic compounds and atomic spectrometric methods for elemental analyses. The major chromatographic methods used have been thin-layer chromatography and high-performance thin-layer chromatography. Atomic absorption spectrometry and inductivity-coupled plasma-optical emission spectrometry have been used for determining certain elements, including metals. These studies have contributed to a better understanding of the chemical composition of larval and adult echinostomes and of the host tissues infected by these digeneans.

10.1 Introduction

This chapter describes certain methods that have been used in the analysis of larval and adult echinostomes. Considered also are studies on the analyses of host tissues infected with either larval or adult echinostomes. Included are the chromatographic methods for determination of different classes of organic compounds and atomic spectrometric methods for elemental analysis. Explanations of the abbreviations used in this chapter are given in Table 10.1. The major chromatographic methods applied have been TLC and HPTLC, with a few papers reporting analyses by GLC and column HPLC; AAS and ICP-OES have been reported for determination of elements, including metals. Earlier studies on the TLC analysis of echinostomes from the 1970s through the mid-1990s were reviewed by Fried and Haseeb (1996). This chapter includes TLC-related publications only from 1995 through 2007. No previous review of analyses by column chromatographic and spectrometric methods is available. Therefore, all references on these topics are covered here.

10.2 Thin-Layer Chromatography Principles and Methods

10.2.1 Principles

TLC is a type of liquid chromatography in which the stationary phase is a layer of sorbent on a glass plate. The plate is developed by placing it in a closed chamber containing a pool of mobile phase, which is a mixture of two or more solvents. As the mobile phase rises through the layer, the compounds move at different rates and are separated, forming the chromatogram. The separated zones are then visualized by their natural color or by application of a detection reagent. Quantification can be performed semiquantitatively by visual comparison of standard and sample zone intensity, or with more accuracy and precision by instrumental densitometry.

The materials, instruments, and techniques of TLC that have been applied to the analysis of echinostomes are described in Sects. 10.2.2–10.2.10, and the references

Table 10.1 The abbreviations used in this review

AAS	Atomic absorption spectrometry
C-18	Silica gel layer chemically bonded with octadicylsilyl groups
DGG	Digestive gland–gonad complex
DGGs	Digestive gland–gonad complexes
DHA	Docosahexaenoic acid
DI	Deionized water
EPA	Environmental Protection Agency
ES	Excretory–secretory
FAAS	Flame atomic absorption spectrometry
FFA	Free fatty acids
FID	Flame ionization detector
FS	Free sterols
GFAAS	Graphite furnace atomic absorption spectrometry
GLC	Gas–liquid chromatography
HPLC	High-performance column liquid chromatography
HPTLC	High-performance thin-layer chromatography
ICP-OES	Inductively coupled plasma-optical emission spectrometry
ICR	Institute of Cancer Research
id	Internal diameter
IXC	Ion exchange chromatography
ME	Methyl esters
MS	Mass spectrometry
NMID	Nonmethylene-interrupted diene
PC	Phosphatidylcholine
PE	Phosphatidylethanolamine
PI	Post-infection
PMA	Phosphomolybdic acid
PMT	Photomultiplier tube
RP	Reversed phase
SCW	Snail-conditioned water
SE	Steryl esters
SEC	Size exclusion chromatography
TG	Triacylglycerols
TLC	Thin-layer chromatography
TMS	Trimethylsilyl
WCOT	Wall coated open tubular

given in these sections are examples rather than comprehensive. Additional details on TLC are available in Sherma and Fried (2003).

10.2.2 Sample Preparation

The classic Folch procedure was used to extract lipids and phospholipids from whole bodies, DGGs, and hemolymph of snails patently infected with *Echinostoma caproni* (Bandstra et al. 2006); feces of BALB/c mice infected with *E. caproni* (Bandstra et al. 2007); the intestinal mucosa, serum, and liver of ICR mice infected with *E. caproni* (Albrecht et al. 1998 and Rivas et al. 1998); adults of *E. caproni*, *E. trivolvis*, and *Zygocotyle lunata* (Lee et al. 1998); cercariae of *E. trivolvis* and *Echinoparyphium* sp. (Muller et al. 1999); *E. trivolvis* redial bodies (Reddy et al. 1997); *E. caproni* adults and excysted metacercariae (Schneck et al. 2004a); and SCW and feces from *Biomphalaria glabrata* infected with *E. caproni* (Schneck et al. 2004b). The Folch procedure involves extraction of the sample with chloroform–methanol (2:1) in a glass homogenizer followed by washing the organic phase with 0.88% KCl. The extract is evaporated to dryness under nitrogen flow, and the residue is dissolved in an appropriate volume of extractant for TLC analysis.

Prior to TLC determination, extraction was performed using acetone in a glass homogenizer for the pigments lutein and beta-carotene from whole bodies or DGGs of *B. glabrata* infected with *E. caproni* and *Helisoma trivolvis* infected with *E. trivolvis* (Evans et al. 2004); by vortexing with 70% aqueous ethanol for free pool amino acids in rediae, cercariae, encysted and excysted metacercariae of *E. caproni* (Ponder et al. 2003a, b); and by homogenization in a vial for sugars in *B. glabrata* infected with *E. caproni* (Wagner et al. 2001).

10.2.3 Layers and Mobile Phases

TLC and HPTLC are carried out on glass plates commercially precoated with the sorbent to be used. HPTLC plates have layers that are composed of particles with a lower mean particle size, have a smaller particle size distribution, and are thinner compared to TLC plates. In general, this leads to better and faster separations for HPTLC compared to TLC.

Neutral and polar lipids have been mostly analyzed on 20 cm × 10-cm HPTLC silica gel plates with diatomaceous earth preadsorbent (or concentrating) zone and 19 scored lanes from Whatman (Florham Park, NJ, USA) (Bandstra et al. 2006). Similar plates from EMD Chemicals, Inc. (Gibbstown, NJ, an affiliate of Merck KGaA, Darmstadt, Germany) with a preadsorbent zone composed of wide pore silica 50,000 have also been used (Bandstra et al. 2007). The preadsorbent zone adjacent to the main analytical layer facilitates manual sample application of relatively high volumes of biological sample extracts, and the lanes serve as a guide for positioning of the initial zones and the densitometer source slit for scanning chromatograms. The usual mobile phase was petroleum ether–diethyl ether–glacial acetic acid (80:20:1) for neutral lipids; SE and ME were better separated from each other and from unidentified nonpolar hydrocarbons using hexane–petroleum ether–diethyl ether–glacial ether (50:25:5:1). Chloroform–methanol–water (65:25:4) was used to separate polar lipids.

Lutein and beta-carotene were determined on unlaned EMD Chemicals, Inc. 20 cm × 10-cm C-18 RP-HPTLC plates with a preadsorbent zone. The mobile phase was petroleum ether–acetonitrile–methanol (1:1:2) (Evans et al. 2004).

Sugars were analyzed on Whatman LK5DF silica gel 20 cm × 20-cm TLC plates containing a preadsorbent zone and 19 lanes. The mobile phase was ethyl acetate–glacial acetic acid–methanol–water (60:15:15:10) (Wagner et al. 2001).

Analysis of amino acids required four layers having different separation mechanisms (Ponder et al. 2003a, b): EMD Chemicals, Inc. silica gel HPTLC with *n*-butanol–acetic acid–water (3:1:1) mobile phase (adsorption), EMD Chemicals, Inc. cellulose HPTLC with the same mobile phase (normal-phase partition), Whatman 20 cm × 20-cm preadsorbent C-18 with *n*-propanol–0.5 M NaCl (4:6) mobile phase (RP), and Polygram Ionex-25 SA-Na sheets (Macherey-Nagel, Bethlehem, PA, USA) with *p*H 3.3 citrate buffer mobile phase (strong acid cation exchange).

10.2.4 Standard and Sample Solution Application

The neutral lipid, phospholipid, carbohydrate, amino acid, and pigment standards were purchased from commercial sources, such as Matreya, Inc. (Pleasant Gap, PA, USA), Nu-Check Prep (Elysian, MN, USA), or Sigma (St. Louis, MO, USA), and dissolved in an appropriate solvent to prepare the standard solutions. Standard solutions and reconstituted sample extracts were generally applied to the layer in 1.00–16.0 μL aliquots. It is important that the sample zone weight applied is within the weight range of the standard calibration graph for quantitative analysis. Drummond (Brommall, PA, USA) Digital Microdispensers (10 and 25 μL) with disposable glass pipets were used to apply the initial zones of samples and standards. Solutions are applied as diffuse vertical streaks to plates containing a preadsorbent zone, and tight band-shaped initial zones are automatically formed at the preadsorbent–analytical sorbent interface. Solutions must be applied as compact spots to layers without a preadsorbent.

10.2.5 Plate Development with the Mobile Phase

Isocratic, linear, ascending development was carried out in a large volume, covered glass TLC or HPTLC chamber (normal chamber or N-chamber) that was lined with a saturation pad (Analtech, Newark, DE, USA) or piece of chromatography paper and pre-equilibrated with the vapors of the mobile phase for 10–15 min before insertion of the plate with the initial zones. Most often used was the Camag (Wilmington, NC, USA) twin trough chamber, a special N-chamber with an inverted V-shaped ridge on the bottom dividing it into two sections that allows development with a very low volume of mobile phase. The ambient temperature and humidity in the laboratory were typically 22°C and 50%, respectively.

10.2.6 Methods for Detection of Separated Zones

After removing the mobile phase from the developed plate by drying inside a fume-hood with cool air from a hair dryer, neutral lipids were detected as blue zones on a yellow background by spraying with PMA detection reagent (5%, w/v, in ethanol) and heating on a plate heater (Camag) for 10 min at 115°C. Polar lipids were detected as gray zones on a white background by spraying with 10%, w/v, aqueous cupric sulfate and heating for 10 min at 140°C (Bandstra et al. 2006) or as brown–black zones on a white background by spraying with a 10% solution of cupric sulfate in 8% phosphoric acid and heating at 140°C for 15 min (Murray et al. 2007). Lutein and beta-carotene appeared as natural yellow zones in daylight (Evans et al. 2004). Sugars were detected as dark purple zones against a light yellow background by spraying with alpha-naphthol–sulfuric acid reagent and heating at 110°C for 5 min (Wagner et al. 2001), and amino acids as purple to blue zones on a pale background by spraying with ninhydrin reagent and heating for 10 min at 110°C (Ponder et al. 2003a, b).

10.2.7 Documentation of Chromatograms

Plates were documented by color photography with a Polaroid camera or by scanning with a videodensitometer (VideoStore, Camag). Pigment zones fade rapidly and must be photographed (and quantified, if required) immediately after plate development.

10.2.8 Identification of Zones

Zones in sample chromatograms have been identified by comparison of R_F values and colors with standard zones developed on adjacent lanes of the same plate. The R_F value is defined as the distance from the origin to the center of the zone divided by the distance from the origin to the mobile-phase front.

Additional reagents can be applied to detect and confirm the identity of various zones in chromatograms. For example, ninhydrin spray reagent was used for the detection of PE and PC, and alpha-naphthol for glycolipids, including cerebrosides (Murray et al. 2007).

10.2.9 Quantification of Chromatograms

In some studies (e.g., Horutz and Fried 1995), lipid amounts were semiquantitatively evaluated by visual comparison of the sizes and intensities of the corresponding zones in adjacent chromatograms.

Accurate and precise quantification was performed using a Camag TLC Scanner II slit scanning densitometer with which standard calibration graphs were calculated by linear regression and sample weights interpolated automatically under computer control. Typical settings of the TLC Scanner II used for measuring the bands formed on the preadsorbent plates by the aliquot scanning method were slit width 4, slit length 4, and scanning speed 4 mm s^{-1}. Scanning wavelengths found to provide optimum quantitative results for the separated and detected standard and sample zones were 610 nm for neutral lipids, 370 nm for polar lipids, 610 nm for amino acids (except 495 nm for histidine), 515 nm for sugars, 448 nm for lutein, and 455 nm for beta-carotene. All wavelengths were provided by the tungsten densitometer source except for 390 nm (deuterium source).

The interpolated weight of analyte in a zone, extract reconstitution volume, volume applied, and sample weight (solid samples) or volume (blood) were used in appropriate equations to calculate the concentration of the analyte in various samples. Up to 15 different samples could be analyzed with four standards on a single plate.

10.2.10 Statistical Analysis of Data

The accuracy and the precision of densitometric quantitative data are high because calibration graphs relating scan areas versus standard weights are established under the same TLC or HPTLC conditions as for sample zones separated on the same plate. Student's *t*-test was used to determine the significance of data based on the mean +/– standard error values of analytes from different sample populations.

10.3 Thin-Layer Chromatography Applications

10.3.1 Introduction

This section describes studies of lipids, amino acids, carbohydrates, and lipophilic pigments in larval and adult echinostomes. Also considered are studies on the effects of larval echinostomes on their intermediate hosts and adult echinostomes on their definitive hosts.

10.3.2 Lipids in Echinostomes

Because lipids may serve as chemoattractants for larval trematodes, Chaffee et al. (1996) used HPTLC to study neutral lipids in SCW from *B. glabrata* snails. A major lipid fraction in SCW at 2 and 4 h after snail incubation contained FAA, and the amounts of these acids was quantified by densitometry. The amount of FFA released at 4 h was significantly greater than at 2 h. Snails also released other lipids into the water. The potential of these neutral lipids to serve as chemoattractants for larval echinostomes remains to be determined. Frazer et al. (1977) used HPTLC to identify neutral lipids and phospholipids in *E. caproni* adults removed from the intestine of experimentally infected ICR mice that had been fed a high fat diet of hen's egg yolk. The control mice were also infected with *E. caproni*, but fed a standard laboratory diet. Significantly greater amounts of TG, FS, PC, and PE were found in the worms from mice on the high fat diet compared with worms from mice on the standard diet at 2 weeks PI. The results of the study suggested that the host diet influenced the lipid content of the worms. Reddy et al. (1997) used HPTLC to study chemoattraction of the rediae of *E. trivolvis* to lipophilic ES products obtained from conspecific rediae. The study showed that such rediae were attracted to ES products and provided evidence for the role of lipophilic compounds serving as pheromones (chemoattractants) in the Trematoda. However, the potential of these neutral lipids to serve as chemoattractants for larval echinostomes remains to be determined.

Lee et al. (1998) used HPTLC to determine chemotaxonomic differences in neutral lipids in three species of adult trematodes, *E. caproni, E. trivolvis,* and *Z. lunata.* Visual observations of the chromatograms revealed some differences between the major and minor neutral lipids present in these worms. It was concluded that HPTLC can be used to distinguish taxonomic differences in adult digeneans. Rivas et al. (1998) used HPTLC to quantify neutral lipids in the intestinal mucosa of mice infected with *E. caproni.* There was a significant increase in FFA and a significant decrease in FS at 2 weeks PI in the mucosa of the infected mice compared to the controls. The increase in FFA in the intestinal mucosa of the infected mice probably resulted from alterations in the host mucosa and the release of worm ES products. Albrecht et al. (1998) used HPTLC to study the effects of *E. caproni* infection on the polar lipid content of the intestinal mucosa of mice experimentally infected with this digenean. The major lipids in both the infected and uninfected mucosa were PC and PE. There was a significant decrease in the concentrations of both PC and PE in the mucosa of infected mice compared to the uninfected controls. Glycolipids were also studied by HPTLC using the same mobile phase as for the analysis of the phospholipids but with detection by spraying with alpha-naphthol–methanol–water reagent followed by sulfuric acid. Cerebrosides and sulfatides, but not sphingolipids, were identified in both the infected and uninfected mucosa. The changes in the lipid content of the host mucosa of the infected animals probably reflected feeding and behavioral activities of the intestinal trematodes. Fried et al. (1998) used HPTLC to study the neutral lipid content of the DGG complex of *H. trivolvis* snails infected with four species of larval trematodes. Two of the species, *Ribeiroia* sp. (an echinostome-like digenean) and *Z. lunata*, contained rediae, and the other two, *Spirorchis* sp. and an armatae xiphidiocercaria, contained sporocysts. HPTLC analysis revealed that the DGGs infected by each species of trematode had their own distinct neutral lipid profile, and the infected snail profiles differed from those of the uninfected snails. HPTLC may provide a useful tool to determine chemotaxonomic differences in snails infected with larval trematodes.

Muller et al. (1999) used HPTLC to determine quantitatively several neutral lipids in the cercariae of two echinostomatid species, *E. trivolvis* and *Echinoparyphium* sp., and also discussed the possible function of lipids in larval trematodes.

Marsit et al. (2000) used HPTLC to analyze the neutral lipids in the rediae, cercariae, and encysted metacercariae of *E. caproni* removed from experimentally infected *B. glabrata* snails. Visual observations showed that the most abundant lipid fraction in all stages was FS. The concentrations of FS were determined in all stages by densitometry.

Schneck et al. (2004a) used HPTLC to analyze neutral lipids in worm incubates isotonic, hypotonic, and hypertonic to the mouse intestinal habitat of *E. caproni* adults. An array of neutral lipids was found in all incubate samples, the most abundant of which was TG. The medium that allowed for the optimal release of TG was a Locke's 0.5× solution. Excysted metacercariae maintained in Locke's solution also released neutral lipids into the medium. Densitometry was used to quantify neutral lipid release into the medium in the incubates of both the adults and the excysted metacercariae. Schneck et al. (2004b) used HPTLC to determine neutral lipids in SCW and feces from *B. glabrata* snails infected with *E. caproni*. Analysis of the SCW showed the presence of FS, FFA, and TG, but significant differences in the concentrations of these lipids from SCW from infected versus uninfected snails were not detected. The FFA fraction was the major neutral lipid detected in the feces, and the amount of this lipid class was significantly lower in the uninfected samples versus the infected samples as determined by quantitative densitometry.

Bandstra et al. (2006) examined the effects of larval echinostome infection on the neutral lipid and phospholipid content of *B. glabrata* patently infected with rediae of *E. caproni*. Uninfected snails were used as controls. The major neutral lipids in whole bodies and DGGs of both groups were FS, FFA, and TG. The major phospholipids were PC and PE. Densitometric analysis showed that the concentration of TG in infected DGGs was significantly less than that of the uninfected snails. Figure 10.1 shows the chromatograms supporting the densitometric analysis.

Bandstra et al. (2007) used HPTLC to determine the neutral lipid profiles in the feces of Balb/c mice each infected with 75 metacercarial cysts of *E. caproni*. There was a significant decrease in the TG fraction (Fig. 10.2) and a significant increase in the FS fraction (Fig. 10.3) in the feces of infected mice compared to uninfected feces. Neutral lipid profiles of infected mice may serve as indicators of intestinal trematode infections.

Murray et al. (2007) used HPTLC to determine the phospholipid and sphingolipid profiles in the feces of Balb/c mice each infected with 75 metacercarial cysts of *E. caproni* versus uninfected controls. No significant changes were found in the phospholipid or sphingolipid profiles in the feces of mice at 1–7 weeks PI compared to matched uninfected controls. The findings of this study suggested that fecal polar lipids are not useful biological markers to determine *E. caproni* infections in mice from uninfected controls.

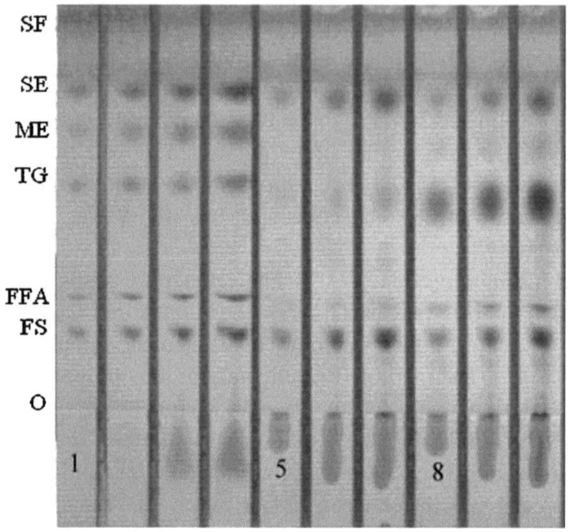

Fig. 10.1 Photograph of chromatograms on a Whatman silica gel HPTLC plate taken in white light with a Camag VideoStore documentation system showing the decrease in triacylglycerols in the infected snail DGGs (lanes 5–7) versus the uninfected snail DGGs (lanes 8–10). Lanes 1–4 contain the 18–4A neutral lipid standard spotted at 2, 4, 8, and 16 μL, respectively. SF = solvent front, SE = steryl esters, ME = methyl esters, TG = triacylglycerols, FFA = free fatty acids, FS = free sterols, O = origin. From Fig. 1 in Bandstra et al. (2006) with the kind permission of Springer Science and Business Media

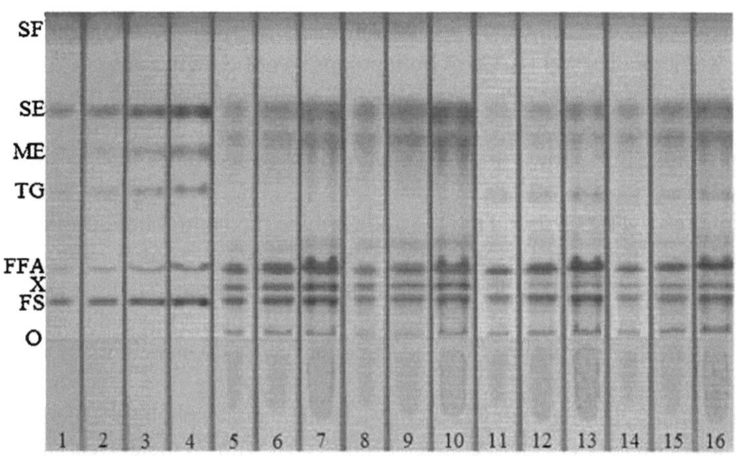

Fig. 10.2 Photograph of chromatograms on an EMD silica gel HPTLC plate taken in white light with a Camag VideoStore documentation system showing the decrease in triacylglycerols in two samples of feces from infected BALB/c mice (lanes 5–10) versus two samples of feces from uninfected BALB/c mice (lanes 11–16) at 5 weeks postinfection. Each sample was applied at 4, 8, and 16 μL. Lanes 1–4 contain the neutral lipid standard spotted at 2, 4, 8, and 16 μL, respectively. See Fig. 10.1 for abbreviations. From Fig. 2 in Bandstra et al. (2007) with the kind permission of the Taylor & Francis Group

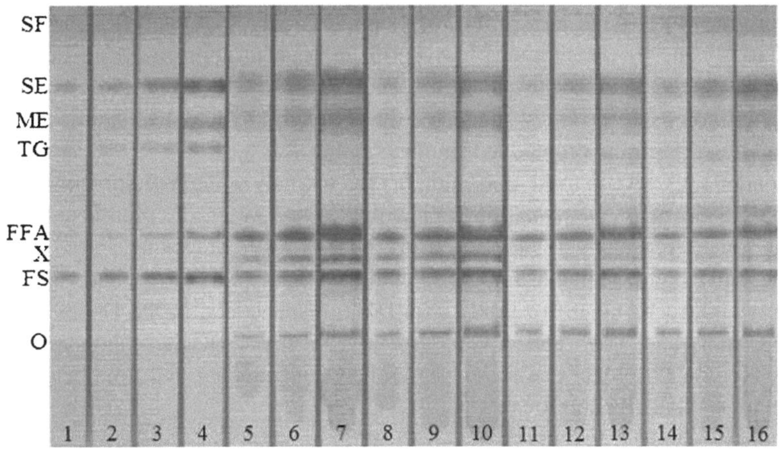

Fig. 10.3 Photograph of chromatograms on an EMD silica gel HPTLC plate taken in white light with a Camag VideoStore documentation system showing the increase in free sterols in two samples of feces from infected BALB/c mice (lanes 5–10) versus two samples of feces from uninfected BALB/c mice (lanes 11–16) at 5 weeks post infection. Each sample was applied at 4, 8, and 16 μL. Lanes 1–4 contain the neutral lipid standard spotted at 2, 4, 8, and 16 μL, respectively. See Fig. 10.1 for abbreviations. From Fig. 1 in Bandstra et al. (2007) with the kind permission of the Taylor & Francis Group

10.3.3 Amino Acids in Echinostomes

Pachuski et al. (2002) used various TLC and HPTLC systems for the analysis of amino acids in *E. caproni* adults removed from the small intestine of experimentally infected laboratory mice, *Mus musculus*. Visual observations of the amino acid chromatograms confirmed the presence of histidine, proline, alanine, leucine, and valine. Quantification of these zones by densitometry gave mean percent values as follows: histidine, 0.09; proline, 0.15; alanine, 0.08; leucine, 0.006; and valine, 0.008. Previous studies on amino acid analyses of trematodes were discussed in the paper.

Ponder et al. (2003a) used TLC and HPTLC to determine the free pool amino acids in the rediae, cercariae, encysted metacercariae, and excysted metacercariae of *E. caproni*. Larvae of each type were pooled and extracted in ethanol, and their free pool amino acids separated using four types of layers with different separation mechanisms. The zones were detected with ninhydrin spray reagent and quantified by densitometry. Qualitative analysis revealed the presence of valine, leucine, lysine, histidine, and alanine in rediae; histidine in cercariae; histidine and alanine in encysted metacercariae; and leucine in excysted metacercariae. Quantitative analysis showed that rediae contained 0.76 ng of lysine/organism and excysted metacercariae contained 0.16 ng of leucine/organism. Ponder et al. (2003b) used TLC and HPTLC to analyze the free pool amino acids in worm incubates isotonic and hypotonic to the intestinal habitat of adult *E. caproni*. Histidine, lysine, alanine, and proline were found in all the worm sample incubates. Quantification of histidine and lysine by densitometry gave a mean concentration of 24 μg histidine/g worm/mL incubate in Locke's solutions and 195 μg lysine/g worm/mL incubate in DI. The findings of the study suggested that *E. caproni* adults leak amino acids for osmoregulation in hypotonic environments.

Ponder et al. (2004) used TLC and HPTLC to analyze the free pool amino acids in the DGGs of *B. glabrata* infected with *E. caproni* versus uninfected control snails. Qualitative analysis showed the presence of histidine, lysine, serine, alanine, valine, and isoleucine in all samples. Quantitative analysis of lysine and valine gave mean weight percentages of 0.007 and 0.002, respectively, in the DGGs of uninfected snails and 0.005 and 0.003, respectively, in the DGGs of infected snails. However, the differences in these values between infected and uninfected snails were not statistically significant.

10.3.4 Carbohydrates in Echinostomes

Wagner et al. (2001) used TLC to analyze the effects of larval trematode parasitism by *E. caproni* on carbohydrates in the whole body of *B. glabrata* snails infected for 10 weeks. The major sugars found in snail whole bodies were glucose, maltose, and raffinose. At 10 weeks PI, parasitism by larval *E. caproni* caused a significant reduction in the concentrations of both maltose and raffinose in the whole bodies of *B. glabrata*. No significant reduction was observed in the concentration of glucose in the infected versus control snails at 10 weeks PI. This paper reviewed earlier studies on the effects of larval trematode parasitism on the carbohydrate content of host snails.

Pachuski et al. (2002) used TLC to analyze the presence of carbohydrates in adults of *E. caproni* removed from the small intestine of experimentally infected laboratory mice. Visual observations of the chromatograms showed the presence of glucose and raffinose. Quantification of the glucose and raffinose zones in adult samples by densitometry gave mean percent values of 0.02 and 0.01, respectively. Previous studies on carbohydrates in trematodes were discussed in this paper.

10.3.5 Lipophilic Pigments in Echinostomes

Evans et al. (2004) used HPTLC to quantify the concentration of beta-carotene and lutein in several planorbid snails infected with larval trematodes. The most important combinations were *H. trivolvis* snails naturally infected with *E. trivolvis* and laboratory-raised *B. glabrata* snails experimentally infected with *E. caproni*. Uninfected snails served as controls. The pigments were extracted from the snail whole bodies and DGGs, separated by RP HPTLC, and quantified by densitometry. Snails infected with echinostomes showed no significant differences in the concentrations of lutein and beta-carotene compared to the uninfected cohorts. Variations in the results of this study compared with other previous studies on this topic reflected intrinsic differences in the larval trematode snail systems used.

10.4 Gas Chromatography

10.4.1 Principles

GLC is a method in which a liquid sample is injected through a rubber septum into a port and vaporized onto the front end of a column. For analyses of echinostomes, isothermal or programmed temperature GLC on a packed column with an inert solid onto which a nonvolatile liquid stationary phase is adsorbed or on a WCOT capillary column has been used. As the inert carrier gas (mobile phase) flows through the column, the sample components move at different rates and reach the detector separated in time. In general, the compound with the lowest boiling point elutes first. The injection port, column, and detector are heated to keep the sample from condensing during the analysis. The FID and the MS detector have been used in echinostome analyses, the latter having the advantage of providing peak detection plus identification.

The chromatogram is a display of the detector signal as a series of peaks at different retention times for the separated compounds. Quantitative analysis is performed by generating a calibration graph of peak area versus weight of a series of standards for each analyte, by dividing the peak area of a compound by the total area of all peaks if the response factors for all analytes are equal, or correcting the detector response using predetermined response factors for each analyte.

See the chapter by Yuwono and Indrayanto (2005) for details of the theory, instrumentation, and techniques of GLC.

10.4.2 Methods and Applications

Barrett et al. (1970) tentatively identified the sterols of four intestinal helminths, including *E. trivolvis* (referred to in their paper as *E. revolutum*), and compared them with sterols in the intestinal contents of their hosts. Analysis of the FS as their TMS and acetate derivatives was carried out at 230°C using a Barber-Colman series 5000 gas chromatograph (no longer commercially available) with 3% QF-1 fluorosilicone and OV-1 methyl silicone packed columns (1.8 m × 0.5 cm id) and an FID. Sterol fractions were isolated prior to GLC by preparative TLC on silica gel H plates, both plain and impregnated with 25% silver nitrate. Cholesterol was found to be the major sterol in *E. trivolvis*, and it did not contain phytosterols as did the other three helminths. None of the helminths synthesized sterols from ^{14}C-labeled acetate or mevalonate, but *E. trivolvis* synthesized cholesterol by dealkylation of the side chain of beta-sitosterol. The evidence suggested that the intestinal helminths used in the study obtained cholesterol only from the host diet or host secretions.

Chitwood et al. (1985) reported the isolation and identification of nine sterols in addition to cholesterol from *E. trivolvis* adults using isothermal capillary column GLC and GLC/MS. Sterols were extracted with chloroform–methanol (2:1), the crude lipid extract was saponified, the sterols were isolated by silicic acid column chromatography, and GLC was on a 12.8 m × 0.32-mm DB-1 dimethylpolysiloxane nonpolar fused silica capillary column (0.25-μm film).

Fried et al. (1990) used GLC to show that the major sterol in the DGG of *H. trivolvis* infected with *E. trivolvis* was cholesterol (about 70%) along with lesser amounts of stigmasterol, beta-sitosterol, campesterol, and desmosterol. No clear distinction was seen in sterols from infected versus uninfected DGG. A Hewlett-Packard 5890 gas chromatograph (Avondale, PA) with a 30 m × 0.32-mm id SPB-1 methylsilicone WCOT column programmed from 200°C to 280°C and an FID was used. Sterols were identified based on retention times compared to standards, and percentage composition was calculated by dividing the peak area of a particular sterol by the total area of all peaks in the chromatogram.

Using the same GLC equipment and conditions, Shetty et al. (1992) studied the sterols in the DGG complex of *B. glabrata* infected with the patent larval stages of *E. caproni* versus uninfected controls. The major sterol present was cholesterol at levels of 59% in the infected snails and 51% in controls. Both populations contained the phytosterols desmosterol, campesterol, stigmasterol, and beta-sitosterol, but the percentage composition was reduced in the infected snails.

Fried et al. (1993a) did GLC studies to determine the fatty acid composition of *B. glabrata* snails experimentally infected with the intramolluskan stages of *E. caproni*. The infection was found to reduce the amounts of saturated fatty acids in whole snail bodies, but increased the amounts of these acids in the DGG complex of the snails. In both whole bodies and DGGs, infection increased markedly the amounts of palmitoleic acid (16:1n – 9) and 20:2 NMID, but reduced the amounts of 16:1n – 7, 20:1n – 11 + 9, and 22:1n – 11 + 13; DHA concentration was markedly decreased in infected whole snail bodies. Lipids were extracted by the Folch method and then treated with 7% BF_3-methanol to convert acyl lipids to ME. GLC of the fatty acid methyl esters was carried out on a Perkin–Elmer (Norwalk, CT) model 8240 gas chromatograph equipped with a 30 m × 0.25-mm id Supelcowax-10 polyethylene glycol fused silica capillary column programmed from 185°C to 230°C and an FID. Peak areas were converted to percent fatty acids by correcting the FID response using a previously developed computer program.

With the same methods for transesterification, GLC separation, peak identification, and quantification, Fried et al. (1993b) studied the fatty acid composition of the DGG complex of *H. trivolvis* infected with the intramolluskan stages of *E. caproni*, rediae freed from the DGG, uninfected DGG, and 40-day-old adult worms grown in golden hamsters. The DGG of infected snails showed significantly higher levels of stearic acid (18:0), hexatrienoic acid (16:3n – 4), and DHA (22:6n – 3) compared to uninfected snails. However, the DGG of uninfected snails showed significantly higher levels of 20:2 NMID and adrenic acid (22:2n – 6) than that of uninfected snails. The profiles of other fatty acids were very similar in infected and uninfected snails. Adult worms showed significantly higher amounts of numerous saturated fatty acids and dienes, and significantly lower amounts of certain monoenes and trienes, compared with the rediae.

10.5 Column Liquid Chromatography

10.5.1 Principles

HPLC involves high-pressure flow of a liquid mobile phase through a metal tube (column) containing the stationary phase. The sample is applied to the column through a loop injector, and separated mixture components are eluted into a detector. HPLC columns have a short length and small id compared to classical LC. Differences of HPLC compared to GLC are that the former is usually performed at ambient temperature, and the HPLC mobile phase affects the selectivity, but the GLC carrier gas does not. Instruments for both HPLC and GLC include a number of components in an integrated system. Mechanisms of separation, depending upon the type of column used, include adsorption, normal and reversed-phase partition, bonded phase, ion exchange, and size exclusion.

See the chapter by Scott (2005) for details of the theory, procedures, and instrumentation of HPLC.

10.5.2 Methods and Applications

White et al. (2005) used ion exchange HPLC to determine that the calcium carbonate content of the shells of *B. glabrata* infected with *E. caproni* was significantly lower than in the shells of uninfected *B. glabrata*. This result, and others with *H. trivolvis*, *Physa* sp., and *Schistosoma mansoni*, disproved the popular hypercalcification thesis, i.e., that larval trematodes induce an increase in the calcium concentrations in shells of their snail hosts. Calcium was extracted from shells with boiling nitric acid, and extracts were analyzed using a DX-120 ion chromatograph (Dionex, Sunnyvale, CA, USA) with an AS40 autosampler, IonPac CG12A guard column (4 mm × 50 mm), IonPac CS12A cation exchange analytical column (4 mm × 250 mm), and conductivity detector with a cation self-regenerating suppressor ultra (100 mA) to suppress background conductivity. The column was eluted isocratically with 20 mM methanesulfonic acid at a flow rate of 1.0 mL min^{-1}.

The snail host signals releasing host-finding responses in miracidia and cercariae of *E. caproni* were analyzed by fractionation of SCW (Haberl et al. 2000).

It was found that miracidia responded to a high molecular weight glycoprotein fraction that could be isolated from SCW by classical, nonhigh-performance IXC and SEC. For the isolation, IXC was carried out on a 2.5-cm id MacroPrep Q (BioRad, Hercules, CA, USA) anion exchange column (unspecified length) eluted with a linear gradient of 0–0.5 M NaCl in Tris buffer. A concentrated and desalted fraction from IXC was then chromatographed on a 66 cm × 1.6-cm Sephacryl S-500 HR SEC column (Pharmacia, Uppsala, Sweden) using 150 mM ammonium acetate buffer as the mobile phase. The results of this study showed for the first time that miracidia and cercariae of the same species may use different signals to identify the same snail host, indicating an independent evolution of host-finding mechanisms in these two larval echinostome stages.

10.6 Inductively Coupled Plasma-Optical Emission Spectrometry and Atomic Absorption Spectrometry

10.6.1 Principles

ICP-OES and AAS are atomic spectrometric methods used to determine elemental analytes in samples. ICP-OES is an emission method in which the source is a very hot plasma of ionized argon, into which samples and standards are introduced. Atoms are ionized and emissions are measured using a sequential, monochromator design or a simultaneous multichannel, polychromator or array-based design to isolate different wavelengths, and one or more PMT detectors. The wavelength is characteristic of a particular element, and the intensity of the emission indicates the concentration of the element in the sample. An article by Sherma (2004) gives further information on the techniques and instrumentation for ICP-OES.

In AAS a liquid sample is aspirated into the flame of an FAAS instrument or pipetted into the graphite tube of a GFAAS instrument at high temperature, and the resulting unexcited gaseous analyte atoms absorb light emitted from a hollow cathode lamp and passed through the flame or furnace. The lamp emits exactly the best wavelength required for the analysis. A PMT detector converts the amount of light reaching it into an electrical signal that is read out as absorption. The analyte concentration in the sample is determined from its amount of absorption using a calibration graph prepared using standards of known concentration. See the paper by Sherma (2002) for further information on the procedures and instrumentation of AAS.

10.6.2 Methods and Applications

Layman et al. (1999) used ICP-OES to study the metallic ions in the intestinal mucosa of ICR mice infected with *E. caproni* and the mucosa of uninfected control mice. Samples were rinsed with ultrapure water and then digested in boiling nitric acid. After dilution, each digested sample was analyzed for 28 elements using a Thermo Jarrell Ash (Franklin, MA, USA) simultaneous reading spectrometer with autosampler. The instrument was calibrated following US EPA Method 610A, which involves a two-point calibration, reagent blanks, and multielement standards. A number of quality control checks were made during analyses to verify the calibration graph, blank, and interelement correction factors that were employed to minimize any interferences between elements in samples. At 2 weeks PI in mice infected with ca. 25 worms/host, five elements were detected above the level of quantification: Ca, K, Mg, Na, and Zn. There were no significant differences using Student's *t*-test ($P > 0.05$) in the concentrations of Ca, K, or Zn in infected versus uninfected mucosa. The concentration of Na was significantly greater, and of Mg was significantly lower, in infected versus uninfected mucosa.

Layman et al. (1996) used ICP-OES, FAAS, and GFAAS to study the metallic ions in the DGG of *H. trivolvis* snails infected with the daughter rediae of *E. trivolvis* and of uninfected DGG. Seven metals were found to be present at or above the detection levels of the methods: Ca, Cu, Fe, Mg, Mn, Na, and Zn. Of these, Na was present in significantly higher amounts in the infected versus uninfected DGG; Mg and Mn occurred in significantly lower amounts in the infected DGG. Twenty-four elements were determined by ICP-OES using the same instrument and methods described earlier. FAAS was performed using a Varian Inc. (Walnut Creek, CA, USA) AA-10 spectrometer to determine Mg, Na, and Zn concentrations sequentially using a different, specific hollow cathode source lamp; the flame and other parameters, such as monochromator wavelength, flame temperature, and sample aspiration rate, were optimized according to the manufacturer's recommendations. Four aqueous standards having analyte concentrations within the linear response range of the instrument and containing the same concentration of nitric acid as in the sample extracts were used for calibration. Each sample, standard, and blank were analyzed using three 10-s integrations. A reagent blank was prepared and its value was subtracted to give the final concentration value. GFAAS was done with a Buck Scientific, Inc. (Norwalk, CT, USA) Model 210 spectrometer with a deuterium lamp background corrector to determine Cu, the concentration of which was below the sensitivity limit of FAAS. Samples, standards, and blanks were placed into the furnace with a 50-μL Eppendorf micropipet, and other conditions were as for FAAS.

10.7 Concluding Remarks

Most of the studies reported here have used TLC and HPTLC for the analyses of larval and adult echinostomes and for tissues infected with these digeneans. These studies have provided considerable information on the qualitative and quantitative analyses of various organic compounds including neutral lipids, phospholipids, glycolipids, amino acids, sugars, and lipophilic pigments such as lutein and beta-carotene. Changes in tissues infected with echinostomes have documented changes in the pathobiochemical effects of the infection on the host. The atomic spectrometry studies have provided new information on elemental analyses of echinostomes and on the pathochemical effects of echinostome infection on host tissues. Less information is available on GLC and HPLC analysis of echinostomes. Further work using these analytical tools may provide much needed information on the molecular species of compound classes, information which is not easily obtained by TLC and HPTLC alone.

Acknowledgment We thank Mr. Robert C. Peoples for assistance in preparing this manuscript.

References

Albrecht, B.K., Fried, B., and Sherma, J. 1998. Effects of *Echinostoma caproni* infection on the phospholipid and sphingolipid content of the intestinal mucosa of ICR mice. Journal of Helminthology 72: 355–357.

Bandstra, S.R., Fried, B., and Sherma, J. 2006. High performance thin layer chromatographic analysis of neutral lipids and phospholipids in *Biomphalaria glabrata* patently infected with *Echinostoma caproni*. Parasitology Research 99: 414–418.

Bandstra, S.R., Murray, K.E., Fried, B., and Sherma, J. 2007. High performance thin layer chromatographic analysis of neutral lipids in the feces of BALB/c mice infected with *Echinostoma caproni*. Journal of Liquid Chromatography and Related Technologies 30: 1437–1445.

Barrett, J., Cain, G.D., and Fairbairn, D. 1970. Sterols in *Ascaris lumbricoides* (Nematoda), *Macracanthorhynchus hirudinaceus* and *Moniliformis dubius* (Acanthocephala), and *Echinostoma revolutum* (Trematoda). Journal of Parasitology 56: 1004–1008.

Chaffee, L.A., Fried, B., and Sherma, J. 1996. Neutral lipids in snail-conditioned water from *Biomphalaria glabrata* (Gastropoda:Planorbidae). Journal of Chemical Ecology 22: 231–235.

Chitwood, D.J., Lusby, W.R., and Fried, B. 1985. Sterols of *Echinostoma revolutum* (Trematoda) adults. Journal of Parasitology 71: 846–847.

Evans, R.T., Fried, B., and Sherma, J. 2004. Effect of diet and larval trematode parasitism on lutein and beta-carotene concentrations in planorbid snails as determined by quantitative high performance thin layer chromatography. Comparative Biochemistry and Physiology 137B: 179–186.

Frazer, B.A., Reddy, A., Fried, B., and Sherma, J. 1997. Effects of diet on the lipid composition of *Echinostoma caproni* (Trematoda) in ICR mice. Parasitology Research 83: 642–645.

Fried, B. and Haseeb, M.A. 1996. Thin layer chromatography in parasitology. In: Practical Thin Layer Chromatography, eds. B. Fried and J. Sherma, pp. 51–70. Boca Raton: CRC.

Fried, B., Frazer, B.A., Lee, M.S., and Sherma, J. 1998. Thin layer chromatography and histochemistry analyses of neutral lipids in *Helisoma trivolvis* infected with four species of larval trematodes. Parasitology Research 84: 369–373.

Fried, B., Holender, E.S., Shetty, P.H., and Sherma, J. 1990. Effects of *Echinostoma trivolvis* (Trematoda) infection on neutral lipids, sterols, and carotenoids in *Helisoma trivolvis* (Gastropoda). Comparative Biochemistry and Physiology 97B: 601–601.

Fried, B., Rao, K.S., Sherma, J., and Huffman, J.E. 1993b. Fatty acid composition of *Echinostoma trivolvis* (Trematoda) rediae and adults of the digestive gland–gonad complex of *Helisoma trivolvis* (Gastropoda) infected with the intramolluscan stages of this echinostome. Parasitology Research 79: 471–474.

Fried, B., Sherma, J., Rao, K.S., and Ackman, R.G. 1993a. Fatty acid composition of *Biomphalaria glabrata* (Gastropoda: Planorbidae) experimentally infected with the intramolluscan stages of *Echinostoma caproni* (Trematoda). Comparative Biochemistry and Physiology 104B: 595–598.

Haberl, B., Korner, M., Spengler, Y., Hertel, J., Kalbe, M., and Haas, W. 2000. Host-finding in *Echinostoma caproni*: miracidia and cercariae use different signals to identify the same snail species. Parasitology 120: 479–486.

Horutz, K. and Fried, B. 1995. Effects of *Echinostoma caproni* infection on the neutral lipid content of the intestinal mucosa of ICR mice. International Journal for Parasitology 25: 653–655.

Layman, L.R., Dory, A.C., Koehnlein, K.M., Fried, B., and Sherma, J. 1996. Effects of *Echinostoma trivolvis* (Trematoda) infection on metallic ions in the host snail *Helisoma trivolvis* (Gastropoda). Parasitology Research 82: 19–21.

Layman, L.R., Mueller, T.J., Koehnlein, K.M., Fried, B., and Sherma, J. 1999. Effects of *Echinostoma caproni* infections on metallic ions in the intestinal mucosa of ICR mice. Journal of Helminthology 73: 367–368.

Lee, M.S., Fried, B., and Sherma, J. 1998. High performance thin layer chromatographic analysis of neutral lipids in *Echinostoma caproni*, *Echinostoma trivolvis*, and *Zygocotyle lunata* (Platyhelminthes: Trematoda). Journal of Planar Chromatography-Modern TLC 11: 105–107.

Marsit, C.J., Fried, B., and Sherma, J. 2000. Neutral lipids in cercariae, encysted metacercariae, and rediae of *Echinostoma caproni*. Journal of Helminthology 74: 365–367.

Muller, E.E., Fried, B., and Sherma, J. 1999. HPTLC determination of neutral lipids in the cercariae of *Echinostoma trivolvis* and *Echinoparyphium* sp. Journal of Planar Chromatography-Modern TLC 12: 306–308.

Murray, K.E., Fried, B., and Sherma, J. 2007. Determination of phospholipid and sphingolipid content in the feces of uninfected BALB/c mice and those infected with *Echinostoma caproni* by high performance silica gel thin layer chromatography. Acta Chromatographica 18: 190–198.

Pachuski, J., Wagner, S.D., Fried, B., and Sherma, J. 2002. Thin layer chromatographic analyses of amino acids and carbohydrates in adults of *Echinostoma caproni*. Comparative Parasitology 69: 202–205.

Ponder, E.L., Fried, B., and Sherma, J. 2003a. Thin layer chromatographic analysis of free pool amino acids in cercariae, rediae, encysted metacercariae and excysted metacercariae of *Echinostoma caproni*. Journal of Liquid Chromatography and Related Technologies 26: 2697–2702.

Ponder, E.L., Fried, B., and Sherma, J. 2003b. Effects of hypotonicity on amino acid release in adult *Echinostoma caproni* as determined by thin layer chromatography. Parasitology Research 89: 242–244.

Ponder, E.L., Fried, B., and Sherma, J. 2004. Free-pool amino acids in *Biomphalaria glabrata* infected with *Echinostoma caproni* as determined by thin layer chromatography. Journal of Parasitology 90: 665–666.

Reddy, A., Frazer, B.A., Fried, B., and Sherma, J. 1997. Chemoattraction of *Echinostoma trivolvis* (Trematoda) rediae to lipophilic excretory–secretory products and thin layer chromatographic analysis of redial lipids. Parasite-Journal de la Societe Francaise de Parasitologie 4: 37–40.

Rivas, F., Sudati, J., Fried, B., and Sherma, J. 1998. HPTLC analysis of neutral lipids in the intestinal mucosa, serum, and liver of ICR mice infected with *Echinostoma caproni* (Trematoda) and of worm excretory–secretory products. Journal of Planar Chromatography-Modern TLC 11: 47–50.

Schneck, J.L., Fried, B., and Sherma, J. 2004a. Effects of tonicity on the release of neutral lipids in *Echinostomoa caproni* adults and observations on lipids in excysted metacercariae. Parasitology Research 92: 285–288.

Schneck, J.L., Bandstra, S.R., Fried, B., and Sherma, J. 2004b. Thin layer chromatographic analysis of neutral lipids in snail conditioned water and feces of *Biomphalaria glabrata* infected with *Echinostoma caproni*. Journal of Liquid Chromatography and Related Technologies 27: 2039–2045.

Scott, R.P.W. 2005. Instrumentation for high performance liquid chromatography. In: Ewing's Analytical Instrumentation Handbook, ed. J. Cazes, pp. 687–726. New York: Marcel Dekker.

Sherma, J. 2002. A field guide to instrumentation: graphite furnace atomic absorption spectrometry (GFAAS). Inside Laboratory Management 6(3): 18–22.

Sherma, J. 2004. A field guide to instrumentation: inductively coupled plasma-optical emission spectrometry and mass spectrometry. Journal of AOAC International 87: 56A–62A.

Sherma, J. and Fried, B. 2003. Handbook of Thin Layer Chromatography, 3rd Edition, (Eds.: J. Sherma and B. Fried). New York: Marcel Dekker.

Shetty, P.H., Fried, B., and Sherma, J. 1992. Effects of *Echinostoma caproni* infection on the sterol composition of the digestive gland–gonad complex of *Biomphalaria glabrata* as determined by gas liquid chromatography. Journal of Helminthology 66: 68–71.

Wagner, S.D., Kim, Y., Fried, B., and Sherma, J. 2001. Thin layer chromatographic analysis of glucose, maltose, and raffinose in *Biomphalaria glabrata* (Gastropoda) infected with *Echinostoma caproni* (Trematoda). Journal of Planar Chromatography-Modern TLC 14: 459–461.

White, M.M., Chejlava, M., Fried, B., and Sherma, J. 2005. Effects of various larval digeneans on the calcium carbonate content of the shells of *Helisoma trivolvis*, *Biomphalaria glabrata*, and *Physa* sp. Parasitology Research 95: 252–255.

Yuwono, M., and Indrayanto, G. 2005. Gas chromatography. In: Ewing's Analytical Instrumentation Handbook, ed. J. Cazes, pp. 727–758. New York: Dekker.

Chapter 11
Effects of Environmental Change on Helminth Infections in Amphibians: Exploring the Emergence of *Ribeiroia* and *Echinostoma* Infections in North America

Pieter T. J. Johnson and Valerie J. McKenzie

Contents

11.1 Introduction 250
11.2 Overview of *Ribeiroia* 253
 11.2.1 History of Frog Deformities and the Link to Trematode Infection 253
 11.2.2 Are Parasite-Induced Malformations a Problem? 259
 11.2.3 Emergence of *Ribeiroia* and Malformations 262
11.3 Overview of Echinostomes 264
 11.3.1 Pathology and Patterns of Infection in Amphibians 265
 11.3.2 Emergence of *Echinostoma* 268
11.4 Possible Causes of Parasite Emergence 269
11.5 Concluding Remarks 273
References 275

Abstract Amphibians have long served as model organisms for studying animal physiology, vertebrate anatomy, and host–parasite interactions. Recently, however, the occurrence of precipitous declines in many amphibian populations and of severe limb malformations in others has catalyzed renewed efforts to understand the effects of parasites on amphibians. In this brief review, we examine the importance of two groups of trematodes that utilize amphibians as intermediate hosts: species in the genus *Ribeiroia* and the broader "echinostome" group which collectively includes the genera *Echinostoma* and *Echinoparyphium*. For each, we specifically explore the pathology resulting from infection, whether the parasite has recently increased in abundance or geographic range, and the biotic and abiotic factors likely to influence infection. Both groups of parasites can induce significant pathology in amphibian hosts. Exposure to *Ribeiroia* cercariae causes substantial increases in mortality and limb malformations in larval amphibians. These malformations, which include missing, malformed and extra limbs, may further reduce survival in amphibians; malformations are extremely rare in adult frogs, even following

P.T.J. Johnson (✉) and V.J. McKenzie
Ecology and Evolutionary Biology Department, University of Colorado,
Ramaley N122, CB 334, Boulder, CO 80309-0334, USA
e-mail: pieter.johnson@colorado.edu

B. Fried and R. Toledo (eds.) *The Biology of Echinostomes*,

DOI: 10.1007/978-0-387-09577-6_11,

years in which they are abundant (>50%) among juvenile frogs. Similarly, the echinostomes, which colonize the kidneys of amphibians, can reduce the survival and increase the incidence of edema and renal failure, particularly in laboratory experiments. Recent surveys of National Wildlife Refuges across the USA suggest that both groups of parasites are widespread and sometimes extremely abundant (~1,000 metacercariae per frog). Infections appear to be most common along major rivers and bird flyways in the northern half of the country. While limited evidence suggests a recent increase in amphibian malformations and *Ribeiroia* infection, the paucity of available historical data precludes a definitive assessment of whether either parasite group has recently emerged. We discuss future approaches to this question and explore contemporary ecological changes known or hypothesized to influence patterns of infection, including changes in land use, increases in nutrient and pesticide runoff, decreases in community diversity and shifts in climate. Considering the documented pathologies of each parasite group, their widespread and often abundant infection patterns, and the ongoing declines observed in amphibian populations, we emphasize the urgent need for further study of *Ribeiroia* and echinostome infections in amphibians.

11.1 Introduction

For more than 50 years, amphibians have served as model hosts for studying the intricate nature of host–trematode interactions (Ingles 1933; Brandt 1936; Smyth and Smyth 1980; Prudhoe and Bray 1982). The first isolation of a trematode from an amphibian occurred in 1737, and a steady number of new species have been recovered over the subsequent centuries (Prudhoe and Bray 1982). The complex life cycles of many amphibians, which involve remarkable shifts in habitat (aquatic to terrestrial), diet (herbivory to carnivory), and morphology (tail to limbs), mirror the complexity of trematode life cycles. Correspondingly, amphibians can act as both intermediate and definitive hosts for a broad diversity of parasites. To date, hundreds of species of digenetic trematodes have been identified that depend on frogs, toads, or salamanders hosts (Prudhoe and Bray 1982). While the bulk of research has focused on amphibian as definitive hosts, amphibians also support larval stages (meso- and metacercariae) of more than 200 trematode species, and this almost certainly represents a considerable underestimate. Because metacercariae are often smaller than adult parasites, distributed throughout diverse regions of the host's body, and difficult to excyst and identify, they have traditionally received less attention than adult parasites. Often they have been ignored altogether.

More recently, renewed attention has focused on the potential importance of larval trematodes within amphibians. This trend is the result of several factors. First, while it was once supposed that trematode metacercariae caused little pathology in their intermediate hosts, waiting almost "patiently" in a resting form, there is growing recognition that this stage may be among the most pathogenic. Because transmission of metacercariae to the definitive host often depends on predation (trophic transmission), there is strong selective pressure on trematodes to increase the susceptibility of intermediate hosts to predation. Examples of parasites that alter the susceptibility of intermediate hosts to predation by definitive hosts are evident across several phyla of parasites, including platyhelminths, nematodes, acanthocephalans, and apicomplexans (Moore 2002). These specific pathologies can include behavioral alteration, physical or developmental modification, and/or significant morbidity all of which can act as signals to definitive hosts of an easy meal. Among the trematodes, for instance, cercariae of *Euhaplorchis californiensis* encyst on the brain of killifishes by the thousands and induce erratic swimming behaviors which attract bird predators (Lafferty and Morris 1996). Recent experimental work by Shaw et al. (submitted) demonstrated that these metacercariae actually cause changes in monoaminergic brain chemical activities in killifish which can directly affect locomotory behavior.

Second, amphibians are now at the forefront of the global biodiversity crisis. With one-third of all species in decline and >100 species extinct or suspected extinct in the last 25 years, amphibians are considered the most imperiled class of vertebrates worldwide (Houlahan et al. 2000; Stuart et al. 2004). The causes of such declines are diverse, including habitat loss, infectious disease, invasive species, and climate change, and may interact through complex mechanisms (Blaustein and Kiesecker 2002). The emergence of the pathogen *Batrachochytrium dendrobatidis*, for example, has led to devastating population losses and extinction of amphibian populations worldwide, and is considered one of the greatest threats to global amphibian diversity (Johnson 2006; Skerratt et al. 2007). Gaining insight into how environmental drivers exacerbate disease threats to amphibians is a critical research area for amphibian conservation. By examining the environmental drivers that influence pathogenic trematode parasites, we can expand this field of study to address other types of amphibian disease as well.

Third and finally, trematode metacercariae have recently been linked to severe pathology in amphibians. Most notably, the digenean *Ribeiroia ondatrae* is a widespread and important cause of limb deformities in amphibians, including missing, extra, and misshapen limbs (Fig. 11.1a; Blaustein and Johnson 2003). These deformities have attracted international attention and generated considerable controversy over the likely causes and implications of the phenomenon (see Kaiser 1997, 1999; Souder 2000). Affected individuals suffer severely restricted mobility, and frequently die soon after metamorphosis. The link between *Ribeiroia* infection and malformations has been substantiated through extensive field surveys and controlled experiments, but the factors that control infection abundance and the apparent increase in deformities remain speculative (see later). Alongside *Ribeiroia*, other trematodes, including members of the echinostomatids (e.g., *Echinostoma* and *Echinoparyphium*), have been suggested to impair amphibian fitness. Because metacercariae aggregate in the kidneys of infected amphibians and may number several thousand within a single frog, these parasites have the potential to inhibit renal function and reduce survival (Fried et al. 1997; Schotthoefer et al. 2003; Skelly et al. 2006; Holland et al. 2007). As both *Ribeiroia* and the echinostomes

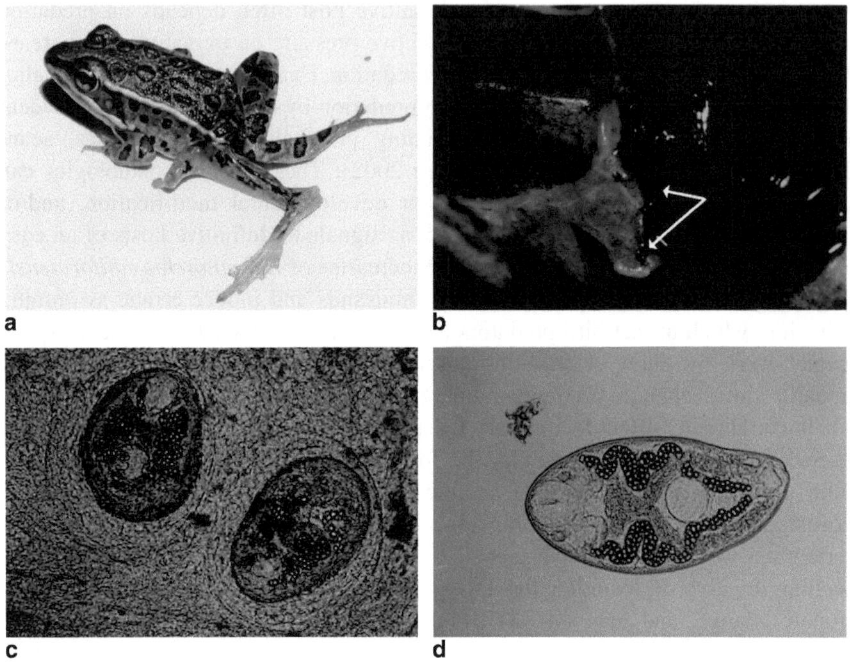

Fig. 11.1 (**a**) Representative malformation associated with *Ribeiroia* infection in a Northern leopard frog (*Rana pipiens*); (**b**) Dorsal view of hind limb and tail resorbtion area of a malformed frog; melanized metacercariae of *Ribeiroia* appear as brown cysts (see arrows); short projection on the left side represents a small supernumerary limb; (**c**) In situ view of two encysted *Ribeiroia* metacercariae within amphibian tissue; (**d**) Excysted metacercariae with visible esophageal diverticula

depend on predation of infected amphibians by vertebrate definitive hosts for transmission, pathology that increases the vulnerability of amphibians to predators could enhance transmission.

In this chapter, our objectives are to (1) review and assess the effects of *Ribeiroia* and *Echinostoma* on their amphibian hosts, (2) evaluate available evidence of an increase in the abundance, distribution, or pathology of each parasite, and (3) explore the environmental factors that influence amphibian infection levels and how they are expected to change in the future. We do not intend this as a comprehensive review for either parasite group, as such detailed information is beyond the scope of this chapter. Throughout, we refer readers instead to other sources for more comprehensive accounts of these groups' taxonomy, physiology, ecology, and pathology. It is our aim that this distillation will help catalyze additional research to address pressing gaps in our knowledge of *Ribeiroia* and *Echinostoma* and their ecological significance. In the sections that follow, we discuss the life cycles and pathology associated with each parasite, assess historical and contemporary evidence of emergence, and evaluate the most likely biotic and abiotic factors to drive changes in infection. Because metacercariae isolated from the kidneys of amphibians are difficult if not impossible to identify on the basis of morphology alone, we suspect many previous investigations (including our own) have inadvertently confused members of the echinostomatids. Accordingly, we make no attempt here to differentiate among these groups, and elect instead to combine our discussions of *Echinostoma trivolvis*, *Echinostoma revolutum*, and *Echinoparyphium* spp. under the broad category of echinostomatids.

11.2 Overview of *Ribeiroia*

Trematodes in the genus *Ribeiroia* (Family Psilostomidae) have complex life cycles involving planorbid snails as first intermediate hosts (genera *Planorbella* and *Biomphalaria*), fish or amphibians as second intermediate hosts, and birds or mammals as definitive hosts (see Table 11.1). While this group's taxonomy will likely require revision, three species are currently recognized: *R. ondatrae*, *R. marini*, and *R. congolensis*. All share a number of morphological similarities, including distinctive esophageal diverticula (Fig. 11.1d; see Johnson et al. 2004 for a full review of this group). These species have been recorded from North and South America (*R. ondatrae*), the Caribbean (*R. marini*), and parts of Africa (*R. congolensis*). Because little is known about the species of *Ribeiroia* that occur outside the US, our focus here is exclusively on *R. ondatrae*, which has recently been linked to severe limb malformations in North American amphibians (Johnson et al. 1999; Blaustein and Johnson 2003; Sutherland 2005). Free-swimming cercariae released from infected snails colonize and infect larval amphibians, wherein they encyst in and around the developing limb buds (Fig. 11.1b,c). The resulting physical and/or chemical disruption of limb growth can lead to dramatic malformations, ranging from complete limb suppression to the induction of multiple extra limbs (Fig. 11.1a). These malformations may benefit the parasite by increasing the vulnerability of infected amphibians to definitive host predation (parasite manipulation). However, some reports suggest that the frequency and severity of these deformities may have increased in recent years (e.g., Johnson et al. 2003; Johnson and Lunde 2005), possibly in association with environmental change. Our goal is to explore the link between *Ribeiroia* and malformations, evaluate the potential importance of parasite-induced malformations for amphibians, and examine available evidence for a recent increase in infection and the resulting pathology.

11.2.1 History of Frog Deformities and the Link to Trematode Infection

Recent interest in amphibian malformations is often traced back to the Ney Pond near Henderson, Minnesota (e.g., Souder 2000). In August 1995, a group of middle school children discovered that nearly half of the emerging leopard frogs (*Rana*

Table 11.1 Hosts involved in the life cycles of *Riberoia ondatrae*, *Echinostoma trivolvis*, and *E. revolutum* in North America

Ribeiroia ondatrae			Echinostomatid species	
Life cycle stage	Common name[1]	Species	Common name	Species
First intermediate hosts	Ramshorn snails	*Biomphalaria obstructa*	Ramshorn snail[2]	*Planorbella trivolvis*
		Helisoma antrosum	Marsh pond snail[11,b]	*Lymnaea elodes*
		Planorbella campanulata		
		Planorbella occidentalis		
		Planorbella subcrenatum		
		Planorbella tenue		
		Planorbella trivolvis		
Second intermediate hosts	**Fish:**		**Planarians**[3]	
	Rock bass	*Ambloplites rupestris*	**Molluscs:**	
	Yellow perch	*Perca flavescens*	Various pulmonate and	*Biomphalaria glabrata*
	Amphibians:		prosobranch snails, mussels[3]	*Planorbella spp.*
	Northern cricket frog[13]	*Acris crepitans*	Ramshorn snails[11,b]	*Lymnaea spp.*
	Long-toed salamander	*Ambystoma macrodactylum*	Pond snails[11,b]	*Physa spp.*
	Blue-spotted salamander	*Ambystoma laterale*	**Fish:**	
	Tiger salamander	*Ambystoma tigrinum*	Various fishes[3]	
	Western toad	*Bufo boreas*	Various fishes[2,b]	
	American toad	*Bufo americanus*	**Amphibians:**	
	Pacific chorus frog	*Pseudacris regilla*	Northern cricket frog[4]	*Acris crepitans*
	Northern red-legged frog	*Rana aurora*	Blue-spotted salamander[13]	*Ambystoma laterale*
	Plains leopard frog[13]	*Rana blairi*	Long-toed salamander[13]	*Ambystoma macrodactylum*
	Cascades frog	*Rana cascadae*	Tiger salamander[5]	*Ambystoma tigrinum*
	American bullfrog	*Rana catesbeiana*	American toad[6]	*Bufo americanus*
	Green frog	*Rana clamitans*	Western toad[13]	*Bufo boreas*
	Colombia spotted frog	*Rana luteiventris*	Woodhouse's toad[13]	*Bufo woodhousei*
	Pickerel frog	*Rana palustris*	Pacific chorus frog[13]	*Pseudacris regilla*
	Northern leopard frog	*Rana pipiens*	Western chorus frog[5]	*Pseudacris triseriata*
	Oregon spotted frog	*Rana pretiosa*	Northern red-legged frog[13]	*Rana aurora*
	Mink frog	*Rana septentrionalis*	American bullfrog[7]	*Rana catesbeiana*
	Southern leopard frog[13]	*Rana sphenocephala*	Plains leopard frog[13]	*Rana blairi*
	Wood frog	*Rana sylvatica*	Green frog[7]	*Rana clamitans*

	Great Basin spadefoot	*Scaphiopus intermontanus*	Columbia spotted frog[13]	*Rana luteiventris*
	Rough-skinned newt	*Taricha granulosa*	Pickerel frog[13]	*Rama palustris*
	California newt	*Tarchica torosa*	Northern leopard frog[7]	*Rana pipiens*
			Mink frog[13]	*Rana septentrionalis*
			Southern leopard frog[13]	*Rana sphenocephala*
			Wood frog[13]	*Rana sylvatica*
			Gray treefrog[12]	*Hyla versicolor*
			Rough-skinned newt[13]	*Taricha torosa*
			Various amphibians[2,b]	
			Reptiles:	
Definitive hosts	**Birds:**	**Birds**	Freshwater turtles[3]	
	Cooper's hawk	*Accipiter cooperi*	Cooper's hawk[8]	*Accipiter cooperi*
	Western grebe	*Aechmophorus occidentalis*	Sharp-shinned hawk[8]	*Accipiter striatus*
	Wood duck	*Aix sponsa*	Northern goshawk[8]	*Accipiter gentilis*
	Mallard	*Anas platyrhynchos*	Northern saw-whet owl[8]	*Aegolius acadicus*
	Anhinga	*Anhinga anhinga*	Boreal owl[8]	*Aegolius funereus*
	Domestic goose	*Anser domesticus*	Common pintail[2]	*Anas acuta*
	Great egret	*Ardea alba*	American widgeon[2]	*Anas americana*
	Great blue heron	*Ardea herodias*	Domestic duck[2]	
	Great-horned owl	*Bubo virginianus*	Domestic duck[2,b]	*Anas borchas*
	Red-tailed hawk	*Buteo jamaicensis*	Cinnamon teal[2]	*Anas cyanoptera septentrionalium*
	Broad-winged hawk	*Buteo platypterus*	Blue-winged teal[2]	*Anas discors*
	Little blue heron	*Egretta caerulea*	Eurasian widgeon[2]	*Anas penelope*
	Redish egret	*Egretta rufescens*	Mallard[2]	*Anas platyrhynchos*
	Louisiana heron	*Egretta tricolor*	Black duck[2]	*Anas rubripes*
	American coot	*Fulica americana*	Australian black duck[2]	*Anas superciliosa rogersi*
	Domestic chicken	*Gallus domesticus*	Emden goose[2]	*Anser sp.*
	Common loon	*Gavia immer*	Graytag goose[2]	*Anser anser*
	Whooping crane	*Grus americana*	Magpie goose[2]	*Anseranus semipalmata*
	Bald eagle	*Haliaeetus leucocephalus*	Short-eared owl[8]	*Asio flammeus*
	Herring gull	*Larus argentatus*	Long-eared owl[8]	*Asio otus*
	California gull	*Larus californicus*	Lesser scaup[2]	*Aythya affinis*
	Common merganser	*Mergus merganser*	Redhead[2]	*Aythya americana*

(continued)

Table11.1 (continued)

Ribeiroia ondatrae			Echinostomatid species	
Life cycle stage	Common name[1]	Species	Common name	Species
	Osprey	*Pandion haliaetus*	Ring-necked duck[2]	*Aythya collaris*
	American white pelican	*Pelecanus erythrorhynchus*	Greater scaup[2]	*Aythya marila*
	Double-crested cormorant	*Phatacrocorax auritus*	Great-horned owl[2]	*Bubo virginianus*
	Red-necked grebe	*Podiceps grisegena*	Red-tailed hawk[8]	*Buteo jamaicensis*
	Pied-billed grebe	*Podilymbus podiceps*	Rough-legged hawk[2]	*Buteo lagopus*
	Mammals:		Broad-winged hawk[8]	*Buteo platypterus*
	Raccoon	*Procyon lotor*	Northern harrier[8]	*Circus cyaneus*
	Muskrat	*Ondatra zibethica*	Muscovy[2]	*Cairina moschata*
	Badger	*Taxidea taxus*	Domestic pidgeon[2,a]	
			Domestic pidgeon[2,b]	*Columba livia*
			Fulvous whistling duck[9]	*Dendrocygna bicolor*
			Peregrine falcon[8]	*Falco peregrinus*
			America kestrel[8]	*Falco sparverius*
			Domestic chicken[2,a]	
			Domestic chicken[2,b]	*Gallus domesticus*
			Blackheaded munia[2,a,b]	*Lonchura ferruginosa*
			Spotted munia[2,b]	*Lonchura punctulata*
			Little cuckoo dove[2,a,b]	*Macropygia ruficeps*
			Common scoter[2]	*Oidemia nigra*
			Long-eared owl[8]	*Otus asio*
			Java sparrow[2,b]	*Padda oryzivora*
			Osprey[2]	*Pandion haliaetus*
			American flamingo[2]	*Phoenicopterus ruber*
			Common grackle[2]	*Quiscalus quiscula*
			Great gray owl[8]	*Strix nebulosa*
			Barred owl[8]	*Strix varia*
			Mourning dove[2]	*Zenaida macroura*
			Mammals:	
			Dog[2,a]	*Canis familiaris*
			Guinea pig[2,a]	*Cavia porcellus*

Opossum[10]	*Didelphis virginiana*
Cat[2,a]	*Felis catus*
Golden hamster[2,a]	*Mesocricetus auratus*
House mouse[2,a]	*Mus musculus*
Muskrat[2]	*Ondatra zibethica*
Rabbit[2,a]	*Oryctolagus cuniculus*
Norway rat[2,a]	*Rattus norvegicus*
Pig[2,a]	*Sus scrofa*
Red fox[2]	*Vulpes vulpes*

Most of the life cycle information presented here for *R. ondatrae* was previously summarized in Johnson et al. (2004) (1) and life cycle information presented here for *Echinostoma* species was previously summarized in Huffman and Fried (1990) (2). Additional references are denoted by superscript numbers following the common names of hosts: (3) Kanev et al. (1995), (4) Beasley et al. (2005), (5) Johnson and McKenzie (unpublished data from Colorado), (6) Ulmer (1970), (7) McAlpine and Burt (1998), (8) Taft et al. (1993), (9) Forrester et al. (1994), (10) Alden (1995), (11) Sorensen et al. (1997), (12) Koprivnikar et al. (2006), (13) Sutherland and Johnson (unpublished)

[a] Indicates a host that was experimentally infected; all others are recorded from natural host infections

[b] Indicates a host species record for *E. revolutum*, while all other hosts listed in the echinostomatids column are records for *Echinostoma trivolvis*

pipiens) around this pond suffered severe limb malformations, including extra, missing, and misshapen limbs. Although some abnormalities are expected in any vertebrate population, this baseline frequency is generally low (0–5%), leading researchers to suspect that something unusual was happening at the pond. The intensity of the media's response to this discovery, coupled with the speed with which images and information were disseminated via the Internet, stimulated a surge of malformation surveys in North American amphibians, ultimately resulting in numerous new reports of deformed amphibians.

Many factors can cause malformations in amphibians (see Ouellet 2000 for a review), and it is likely that the current "phenomenon" is actually the product of multiple etiological factors. Nevertheless, growing evidence supports a link between *Ribeiroia* infection and high-frequency cases of amphibian limb deformities. The hypothesis that trematodes could cause amphibian malformations was first suggested by Sessions and Ruth (1990), who noted high concentrations of unidentified metacercarial cysts around the deformed limbs of Pacific treefrogs (*Pseudacris regilla*) and California long-toed salamanders (*Ambystoma macrodactylum croceum*) in California. More recently, a series of laboratory and field experiments have definitively tested this hypothesis in a variety of frog, toad, and salamander species. These experiments revealed that exposure to realistic levels of *Ribeiroia* cercariae causes elevated mortality and severe malformations (up to 100%) in larval amphibians, including extra limbs, skin webbings, bony triangles, and missing limbs in both the fore- and hind limbs (Johnson et al. 1999; 2001a; 2006; Kiesecker 2002; Stopper et al. 2002; Schotthoefer et al. 2003; Johnson and Hartson 2008). Importantly, however, the types and severity of malformations vary with parasite dosage (number of cercariae), the timing of exposure, and the amphibian species being exposed. More parasites generally translate into higher pathology, but only during certain periods of development (e.g., early larval stages) and in species that are susceptible to infection (Bowerman and Johnson 2003; Schotthoefer et al. 2003; Johnson and Hartson 2008). Field data further support the link between infection and malformations, and *Ribeiroia* has been connected to deformed amphibians in the Midwest, the Northeast, and especially the Western USA (Johnson et al. 2001b, 2002; Kiesecker 2002; Lannoo et al. 2003; Sutherland 2005; Johnson and Hartson 2008). Once again, higher average levels of *Ribeiroia* infection within a wetland are associated with higher frequencies of malformations, which can range from <5% to as high as >90% in metamorphosing amphibians.

Collectively, results of this growing body of research indicate that *Ribeiroia* is an important and widespread cause of amphibian deformities in the USA. It has even been recovered from many of the early reported "hotspot" malformation sites, such as the original Ney Pond and the enigmatic "CWB" pond, also in Minnesota (e.g., see Hoppe 2005; Sutherland 2005; Sutherland and Johnson, unpublished). This does not imply, however, that *Ribeiroia* infection explains all malformations in amphibians. *Ribeiroia* is unlikely to causes cases of predominantly or exclusively missing-legged frogs, as have been reported in some areas (Skelly et al. 2007). While *Ribeiroia* can and does cause missing and partially missing limbs in amphibians, these usually occur alongside a broad variety of other abnormality types, such as skin webbings and bony triangles, such that parasite-induced outbreaks of *only* missing legged amphibians are rare. Extra limbs and digits may be a common form of pathology in select species (most notably *Pseudacris regilla*), but are uncommon or absent in many other species exhibiting parasite-induced malformations (Johnson et al. 2001a, 2002; Johnson and Hartson 2008).

Over the last 10 years, we have accumulated records of *Ribeiroia* and malformations from a broad range of locations and amphibian species in the USA. Much of this work has been in cooperation with scientists from the US Fish and Wildlife Service, who have collected and examined amphibians from National Wildlife Refuges across the country (Johnson et al., in press). Necropsies of animals from many of these refuges have revealed interesting patterns in the abundance and distribution of *Ribeiroia* and echinostomatids (see later). For example, within the National Wildlife Refuge survey, *Ribeiroia* was recovered from eight species of amphibians across 13 states, representing 14 of the 37 examined refuges. Many of these represent new host or locality records. The distributional pattern of *Ribeiroia* and of malformations corresponds roughly with the major flyways of migratory birds in the USA: the Pacific, the Mississippi, and the Atlantic flyways (Fig. 11.2). Intriguingly, however, records of *Ribeiroia* in amphibians are conspicuously absent from the southern USA below around 37° latitude. This is somewhat surprising given the diversity of birds in these areas, and in some cases the number of *Ribeiroia* records from birds (e.g., Forrester and Spalding 2003). It is not yet clear to what extent this pattern is truly representative of the parasite's distribution or owes instead to incomplete sampling, but it is interesting to note that deformed amphibians have also predominantly been recorded in the northern half of the USA (http://frogweb.nbii.gov/narcam/). These results also illustrate the broad range in infection levels exhibited by amphibians. While many populations exhibit low average infection abundances, other amphibians may support up to 1,000 metacercariae in a single frog, generally with a high corresponding frequency of severe malformations in the population (Fig. 11.3).

11.2.2 Are Parasite-Induced Malformations a Problem?

One of the most challenging and pressing questions to address is whether *Ribeiroia* infection and the resulting malformations represent a threat to affected amphibian populations. The answer to this question depends, in part, on whether infection and malformations are increasing (see below). Nevertheless, several compelling pieces of evidence suggest that *Ribeiroia*-induced pathology is problematic. First, based on laboratory exposure studies, even low numbers of *Ribeiroia* cercariae (10–50) cause high mortality in many amphibian species. For example, experiments with *P. regilla*, *B. americanus*, *B. boreas*, and *R. pipiens* have shown that exposure to as few as 25 cercariae causes 30–95% mortality in larval amphibians, often in <1 week. Invading cercariae penetrate amphibian tissue directly, with each parasite incurring an injury in its host. The combined effects of those injuries may be substantial

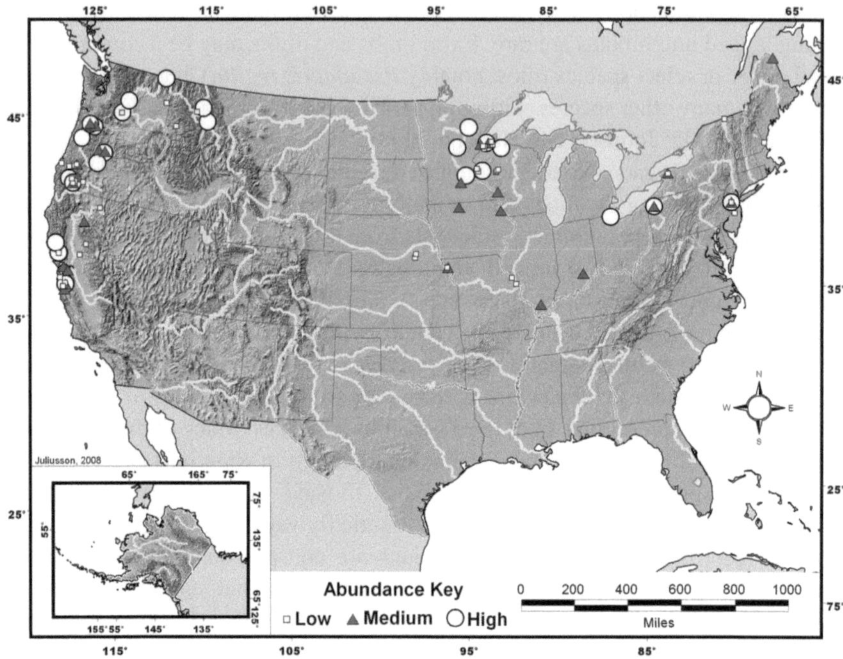

Fig. 11.2 Geographic distribution of *Ribeiroia* as determined from amphibian necropsies. Data represent a compilation of samples from USFWS National Wildlife Refuges and additional sampling on private lands. The size of each circle reflects the average infection abundance recorded in the sample (usually determined from a sample of 10 amphibians). In total, 16 amphibian species from 107 sites distributed across 20 states are included (1999–2007). Abundance values as follows: low (1-10 metacercariae per amphibian), medium (11–30 metacercariae), and high (31–135 metacercariae). Infection intensity for individual frogs ranged form 1 to 960

in early stage larvae. At wetlands that support *Ribeiroia*, one of us (PTJJ) has routinely observed dead and dying amphibian larvae with microhemorrhaging around the limbs characteristic of heavy cercarial penetration. In most cases, however, such mortality is difficult to observe and quantify given the rate of predation and decomposition in nature.

Second, many of the malformations induced by *Ribeiroia*, which typically affect the hind limbs, are almost certainly detrimental to amphibian survival. Missing, extra, or severely malformed limbs interfere with an animal's ability to swim, jump, obtain food, and, especially, to avoid predators. Following years in which >50% of metamorphic amphibians exhibited severe limb deformities, fewer than 2% of returning adults exhibited abnormalities, strongly suggesting that abnormal individuals are less likely to reach sexual maturity than their normal counterparts (Johnson et al. 2001b; Lunde and Johnson, unpublished). Those abnormalities that were observed in adults tended to be very minor. At a malformation hotspot in Minnesota, American toads (*B. americanus*) completely failed

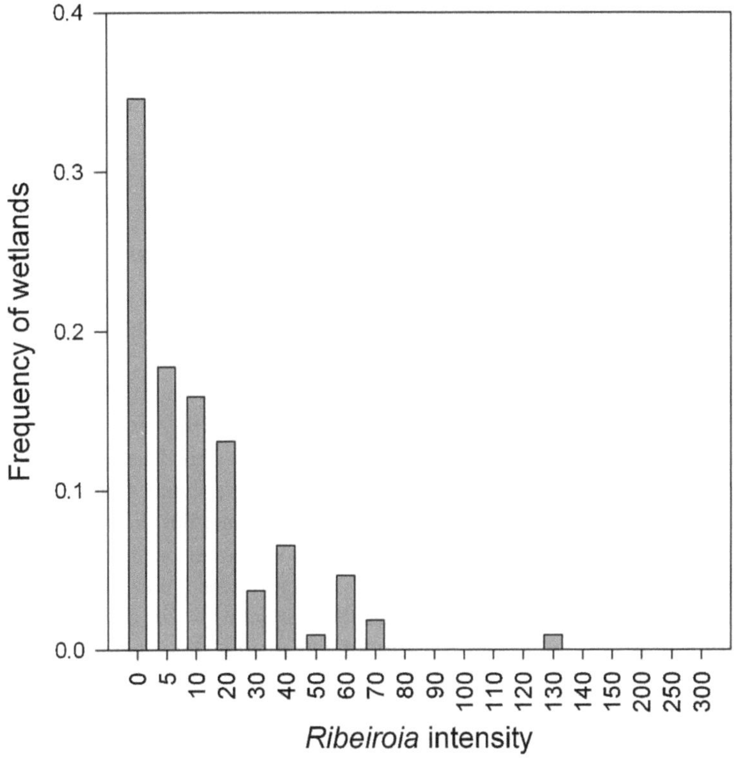

Fig. 11.3 Frequency distribution of *Ribeiroia* infection intensity among wetlands that support the parasite. Values represent the average level of infection within a wetland, as determined from the necropsy of ten individual amphibians. Data are combined among amphibian species and among years

to breed following two years of high-frequency (up to 80%) malformations in metamorphs (Johnson and Hartson 2008). Similarly, Hoppe (2005) noted the disappearance of several anuran species from a Minnesota wetland ("CWB") with high levels of *Ribeiroia* infection following several years of severe malformations. At Jette Pond, Montana, which boasts the longest history of malformations in the USA (1950s to present), three amphibian species have declined or disappeared, while *Ribeiroia* and malformations appear to have increased (see Johnson et al. 2003).

In all these cases, however, the role of infection and malformations in explaining such declines remains correlative and inconclusive. Definitive evidence linking *Ribeiroia* and population-level effects in affected amphibians remains lacking. A combination of long-term monitoring, mark-recapture data, and ecosystem-level manipulations (e.g., parasite removal or addition) is needed to address this issue.

Considering the widespread and abundant occurrence of *Ribeiroia*, as well as the diversity of amphibian species affected by malformations, we argue that population effects of conservation importance are likely, particularly when infection occurs in combination with additional stressors (e.g., habitat loss, pollution, introduced species, etc.).

11.2.3 Emergence of *Ribeiroia* and Malformations

The current and future importance of malformations for amphibian conservation depends on whether *Ribeiroia* represents an emerging infection. Owing to a lack of reliable historical data on infection patterns, this remains one of the most difficult questions to address adequately. For example, prior to 1999, we are aware of no published records documenting *Ribeiroia* infection within free-living amphibian populations. Beaver (1939) and Riggin (1956) each reported conducting experimental infections with *Ribeiroia* and larval amphibians, but did not comment on naturally occurring infections. Since 1999, however, *Ribeiroia* metacercariae have been recorded in more than 20 amphibian species, sometimes at relatively high abundance (Table 11.1; Fig. 11.3). Rather than signaling a recent host shift toward amphibians, this information most likely reflects a recording bias: trematode metacercariae are notoriously difficult to identify, particularly among preserved or frozen hosts, and many parasitologists focus on adult parasites. There are numerous records of *Ribeiroia* from definitive hosts between 1931 and 1999, collectively representing 40 species of birds and 4 species of mammals (Table 11.1). Thus, while unlikely to be an introduced parasite into the US, we must ask whether the abundance and distribution of *Ribeiroia* or the severity of amphibians' response to infection have recently changed. Several studies have reported an increase in the baseline frequency of abnormalities in amphibian populations (Hoppe 2000; Gray 2000; McCallum and Trauth 2003), but the role of changes in parasitic infection in these instances is unknown.

Alongside changes in baseline levels of abnormalities, an important question to address is whether the abundance or severity of "mass malformations" accounts, which involve deformities among >5% of the amphibian population (Johnson et al. 2003), have recently increased. Reports of mass malformations in amphibians occur in the historical literature, but they are exceptionally rare. Johnson et al. (2003) coupled examinations of malformed voucher specimens with contemporary resurveys to evaluate the role of *Ribeiroia* in historical mass malformation accounts. Their results indicate that *Ribeiroia* likely explained six of eight such historical cases of mass malformations in the USA, dating back to the 1940s. This suggests that parasite-induced malformations are not a new phenomenon, but does not address whether they have increased in number of cases, severity, or geographic distribution. Nevertheless, the authors suggested that several lines of evidence support the notion that parasite-induced malformations have increased. First, despite a long and intensive history of amphibian surveys, historical cases of mass malformations

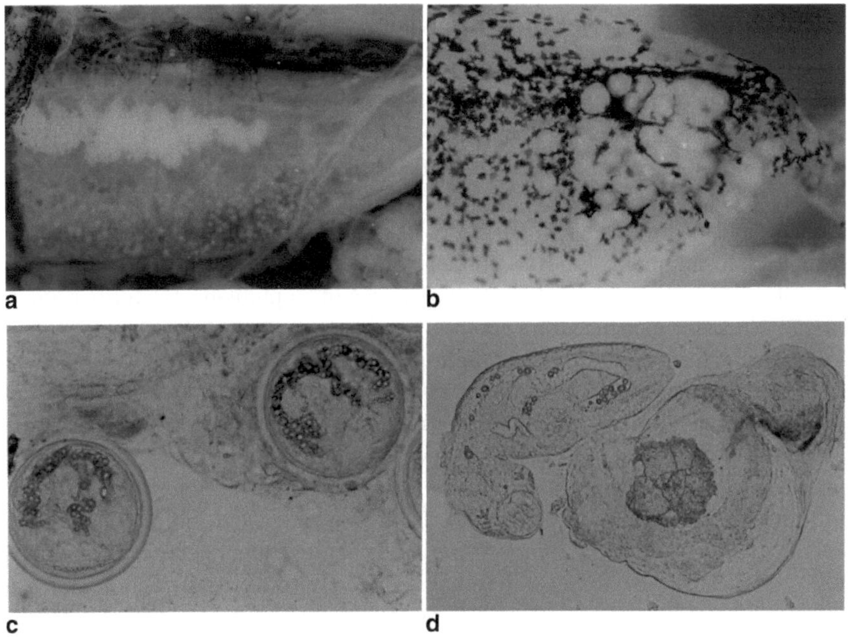

Fig. 11.4 (**a**) In situ view of echinostome infection within an amphibian kidney; metacercariae are visible as white cysts along the lower half of the kidney; (**b**) Close-up view of metacercariae within the kidney; (**c**) isolated encysted metacercariae; (**d**) excysted metacercaria next to its cyst (note collar spines)

in North American amphibians are extremely rare (less than 10 between 1947 and 1990). In contrast, more than 50 mass malformation sites associated with *Ribeiroia* have been recorded since 1996 (Fig. 11.3). Second, most mass malformations accounts (recent and historical) occurred in artificial or highly modified wetlands, such as farm ponds, cattle ponds, impoundments, or retention wetlands. *Ribeiroia* and its snail hosts (*Planorbella* spp.) often thrive in the nutrient-rich conditions of these environments, wherein they may be more likely to achieve elevated abundance. While *Ribeiroia* also occurs in natural wetlands, it typically exhibits a higher abundance that may be more likely to induce mortality and malformations in artificial wetlands (Johnson et al. 2002; 2003; Lannoo et al. 2003; Johnson and Lunde 2005). It is tempting to speculate that, prior to the widespread replacement of natural wetlands with artificial systems, which often differ dramatically in size, biota, hydroperiod, and water chemistry, high levels of *Ribeiroia* infection and the resulting malformations were less common than they are today. Again, however, extensive long-term data on sites with and without *Ribeiroia* are needed to identify the temporal trends in malformation frequency.

11.3 Overview of Echinostomes

Here, we focus on the echinostomes that utilize amphibians as intermediate hosts in their life cycles and compare and contrast them with *Ribeiroia*. While *Ribeiroia* infection in amphibians has been the subject of extensive field observations and ecological studies (see earlier), species of *Echinostoma* have proven excellent model organisms for laboratory studies of host–parasite relationships, ranging from physiology to immunology (Huffman and Fried 1990; Toledo et al. 2007). Given the ecological similarities of these two parasites in amphibians, we further aim to critically examine evidence for a possible rise in echinostome infection in response to the same drivers that influence *Ribeiroia*.

In North America, *Echinostoma trivolvis* is the most commonly reported echinostome that infects amphibians as second intermediate hosts (Fried and Graczyk 2004). Several other echinostomes that occur in North America and infect amphibians as intermediate hosts include *E. revolutum* and *Echinoparyphium* spp., but these are less commonly reported. Both molecular and morphological data support *Echinostoma* and *Echinoparyphium* as being distinct genera (Kostadinova et al. 2003). However, because they have relatively few recognizable features to separate them, especially as metacercariae, some workers believe that misidentification may be common (Sorensen et al. 1997; Fried et al. 1998). *Echinostoma revolutum* has a more prevalent distribution in Europe, Asia, and Africa but has been confirmed in the midwestern USA by Sorensen et al. (1997) using morphological traits and in Connecticut by Holland et al. (2007) using molecular techniques. The overall distribution of *E. revolutum* is not well understood, probably due, in part, to misclassification as *E. trivolvis* in some regions. *Echinoparyphium* species are known from snail hosts in North Carolina (Fried et al. 1998), Michigan, and California (Najarian 1954; Kanev et al. 1998), but they are also suspected to be frequently misidentified as *E. trivolvis* (Fried et al. 1998). Because of such difficulties in identification, we make no attempt here to differentiate among these groups and elect instead to combine our discussions of *Echinostoma trivolvis*, *Echinostoma revolutum*, and *Echinoparyphium* species under the broad category of echinostomes.

Similar to *Ribeiroia*, echinostomes require three hosts to complete their life cycles (Table 11.1). *Ribeiroia ondatrae* infects several genera of planorbid snails as first intermediate hosts, whereas echinostomes appear to be more specialized at this life cycle stage. *Echinostoma trivolvis* has only been reported from *Planorbella trivolvis*; trial infections with other planorbid species have failed, and miracidia from a Pennsylvania strain were infective to *P. trivolvis* from Pennsylvania, but not from Colorado (Fried et al. 1987). Similarly, miracidia of *Echinostoma revolutum* from Indiana infected *Lymnaea elodes*, but not *Lymnaea stagnalis, Physa gyrina*, or *Planorbella trivolvis* (Sorensen et al. 1997) (although Holland et al. 2007 reported *E. revolutum* from *P. trivolvis*). *Echinoparyphium flexum* infects *Lymnaea palustris* in North America (Najarian 1954), whereas *Echinoparyphium rubrum* infects *Physa gyrina* in Michigan and *Physa occidentalis* in California (Kanev et al. 1998).

With respect to second intermediate and definitive hosts, echinostomes are adept generalists. In addition to amphibians and fish, the echinostomes employ a diversity of snails as second intermediate hosts, as well as a few planarians and reptiles (see Table 11.1). Echinostome cercariae emerging from an infected snail thus have many potential second intermediate hosts to infect: they can (a) reinfect the same snail by encysting in the renal organs as metacercariae, (b) infect other snails of any number of species, or (c) encyst in the kidneys of fish, amphibians (Fig. 11.4), or reptiles (Fried and Graczyk 2004). Less commonly, encysted echinostomes can also be found within the body cavity or near the tail resorption site of metamorphosing amphibians (J. Koprivnikar, personal communication; Johnson, unpublished data). Such a range of second intermediate hosts increases the prospects in the trophic food web to cover many common prey animals taken by aquatic feeding birds. The use of snails as second intermediate hosts also simplifies the number of hosts required to complete the life cycle from three (e.g., *Ribeiroia*) to two (e.g., echinostomes). As a result, echinostomes are often more common among dabbling ducks (Anatidae) than is *Ribeiroia*, as these ducks are common consumers of gastropods (Benoy et al. 2002). Both *Ribeiroia* and echinostomes utilize a broad diversity of predatory birds, ranging from herons to owls to raptors, each of which commonly consume amphibians and fish. The hosts listed in Table 11.1 should not be regarded as complete, since many regions of North America have not been studied with regard to these parasites and specific training is needed to differentiate these parasite species, particularly at the larval stages.

11.3.1 Pathology and Patterns of Infection in Amphibians

Although echinostomes are relatively uncommon in the historical literature on amphibian parasites (e.g., Beaver 1937; Najarian 1952), many researchers have observed these parasites in amphibians in recent decades. Muzzall et al. (2001) reported "echinostomid-like metacercariae" from green frogs (*R. clamitans*) in southern Michigan. Skelly et al. (2006) and Holland et al. (2007) reported *Echinostoma revolutum* from green frogs in Connecticut. King et al. (2007) found echinstomes to be widely prevalent in Northern leopard frogs (*Rana pipiens*) in southern Quebec, while Koprivnikar et al. (2007a) reported echinostomes from snails in southern Ontario. In our own work with the National Wildlife Refuge survey, we recorded echinostomes in more than one-third of all examined amphibians ($n > 1{,}500$) from across the USA, including 16 species from 25 states. Again, many of these represent new host and/or locality records. All the 37 surveyed refuges supported at least one amphibian or wetland with infections by echinostomes. As found with *Ribeiroia*, the abundance of *Echinostoma* was greatest along the major bird flyways and in the temperate regions (Fig. 11.5). Echinostome infections exhibited a remarkable variation in intensity, ranging from 1 metacercariae per frog up to 2,750 metacercariae between the two kidneys (Fig. 11.6).

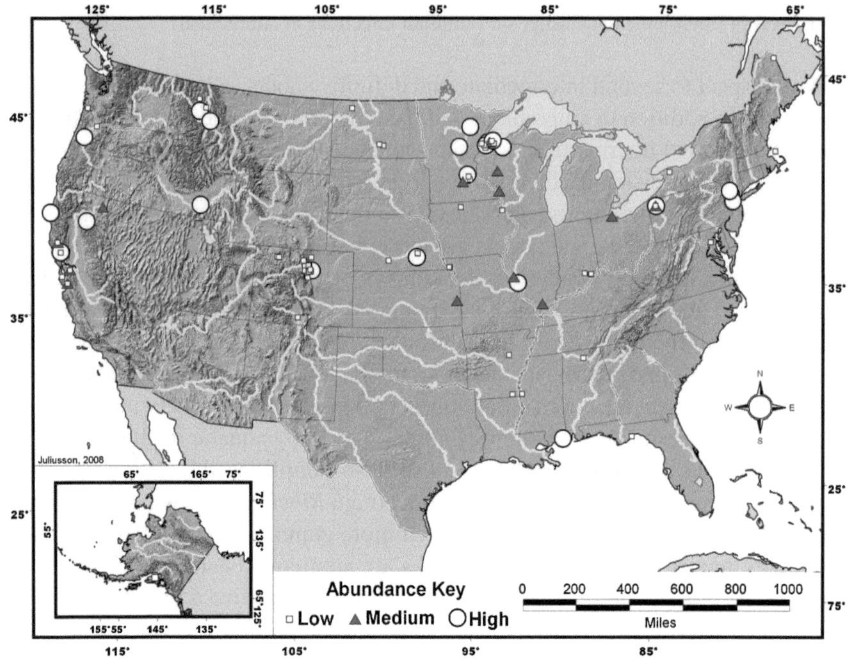

Fig. 11.5 Geographic distribution of echinostomes as determined from amphibian necropsies. Data represent a compilation of samples from USFWS National Wildlife Refuges and additional sampling on private lands. The size of each circle reflects the average infection abundance recorded in the sample (usually determined from a sample of 10 amphibians). In total, 16 amphibian species from 231 sites distributed across 30 states are included (1999–2007). Abundance values as follows: low (1–50 metacercariae per amphibian), medium (51–100 metacercariae), and high (101–1,960 metacercariae). Infection intensity for individual amphibians ranged from 1 to 2,750

Fig. 11.6 Frequency distribution of echinostome infection intensity among wetlands that support the parasite. Values represent the average level of infection within a wetland, as determined from the necropsy of ten individual amphibians. Data are from 231 wetlands combined among amphibian species and among years (1999–2007). Average infection intensity ranged from 1 to 1,960 metacercariae per amphibian

Within larval and postmetamorphic amphibians, the echinostomes encyst in the kidneys (Fig. 11.4). Whether such infection increases the susceptibility of amphibian hosts to definitive host predators is unknown. However, infection of vital organs such as kidneys can cause significant pathology at high parasite intensity (Martin and Conn 1990). Beaver (1937) was the first to observe the edema experienced by infected tadpoles and noted the pattern by which *E. trivolvis* cercariae systematically crawl across the body of a tadpole heading toward the cloaca, wherein they enter the host. Inside the cloaca, migrating cercariae enter the mesonephric ducts, crawl toward the kidneys and form a cyst wall, which becomes infective to definitive hosts within six hours (Fried et al. 1997). Fried et al. (1997) found that experimentally infected tadpoles (*Rana pipiens*) suffered edema, intensity-dependent mortality, and inhibited growth. Schotthoefer et al. (2003) made a significant contribution to the understanding of echinostome pathology in amphibians by considering the role of developmental stage in determining the response of tadpoles to parasite exposure. Early stage tadpoles (stage 25, Gosner 1960) exhibited the highest mortality following *E. trivolvis* exposure, suggesting that the developmental stage of the kidneys plays an important role in host response to infection. Early stage tadpoles only have small pronephroi (or "head kidneys"), and Schotthoefer et al. (2003) argued that these tiny, fragile organs are more susceptible to developmental disruption caused by encysting metacercariae relative to the fully developed mesonephroi of later stage tadpoles. They further noted that tadpoles became progressively less susceptible to infection with continued development toward metamorphosis. Working with *E. revolutum* infection in tadpoles of *Rana clamitans*, Holland et al. (2007) presented similar results linking host mortality and the loss of renal function. Corresponding histopathology of infected kidneys revealed signs of edema and granulomas surrounding metacercariae. Intriguingly, Thiemann and Wassersug (2000a) suggested that echinostome parasites have evolved a mechanism to reduce pathology and mortality risk within amphibian hosts: the bulk of cercariae colonize only one kidney within an individual host, thereby reducing the risk that overinfection leads to renal failure in both kidneys (vertebrates can survive with only one kidney). In support of this hypothesis, the authors noted a significant right side bias in echinostome infection in tadpoles of *Rana sylvatica* and *R. clamitans*. The generality of this infection pattern and the mechanisms producing it remain unknown.

In summary, the pathology of echinostomes in amphibians can be significant and, like *Ribeiroia*, varies with the developmental stage of the amphibian host and the level of cercarial exposure. However, the dosages of echinostome cercariae necessary to cause pathology are often greater than those observed for *Ribeiroia*. For example, while high mortality due to echinostome exposure has only been observed among early stage amphibian larvae (see Schotthoefer et al. 2003), even following relatively high cercarial exposures (100 cercariae), low-to-moderate dosages of *Ribeiroia* cercariae (e.g., 12–50) can induce high mortality across a wide range of developmental stages (Johnson and Hartson 2008; Johnson et al. 1999, 2001b; Schotthoefer et al. 2003). On the other hand, echinostome trematodes are extremely widespread in wetland habitats and are frequently abundant within amphibian hosts. Thus, while it is difficult to project the population level impacts

Fig. 11.7 Kidneys of a recently metamorphosed Pacific chorus frog (*Pseudacris regilla*) with a heavy echinostome infection. Note the large proportion of renal tissue with metacercariae (visible as small white cysts)

of echinostome parasites in amphibians, the combination of their high prevalence, high abundance, and pathogenic effects in laboratory studies suggest that, in some wetlands, their effects may be substantial. For example, Beasley et al. (2005) studied factors related to the declines of Northern cricket frogs (*Acris crepitans*) in Illinois and found all eight sites examined for parasites to have echinostomes present in juvenile frogs. In three of the eight sites, the prevalence of echinostomes in tadpoles and juvenile frogs was between 69% and 100%. In one of these sites, the majority of infected frogs had high-intensity infections where more than 50% of the kidney tissue was occupied by echinostome metacercariae as determined by histological sections (e.g., Fig. 11.7). This level of infection translates to individual tadpoles being exposed to multiple thousands of cercariae during their development in the aquatic environment. The following year, the authors observed a decrease in cricket frog recruitment in sites that had exhibited high echinostome infections, leading them to argue that high mortality due to infections was likely an important factor. Taken together with similar accounts for populations heavily impacted by *Ribeiroia* (discussed earlier), we feel these parasites warrant more attention in regions where they are highly abundant.

11.3.2 Emergence of Echinostoma

Finally, we ask whether echinostome infections have recently exhibited changes in abundance, in host usage, or in geographic distribution. Given the limited data available, we can only begin to address these questions here, and our goal is to suggest avenues of research that will advance this intriguing research area. Historical records of echinostome infections in amphibians are few (Beaver 1937; Najarian 1952). However, it is not well understood to what extent previous researchers examined the kidneys for metacercariae. For example, Andrews et al. (1992) provide an extensive checklist of 51 species of digeneans reported from bullfrogs (*R. catesbeiana*) in North America, including many metacercariae and mesocercariae, but no echinostomes. It is difficult to use the rarity of historical records as a reliable indicator of past prevalence or abundance of echinostomes, similar to the limitations of the parasitology data for evaluating the historical abundance of *Ribeiroia*. But a few recent studies that have examined amphibian kidneys provide a strong signal that we should address this question more carefully. Beasley et al. (2005) observed extremely heavy echinostome infections in Northern cricket frogs where more than half the kidney tissues of tadpoles consisted of metacercariae. Similarly, Skelly et al. (2006) reported echinostome infections of up to 1,648 metacercarial cysts in a single green frog (*R. clamitans*) in Connecticut. There is no doubt that these high-intensity infections can cause extreme pathology and death, especially given that experimental infections of 100 cercariae led to high mortality in tadpoles (Schotthoefer et al. 2003). To more accurately address whether amphibians are experiencing more frequent and heavier echinostome infections, a comprehensive study of vouchered amphibians in museum collections is in order, particularly for sites with a long collection history. Though the current evidence for increasing levels of echinostomes is thin, it is noteworthy that, among our recent surveys, this group of parasites is the most common larval trematode encountered, often with 100% prevalence within the amphibian population and individual infections numbering in the several hundred.

11.4 Possible Causes of Parasite Emergence

Growing attention has recently focused on the question of how environmental change affects host–macroparasite interactions (Bradley and Altizer 2007; McKenzie and Townsend 2007; Johnson and Carpenter 2008). Because many of these parasites have complex or multihost life cycles, and different hosts vary in their responses to environmental changes, addressing this question is neither simple nor does it have a single answer. Ultimately, the response of a parasite will vary among types of parasites and types of environmental change, and the product of host abundance, host physiological condition, spatial and temporal interactions among hosts, the abundance and composition of the parasite community, and the effect of environmental conditions on the parasite's free-living stages (e.g., Lafferty and Holt 2003).

Investigations into this area have fallen largely into two distinct categories. On one hand, the diversity of macroparasite infection in a host species and community can be used as a biological indicator. Because of the complexity of helminth life cycles and their dependency on multiple interacting species across several trophic levels, the diversity of parasites is often indicative of a species-rich community of hosts (e.g., Hechinger et al. 2007). Simply stated, ecosystems with low levels of disturbance can be expected to support more diverse host and parasite communities (Hudson et al. 2006). For example, Huspeni and Lafferty (2004) found that estuarine environments surrounded by development supported half as many parasite species as estuaries surrounded by natural areas. This difference was attributed to the reduced diversity of birds and other free-living species within impacted estuaries.

On the other hand, a second important issue is what controls the abundance of a macroparasite species within a community, particularly for parasites that cause severe pathology in their hosts. Many macroparasite infections are relatively benign at low levels. As infection abundance increases, however, so too does the likelihood of pathology and disease ("intensity-dependent" pathology). Thus, while the diversity of macroparasite species may decrease with certain forms of environmental change, the abundance of certain pathogenic species may increase, potentially increasing the risk of disease outcomes. In fact, the loss of many parasite species from a community following a disturbance may reduce competition among the remaining parasites, further contributing to an increase in parasite abundance. Parasite species richness and parasite species even-ness may therefore be uncorrelated or even negatively correlated, such that select parasites become over-represented in species-poor communities. We focus here on how environmental change can increase the abundance of pathogenic parasites.

With respect to *Ribeiroia* and echinostome infections in amphibians, several forms of environmental change have been shown or suggested to enhance infections. Collectively such changes alter the abundance or distribution of hosts, their susceptibility to infection, or the abundance and productivity of the parasite. Many of the mechanisms discussed as follows are not mutually exclusive or independent, and may interact to additively or synergistically influence patterns of infection:

1. Changes in land use and wetland characteristics

- Changes in patterns of trematode infection within amphibians ultimately result from shifts in surrounding land use and wetland characteristics. This forms the foundational context for the specific mechanisms discussed later. One particularly important form of land use change involves the destruction of natural wetlands and their replacement by artificial systems. As natural wetlands continue to be destroyed and altered, amphibians rely increasingly on artificial or highly modified wetlands to complete their breeding cycles, including farm ponds, stock ponds, retention systems, and other impoundments (Knutson et al. 2004). These artificial environments often exhibit strikingly different biotic and abiotic characteristics relative to their natural counterparts, and we discuss some of the important changes later.

2. Nutrient pollution

- Nutrient runoff from fertilizers, livestock, and erosion often leads to eutrophication in wetlands surrounded by agrarian or urbanized landscapes. The resulting increase in algal growth in freshwater environments may differentially favor some groups of freshwater snails, including the planorbids (Chase 2003; Johnson and Chase 2004). Johnson et al. (2002) and Johnson and Chase (2004) reported field correlations among nutrient levels, snail host density, and *Ribeiroia* infection in amphibians. Skelly et al. (2006) suggested a similar pattern may apply to echinostome infections in amphibians from urban wetlands. They found that, while variable, the abundance of snails and echinostome infections in frogs were highest in urban wetlands. Recently, the relationship between nutrient inputs and trematode infection has been tested experimentally using outdoor mesocosms. By manipulating nutrient levels and the input of *Ribeiroia* eggs (e.g., bird activity), Johnson et al. (2007) found that eutrophication enhanced *Ribeiroia* infection in green frog larvae through two related mechanisms: nutrient-mediated increases in algal growth enhanced both the density of *Ribeiroia* infected snails and the per-snail production of *Ribeiroia* cercariae. Collectively, these changes caused a three- to five-fold increase in metacercerial abundance in co-occurring tadpoles.

- Importantly, however, agriculture and urban development can also affect definitive host activity. Forms of land use that depress activity of definitive hosts could reduce the abundance of parasite inputs (e.g., eggs) into aquatic systems, even while the density and susceptibility of snail hosts are maximized. Clearly this outcome depends on the type and severity of development and the particular needs of the definitive hosts. In some cases, eutrophication has been linked to increased colonization by vertebrate definitive hosts (e.g., Esch 1971; Wisniewiski 1958; Zander et al. 2002), whereas in others the loss of forested areas following development is associated with decreases in bird and mammal activity (e.g., King et al. 2007).

3. Pesticide contamination

- In addition to nutrients, wetlands surrounded by agriculture, industrial development, or residential areas often support a diverse mixture of pesticides and other contaminants. Because some contaminants can suppress immune function in amphibians, their exposure may increase the susceptibility of amphibian larvae to infection by trematode cercariae. Changes in tadpole behavior following contaminant exposure, such as reduced swimming activity, could further increase infection by parasites (Thiemann and Wassersug 2000b; Taylor et al. 2004). Kiesecker (2002) reported that amphibians in pesticide-contaminated ponds in Pennsylvania exhibited more *Ribeiroia* metacercariae and higher malformations than those in pesticide-free ponds. In complementary experiments, he showed that exposure to atrazine and malathion increased trematode infection success in wood frog (*Rana sylvatica*) larvae, possibly by reducing immune function (measured by eosinophil count). Other studies have also suggested a link between pesticide exposure and infections of amphibians by viruses (Forson and Storfer 2006) and other trematodes (Koprivnikar et al. 2007a). With respect to *Ribeiroia* and *Echinostoma*, however, more data are needed on the effects of contaminants on the parasites' free-living stages (e.g., miracidia and cercariae) and on snail hosts, including especially infected snails, which can be more sensitive than uninfected individuals (see also Koprivnikar et al. 2007b). Without such information, it is difficult to evaluate the ecological relevance of pesticide-parasite-amphibian interactions.

4. Loss of biodiversity

- The dilution effect hypothesis suggests that with decreases in host diversity, the abundance and pathology of some parasites will increase (e.g., Keesing et al. 2006). This hypothesis holds when more diverse communities support a greater relative abundance of low competency hosts that either resist infection or do a poor job of supporting parasites once infected. In this manner, diverse communities lead to more "wasted" transmission events, in which infectious stages of the parasites (e.g., cercariae) are eliminated by infecting less susceptible hosts. While most of this hypothesis has been developed with a focus on vector-borne infections (e.g., Lyme disease, West Nile virus), increasing evidence suggests that it may also apply to complex life cycle parasites with free-living infectious stages. Experimental evidence supports this principle in relation to infections of first intermediate hosts (e.g., snails) and second intermediate hosts (e.g., amphibians). Johnson et al. (2008) found that the addition of "decoy" snails, which were not susceptible to trematode miracidia, dramatically reduced infection in susceptible snails, even when the density of susceptible snail hosts remained constant among treatments. This ultimately caused a 40% reduction in snail infection prevalence and a 60% reduction in cercarial production. Similar results involving snail-miracidia infections have been reported by Chernin (1968), Mone and Combes (1986), and Kopp and Jokela (2007).
- With respect to amphibian infections, heterospecific amphibian communities containing larvae of both American toads (*B. americanus*) and gray treefrogs (*Hyla versicolor*) supported 40–60% fewer *Ribeiroia* metacercariae than did monospecific communities with *B. americanus* alone (Johnson et al. 2008). All communities were initially exposed to the same number of *Ribeiroia* cercariae. However, larval *H. versicolor* are largely resistant to infection, and cercariae that colonized these hosts were killed by the immune system. This led to a reduction in the total abundance of surviving metacercariae. Because the presence of *H. versicolor* also reduced infections in co-occurring toads, toads from the heterospecific treatments exhibited higher survival and fewer malformations than toads raised alone or with another toad (Johnson et al. 2008). Taken together, these results suggest that ongoing losses of aquatic biodiversity, including snails and especially amphibians, may influence patterns of trematode transmission and pathology. Artificial or modified wetlands often support lower community diversity than natural systems, which may therefore exacerbate infection levels via the dilution effect. It is important to recognize, however, that the applicability of the dilution effect to *Ribeiroia* and *Echinostoma* infections depends critically on the patterns of community assembly and community disassembly (Ostfeld and LoGiudice 2003). If, as has been suggested on coevolutionary grounds by Ostfeld and colleagues, low-diversity communities are dominated by highly competent hosts, with less competent hosts (dilution or decoy hosts) occurring with increasing diversity, then transmission will likely be maximized in species-poor communities. If, however, necessary hosts in the life cycles of either parasite are absent in low-diversity communities, transmission will be interrupted. Thus far, insufficient information is available on the (a) competency

of snail and amphibian hosts to each parasite and (b) the patterns of community disassembly in wetlands to assess the generality of the dilution effect under field conditions. Finally, while less commonly included in discussions of the dilution effect, it is important to recognize that nonhost species such as predators can also influence the abundance of infectious parasites in the environment through predation (see Schotthoefer et al. 2007; Thieltges et al. 2008).

5. Climate change

- Finally, forecasted changes in climate are expected to influence patterns of trematode infection and pathology in amphibians. Increases in minimum temperatures and the length of the growing season are likely to affect both the timing and maximal levels of infection. Despite growing interest in the effects of climate change on human and wildlife diseases (e.g., Harvell et al. 2002), the full complexity of host–parasite–climate interactions remains poorly understood. The pathology induced in amphibians by *Ribeiroia* and *Echinostoma* is stage dependent, such that early stage tadpoles are most vulnerable to mortality and malformations with progressively reduced vulnerability as development progresses toward metamorphosis (Bowerman and Johnson 2003; Schotthoefer et al. 2003; Holland et al. 2007). As a result, changes in the timing of cercarial release from snails could sharply affect pathology in amphibians. Because trematode maturation and cercarial production in snails are controlled largely by temperature (Poulin 2006), whereas amphibian breeding is often triggered by rainfall, forecasted increases in temperature and growing season duration could enhance amphibian exposure to trematodes early in their development when larvae are most vulnerable. Moreover, because of the high metabolic efficiency of trematodes relative to other invertebrates and vertebrates, any acceleration in amphibian growth rate will likely be overshadowed by the increase in trematode development (Poulin 2006). An extended growth season could also allow for a transition from a single snail infection cycle per year to multiple infection cycles per year, as found in wetlands exposed to thermal effluent (Marcogliese 2001). Models built by Kutz et al. (2005) for a protostrongylid nematode with a similarly complex life cycle suggested that even small increases in temperature (as little as 1°C) could increase infection from a 2-year cycle to a 1-year cycle. Limited empirical or experimental data are available to evaluate the effects of climate on *Ribeiroia* and *Echinostoma* infections, but we suspect the magnitude of such effects would be greatest in temperate regions where climate changes are expected to be pronounced and interactions between trematodes and amphibians are currently constrained by temperature.

11.5 Concluding Remarks

Emerging diseases include those that have expanded in prevalence, geographic distribution, host range, or pathology (e.g., CDC 1994; Friend et al. 2001). The causes of emergence are complex and may involve changes in the evolutionary and ecological relationships between host(s) and pathogen. In many cases, definitive evidence of emergence is elusive, as historical baseline data for most pathogens (especially those of nonhuman hosts) are unavailable. Here, we explored available evidence to evaluate whether infections by *Ribeiroia* and *Echinostoma* in amphibians have recently increased. Owing to a lack of published historical information on these parasites in amphibians, this question remains difficult to answer definitively. Both parasites have only infrequently been recorded in amphibians historically (e.g., http://www.nhm.ac.uk/research-curation/projects/host-parasites/database/). While the rarity of published records of amphibians supporting either parasite is notable, particularly in contrast to their relative abundance in the last decade, significant biases exist in what parasites have been recorded in amphibians (e.g., adult worms over metacercariae). The apparent increase in records of mass amphibian malformations associated with *Ribeiroia* – a relatively obvious and severe form of pathology – is also suggestive of a recent increase in infection, but it is difficult to quantitatively compare historical and contemporary surveillance efforts to assess how large such a change might actually be (see Johnson et al. 2003). We suggest that examination of vouchered museum specimens, in combination with rigorous and long-term monitoring programs, holds the greatest promise in providing an adequate test of the emergence hypothesis.

Several characteristics of trematodes in the genera *Ribeiroia* and *Echinostoma* suggest they will be strongly influenced by environmental changes and may therefore be predisposed toward emergence. Despite their complex life cycles, both parasites can be considered relative generalists. Each can utilize an enormous diversity of second intermediate and definitive hosts, including a wide range of fishes, amphibians, birds, and mammals. Members of *Echinostoma* (but not *Ribeiroia*) can even infect snails as second intermediate hosts, which could be one contributing factor explaining the near ubiquity of echinostomes in amphibian habitats. This suggests that these parasites, which already occur in many wetland environments, will be influenced by changes in the abundance and composition of the host community. Importantly, *Ribeiroia* and *Echinostoma* can infect many species that thrive in disturbed or human-altered conditions, such as rats, raccoons, and some herons. The same principle apparently holds for their use of first intermediate hosts. Members of the genera *Physa* and *Planorbella* are common components of pond habitats, often achieving high densities in eutrophic, artificial wetlands (see Johnson et al. 2007). Thus, rather than eliminating these parasites, environmental changes could enhance their levels of infection by promoting such human-associated hosts.

Finally, because amphibians represent a vehicle of transmission between snail intermediate hosts and vertebrate definitive hosts, both *Ribeiroia* and *Echinostoma* may benefit by inducing pathology that enhances the predation risk of infected amphibians. This is particularly true for *Ribeiroia*, which can cause high levels of direct mortality among infected amphibian larvae, and surviving animals often develop severe malformations that are detrimental to long-term survival. Cercariae of *Echinostoma* are less pathogenic than those of *Ribeiroia*, but they can nevertheless cause impaired renal function, edema, and mortality at high dosages or during certain stages of amphibian development. The echinostomes are also incredibly widespread across the landscape and often locally abundant in amphibians, suggesting their collective effects may be considerable. Because trematodes exist in a metacommunity dynamic, with mobile definitive hosts connecting wetlands that support the parasites, high amphibian mortality or population extirpation within a single pond is unlikely to dramatically affect parasite transmission (but it will take a larger toll on the amphibian hosts). Thus, a high abundance of *Ribeiroia* or *Echinostoma* has the potential to sharply affect the viability of amphibians, which are already threatened by numerous forms of environmental changes and are now considered the most threatened class of vertebrates worldwide.

Taken together, these characteristics suggest that current and future research should focus on evaluating ongoing changes in infection by these trematodes. While the exact trajectory of how their infections have changed in host range, geography, and abundance over time remains unavailable, environmental changes such as nutrient runoff, pesticide contamination, biodiversity loss and climate change all have demonstrated potential to enhance infection and/or pathology in amphibians. These changes occur through increases in snail host abundance, changes in parasite production, decreases in amphibian infection resistance, or by altering the transmission success among hosts. Based on the distributional data collected for each parasite in the USA, preliminary results are suggestive of higher infection along bird migratory routes and at more temperate latitudes. This latter trend is perhaps of particular concern, as temperate regions also tend to (a) support lower biodiversity, which may enhance transmission through the dilution effect, (b) be associated with high agricultural activity, leading to higher inputs of fertilizers and pesticides, and (c) be sharply affected by forecasted increases in temperature, which are expected to reduce winter duration and increase the number or intensity of parasite cycles per year. This "perfect storm" convergence underscores the importance of understanding the individual and combined effects of environmental changes on host–parasite dynamics, and of incorporating parasites into monitoring efforts aimed at amphibian conservation and restoration. We further advocate the need for greater molecular and morphological resolution of these groups, long-term studies of parasite abundance (preferably in conjunction with museum vouchers), and field-based manipulations to quantitatively assess the roles of multiple interacting drivers on parasite-induced pathology in amphibians.

Acknowledgments We wish to thank the editors of this volume, R. Toledo and B. Fried, for inviting us to contribute, J. Koprivnikar for helpful comments on an earlier draft of this chapter, and all the participants in the US Fish and Wildlife Service's Amphibian Abnormality Survey which contributed greatly to the development of ideas presented here. Finally, we gratefully acknowledge contributions from the late Dr. Daniel Sutherland, whose pioneering work in the area of amphibian parasitology remains an inspiration to us all.

References

Alden, K. J. 1995. Helminths of the opossum, *Didelphis virginiana*, in southern Illinois, with a compilation of all helminths reported from this host in North America. Journal of the Helminthological Society of Washington 62: 197–208.

Andrews, K. D., Lampley, R. L., Gillman, M. A., Corey, D. T., Ballard, S. R., Blasczyk, M. J., and Dyer, W. G. 1992. Helminths of *Rana catesbeiana* in southern Illinois with a checklist of helminths in bullfrogs of North America. Transactions of the Illinois State Academy of Science 85: 147–172.

Beasley, V. R., Faeh, S. A., Wikoff, B., Staehle, C., Eisold, J., Nichols, D., Cole, R., Schotthoefer, A. M., Greenwell, M., and Brown, L. E. 2005. Risk factors and the decline of the cricket frog, *Acris crepitans*: evidence for involvement of herbicides, parasitism, and habitat modifications. In: Lannoo, M. J., ed., Amphibian Declines: The Conservation Status of United States Species, pp. 75–87, Chicago: University of Chicago Press.

Beaver, P. C. 1937. Experimental studies on *Echinostoma revolutum* (Froelich) a fluke from birds and mammals. Illinois Biological Monographs 15: 1–96.

Beaver, P. C. 1939. The morphology and life history of *Psilostomum ondatrae* Price 1931 (Trematoda: Psilostomidae), Journal of Parasitology 25: 383–393.

Benoy, G. A., Nudds, T. D., and Dunlop, E. 2002. Patterns of habitat and invertebrate diet overlap between tiger salamanders and ducks in prairie potholes. Hydrobiologia 481: 47–59.

Blaustein, A. R., and Johnson, P. T. J. 2003. The complexity of deformed amphibians. Frontiers in Ecology and the Environment 1: 87–94.

Bowerman, J., and Johnson, P. T. J. 2003. Timing of trematode-related malformations in Oregon spotted frogs and Pacific treefrogs. Northwestern Naturalist 84: 142–145.

Bradley, C. A., and Altizer, S. 2007. Urbanization and the ecology of wildlife diseases. Trends in Ecology and Evolution 22: 95–102.

Brandt, B. B. 1936. Parasites of certain North Carolina salienta. Ecological Monographs 6: 491–532.

Centers for Disease Control and Prevention 1994. Addressing emerging infectious disease threats: a prevention strategy for the United States. Atlanta: US Department of Health and Human Services, Public Health Service.

Chase, J. M. 2003. Strong and weak trophic cascades along a productivity gradient. Oikos 101: 187–195.

Chernin, E. 1968. Interference with the capacity of *Schistosoma mansoni* miracidia to infect the molluscan host. Journal of Parasitology 54: 509–516.

Esch, G. W. 1971. Impact of ecological succession on parasite fauna in centrarchids from oligotrophic and eutrophic ecosystems. American Midland Naturalist 86: 160–168.

Forrester, D. J., and Spalding, M. G. 2003. Parasites and Diseases of Wild Birds in Florida. Gainesville: University Press of Florida.

Forrester, D. J., Kinsella, J. M., Mertins, J. W., Price, R. D., and Turnbull, R. E. 1994. Parasitic helminths and arthropods of fulvous whistling-ducks (*Dendrocygna bicolor*) in southern Florida. Journal of the Helminthological Society of Washington 61: 84–88.

Forson, D. D. and A. Storfer (2006). Atrazine increases ranavirus susceptibility in the tiger salamander, *Ambystoma tigrinum*. Ecological Applications 16: 2325–2332.

Fried, B., and Graczyk, T. K. 2004. Recent advances in the biology of *Echinostoma* species in the "revolutum" group. Advances in Parasitology 58: 139–195.

Fried, B., Scheuermann, S., and Moore, J. 1987. Infectivity of *Echinostoma revolutum* miracidia for laboratory-raised pulmonate snails. Journal of Parasitology 73: 1047–1048.

Fried, B., Pane, P. L., and Reddy, A. 1997. Experimental infection of *Rana pipiens* tadpoles with *Echinostoma trivolvis* cercariae. Parasitology Research 83: 666–669.

Fried, B., Frazer, B. A., and Kanev, I. 1998. Comparative observations on cercariae and metacercariae of *Echinostoma trivolvis* and *Echinoparyphium* sp. Journal of Parasitology 84: 623–626.

Friend, M., McLean, R. G., and Dein, F. J. 2001. Disease emergence in birds: challenges for the twenty-first century. The Auk 118: 290–303.

Gosner, K. L. 1960. A simplified table for staging anuran embryos and larvae with notes of identification. Herpetologica 16: 183–190.

Gray, R. H. 2000. Historical occurrence of malformations in the cricket frog, *Acris crepitans*, in Illinois. Transactions of the Illinois State Academy of Sciences 93: 279–284.

Harvell, C. D., Mitchell, C. E., Ward, J. R., Altizer, S., Dobson, A. P., Ostfeld, R. S. and M. D. Samuel (2002). Climate warming and disease risks for terrestrial and marine biota. Science 296: 2158–2162.

Hechinger, R. F., Lafferty, K. D., Huspeni, T. C., Brooks, A. J., and Kuris, A. M. 2007. Can parasites be indicators of free-living diversity? Relationships between species richness and the abundance of larval trematodes and of local benthos and fishes. Oecologia 151: 82–92.

Holland, M. P., Skelly, D. K., Kashgarian, M., Bolden, S. R., Harrison, L. M., and Cappello, M. 2007. Echinostome infection in green frogs (*Rana clamitans*) is stage and age dependent. Journal of Zoology 271: 455–462.

Hoppe, D. M. 2000. History of Minnesota frog abnormalities: do recent findings represent a new phenomenon? In: Kaiser, H., Casper, G. S., and Bernstein, N., eds., Investigating Amphibian Declines. Journal of the Iowa Academy of Science 107: 86–89.

Hoppe, D. M. 2005. Malformed frogs in Minnesota: history and interspecific differences. In: Lannoo, M. J., ed., Amphibian Declines: The Conservation Status of United States Species, pp. 75–87. California: University of California Press.

Houlahan, J. E., Findlay, C. S., Schmidt, B. R., Meyer, A. H., and Kuzmin, S. L. 2000. Quantitative evidence for global amphibian population declines. Nature 404: 752–755.

Hudson, P. J., Dobson, A. P., and Lafferty, K. D. 2006. Is a healthy ecosystem one that is rich in parasites? Trends in Ecology and Evolution 21: 381–385.

Huffman, J. E., and Fried, B. 1990. *Echinostoma* and Echinostomiasis. Advances in Parasitology 29: 215–269.

Huspeni, T. C., and Lafferty, K. D. 2004. Using larval trematodes that parasitize snails to evaluate a saltmarsh restoration project. Ecological Applications 14: 795–804.

Ingles, L. 1933. The specificity of frog flukes. Science 78: 168.

Johnson, P. T. J. 2006. Amphibian diversity: decimation by disease. Proceedings of the National Academy of Sciences of the United States of America 103: 3011–3012.

Johnson, P. T. J., and Chase, J. M. 2004. Parasites in the food web: linking amphibian malformations and aquatic eutrophication. Ecology Letters 7: 521–526.

Johnson, P. T. J., and Hartson, R. B. 2008. All hosts are not equal: explaining differential patterns of malformations in an amphibian community (submitted).

Johnson, P. T. J., and Lunde, K. B. 2005. Parasite infection and limb malformations: a growing problem in amphibian conservation. In: Lannoo, M. J., ed., Amphibian Declines: The Conservation Status of United States Species, pp. 124–138. California: University of California Press.

Johnson, P. T. J., Lunde, K. B., Ritchie, E. G., and Launer, A. E. 1999. The effect of trematode infection on amphibian limb development and survivorship. Science 284: 802–804.

Johnson, P. T. J., Lunde, K. B., Haight, R. W., Bowerman, J., and Blaustein, A. R. 2001a. *Ribeiroia ondatrae* (Trematoda: Digenea) infection induces severe limb malformations in western toads (*Bufo boreas*). Canadian Journal of Zoology 79: 370–379.

Johnson, P. T. J., Lunde, K. B., Ritchie, E. G., Reaser, J. K., and Launer, A. E. 2001b. Morphological abnormality patterns in a California amphibian community. Herpetologica 57: 336–352.

Johnson, P. T. J., Lunde, K. B., Thurman, E. M., Ritchie, E. G., Wray, S. N., Sutherland, D. R., Kapfer, J. M., Frest, T. J., Bowerman, J., and Blaustein, A. R. 2002. Parasite (*Ribeiroia ondatrae*) infection linked to amphibian malformations in the western United States. Ecological Monographs 72: 151–168.

Johnson, P. T. J., Lunde, K. B., Zelmer, D. A., and Werner, J. K. 2003. Limb deformities as an emerging parasitic disease in amphibians: evidence from museum specimens and resurvey data. Conservation Biology 17: 1724–1737.

Johnson, P. T. J., Sutherland, D. R., Kinsella, J. M., and Lunde, K. B. 2004. Review of the trematode genus *Ribeiroia* (Psilostomidae): Ecology, life history, and pathogenesis with special emphasis on the amphibian malformation problem. Advances in Parasitology 57: 191–253.

Johnson PTJ, Preu ER, Sutherland DR, Romansic J, Han B, Blaustein AR. 2006. Adding infection to injury: synergistic effects of predation and parasitism on salamander limb malformations. Ecology 87: 2227–2235.

Johnson, P. T. J., Chase, J. M., Dosch, K. L., Hartson, R. B., Gross, J. A., Larson, D. J., Sutherland, D. R., and Carpenter, S. R. 2007. Aquatic eutrophication promotes pathogenic infection in amphibians. Proceedings of the National Academy of Sciences of the United States of America 104: 15781–15786.

Johnson, P. T. J. and S. R. Carpenter (2008). Influence of eutrophication on disease in aquatic ecosystems: patterns, processes, and predictions. Chapter 4 (pp. 71–99) in Ostfeld, R. S., F. Keesing, and V. T. Eviner, *Infectious disease ecology: effects of ecosystems on disease and of disease on ecosystems*. Princeton University Press, Princeton, NJ.

Johnson, P. T. J., Hartson, R. B., Larson, D. J., and Sutherland, D. R. 2008. Linking biodiversity loss and disease emergence: amphibian community structure determines parasite transmission and pathology (in preparation).

Johnson, PTJ, Hartson, RB, Larson, DJ, Sutherland, DR. 2008. Diversity and disease: community structure drives parasite transmission and host fitness. Ecology Letters (in press, October).

Johnson, PTJ, Reeves, MK, Krest, SK, and Pinkney, AE. A decade of deformities: advances in our understanding of amphibian malformations and their implications. In Sparling, Linder, Bishop, Krest (eds), Ecotoxicology of Amphibians and Reptiles, 2nd edtion. SETAC Press, Pensacola FL (in press).

Kaiser, J. 1997. Ecology – deformed frogs leap into spotlight at health workshop. Science 278: 2051–2052.

Kaiser, J. 1999. Frog declines – deformities – a trematode parasite causes some frog deformities. Science 284: 731.

Kanev, I., Fried, B., Dimitrov, V., and Radev, V. 1995. Redescription of *Echinostoma trivolvis* (Cort, 1914) (Trematoda, Echinostomatidae) with a discussion on its identity. Systematic Parasitology 32: 61–70.

Kanev, I., Sorensen, R., Sterner, M., Cole, R., and Fried, B. 1998. The identification and characteristics of *Echinoparyphium rubrum* (Cort, 1914) comb. new (Trematoda, Echinostomatidae) based on experimental evidence of the life cycle. Acta Parasitologica 43: 181–188.

Keesing, F., Holt, R. D., and Ostfeld, R. S. 2006. Effects of species diversity on disease risk. Ecology Letters 9: 485–498.

Kiesecker, J. M. 2002. Synergism between trematode infection and pesticide exposure: a link to amphibian limb deformities in nature? Proceedings of the National Academy of Sciences of the United States of America 99: 9900–9904.

King, K. C., McLaughlin, J. D., Gendron, A. D., Pauli, B. D., Giroux, I., Rondeau, B., Boily, M., Juneau, P., and Marcogliese, D. J. 2007. Impacts of agriculture on the parasite communities of northern leopard frogs (*Rana pipiens*) in southern Quebec, Canada. Parasitology 134: 2063–2080.

Knutson, M. G., Richardson, W. B., Reineke, D. M., Gray, B. R., Parmelee, J. R., and Weick, S. E. 2004. Agricultural ponds support amphibian populations. Ecological Applications 14: 669–684.

Kopp, K., and Jokela, J. 2007. Resistant invaders can convey benefits to native species. Oikos 116: 295–301.

Koprivnikar, J., Baker, R. L., and Forbes, M. R. 2006. Environmental factors influencing trematode prevalence in grey tree frog (*Hyla versicolor*) tadpoles in southern Ontario. Journal of Parasitology 92: 997–1001.

Koprivnikar, J., Baker, R. L., and Forbes, M. R. 2007a. Environmental factors influencing community composition of gastropods and their trematode parasites in southern Ontario. Journal of Parasitology 93: 992–998.

Koprivnikar, J., Forbes, M. R., and Baker, R. L. 2007b. Contaminant effects on host–parasite interactions: atrazine, frogs and trematodes. Environmental Toxicology and Chemistry 26: 2166–2170.

Kostadinova, A., Herniou, E. A., Barrett, J., and Littlewood, D. T. J. 2003. Phylogenetic relationships of *Echinostoma* Rudolphi, 1809 (Digenea: Echinostomatidae) and related genera re-assessed via DNA and morphological analyses. Systematic Parasitology 54: 159–176.

Kutz, S. J., Hoberg, E. P., Polley, L., and Jenkins, E. J. 2005. Global warming is changing the dynamics of Arctic host–parasite systems. Proceedings of the Royal Society B-Biological Sciences 272: 2571–2576.
Lafferty, K. D. and Morris, A. K. 1996. Altered behavior of parasitized killifish increases host susceptibility to predation by bird final hosts. Ecology 77: 1390–1397.
Lafferty, K. D., and Holt, R. D. 2003. How should environmental stress affect the population dynamics of disease? Ecology Letters 6: 654–664.
Lannoo, M. J., Sutherland, D. R., Jones, P., Rosenberry, D., Klaver, R. W., Hoppe, D. M., Johnson, P. T. J., Lunde, K. B., Facemire, C., and Kapfer, J. M. 2003. Multiple causes for the malformed frog phenomenon. In: ATSM STP 1443, Linder, G., Little, E., Krest, S., and Sparling, D., eds., Multiple Stressor Effects in Relation to Declining Amphibian Populations. West Conshoshocken: ASTM International.
Marcogliese, D. J. 2001. Implications of climate change for parasitism of animals in the aquatic environment. Canadian Journal of Zoology 79: 1331–1352.
Martin, T. R., and D. B. Conn. 1990. The pathogenicity, localization and cyst structure of echinostomatid metacercariae (Trematoda) infecting the kidneys of the frogs *Rana clamitans* and *Rana pipiens*. Journal of Parasitology 76: 414–419.
McAlpine, D. F., and Burt, M. D. B. 1998. Helminths of bullfrogs, *Rana catesbeiana*, green frogs, *R. clamitans*, and leopard frogs, *R. pipiens* in New Brunswick. Canadian Field-Naturalist 112: 50–68.
McCallum, M. L., and Trauth, S. E. 2003. A forty-three year museum study of northern cricket frog (*Acris crepitans*) abnormalities in Arkansas: upward trends and distributions. Journal of Wildlife Diseases 39: 522–528.
McKenzie, V. J. 2007. Human land use and patterns of parasitism in tropical amphibian hosts. Biological Conservation 137: 102–116.
McKenzie, V. J., and Townsend, A. R. 2007. Parasitic and infectious disease responses to changing global nutrient cycles. Ecohealth 4: 384–396.
Mone, H., and Combes, C. 1986. Experimental analysis of the decoy effect exerted by nontarget mollusks on the *Biomphalaria glabrata* (Say, 1818) – *Schistosoma mansoni* Sambon, 1907 host–parasite system. Acta Oecologica-Oecologia Applicata 7: 281–286.
Moore, J. 2002. Parasites and the Behavior of Animals. New York: Oxford University Press.
Muzzall, P. M., Gillilland, M. G., Summer, C. S., and Mehne, C. J. 2001. Helminth communities of green frogs *Rana clamitans* Latreille, from southwestern Michigan. Journal of Parasitology 87: 962–968.
Najarian, H. H. 1952. The metacercariae of *Echinoparyphium flexum* (Linton) Dietz 1909 in frog kidneys. Journal of Parasitology 38: S38.
Najarian, H. H. 1954. Developmental stages in the life cycle of *Echinoparyphium flexum* (Linton, 1892) Dietz, 1910 (Trematoda, Echinostomatidae). Journal of Morphology 94: 165–197.
Ostfeld, R. S., and LoGiudice, K. 2003. Community disassembly, biodiversity loss, and the erosion of an ecosystem service. Ecology 84: 1421–1427.
Ouellet, M. 2000. Amphibian deformities: current state of knowledge. In: Sparling, D. W., Linder, G., and Bishop, C. A., eds., Ecotoxicology of Amphibians and Reptiles, pp. 617–661. Pensacola: Society of Environmental Toxicology and Chemistry.
Poulin, R. 2006. Global warming and temperature-mediated increases in cercarial emergence in trematode parasites. Parasitology 132: 143–151.
Prudhoe, S., and Bray, R. A. 1982. Platyhleminth parasites of the Amphibia. Oxford: Oxford University Press.
Riggin, G. T. 1956. A note on *Ribeiroia ondatrae* (Price, 1931) in Puerto Rico. Proceedings of the Helominthological Society 32: 28–29.
Schotthoefer, A. M., Cole, R. A., and Beasley, V. R. 2003. Relationship of tadpole stage to location of echinostome cercariae encystment and the consequences for tadpole survival. Journal of Parasitology 89: 475–482.

Schotthoefer, A. M., Labak, K. M., and Beasley, V. R. 2007. *Ribeiroia ondatrae* cercariae are consumed by aquatic invertebrate predators. Journal of Parasitology 93: 1240–1243.

Sessions, S. K., and Ruth, S. B. 1990. Explanation for naturally-occurring supernumerary limbs in Amphibians. Journal of Experimental Zoology 254: 38–47.

Shaw, J.C., Korzan, W.J., Carpenter R.E., Kuris, A.M., Lafferty, K.D., Summers, C.H., Øverli, Ø (*In Press*). Parasite manipulation of brain monoamines in California killifish (*Fundulus parvipinnis*) by the trematode Euhaplorchis californiensis. Proceedings of the Royal Society B.

Skelly, D. K., Bolden, S. R., Holland, M. P., Friedenburg, L. K., Friedenfelds, N. A., and Malcom, T. R. 2006. Urbanization and disease in amphibians. In: Collinge, S. K., and Ray, C., eds., Disease Ecology: Community Structure and Pathogen Dynamics, pp. 153–167. Cary, NC: Oxford University Press.

Skelly, D. K., Bolden, S. R., Freidenburg, L. K., Freidenfelds, N. A., Levey, R. 2007. *Ribeiroia* infection is not responsible for Vermont amphibian deformities. Ecohealth 4: 156–163.

Skerratt, L. F., Berger, L., Speare, R., Cashins, S., McDonald, K. R., Phillott, A. D., Hines, H. B., and Kenyon, N. 2007. Spread of chytridiomycosis has caused the rapid global decline and extinction of frogs. Ecohealth 4: 125–134.

Smyth, J. D., and Smyth, M. M. 1980. Frogs as host–parasite systems I. An introduction to parasitology through the parasites of *Rana temporaria, R. esculenta* and *R. pipiens*. London: Macmillan.

Sorensen, R. E., Kanev, I., Fried, B., and Minchella, D. J. 1997. The occurrence and identification of *Echinostoma revolutum* from North American *Lymnaea elodes* snails. Journal of Parasitology 83: 169–170.

Souder, W. 2000. A Plague of Frogs. New York: Hyperion.

Stopper, G. F., Hecker, L., Franssen, R. A., and Sessions, S. K. 2002. How trematodes cause limb deformities in amphibians. Journal of Experimental Zoology 294: 252–263.

Stuart, S. N., Chanson, J. S., Cox, N. A., Young, B. E., Rodrigues, A. S. L., Fischman, D. L., and Waller, R. W. 2004. Status and trends of amphibian declines and extinctions worldwide. Science 306: 1783–1786.

Sutherland, D. R. 2005. Parasites of North American Frogs. In: Lannoo, M. J., ed., Amphibian Declines: The Conservation Status of United States Species, pp. 109–123. California: University of California Press.

Taft, S. J., Suchow, K., and Vanhorn, M. 1993. Helminths from some Minnesota and Wisconsin raptors. Journal of the Helminthological Society of Washington 60: 260–263.

Taylor, C. N., Oseen, K. L., and Wassersug, R. J. 2004. On the behavioural response of *Rana* and *Bufo* tadpoles to echinostomatoid cercariae: implications to synergistic factors influencing trematode infections in anurans. Canadian Journal of Zoology 82: 701–706.

Thieltges, D. W., Jensen, K. T., and Poulin, R. 2008. The role of biotic factors in the transmission of free-living endohelminth stages. Parasitology 135: 407–426.

Thiemann, G. W., and Wassersug, R. J. 2000a. Biased distribution of trematode metacercariae in the nephric system of *Rana* tadpoles. Journal of Zoology 252: 534–538.

Thiemann, G. W., and Wassersug, R. J. 2000b. Patterns and consequences of behavioural responses to predators and parasites in *Rana* tadpoles. Biological Journal of the Linnean Society 71: 513–528.

Toledo, R., Muñoz-Antoli, C., and Fried, B. 2007. The use of echinostomes to study host–parasite relationships between larval trematodes and invertebrate and cold-blooded vertebrate hosts. Parasitology Research 100: 1177–1185.

Ulmer, M. J. 1970. Studies on the helminth fauna of Iowa I: trematodes of amphibians. American Midland Naturalist 83: 38–64.

Wisniewiski, W. L. 1958. Characterization of the parasitofauna of a eutrophic lake. Acta Parasitologica Polonica 6: 1–64.

Zander, C. D., Koçogluk, Ö, Skroblies, M., and Strohbach, U. 2002. Parasite populations and communities from the shallow littoral of the Orther Bight (Fehmarn, SW Baltic Sea). Parasitology Research 88: 734–744.

Chapter 12
Concurrent Infections of Echinostomes with Helminths and Protozoans

Gregory S. Noland and Thaddeus K. Graczyk

Contents

12.1 Introduction 282
12.2 Echinostome Coinfection with Heterologous Echinostomes 282
12.3 Echinostome Coinfection with Other Helminths 284
12.3.1 Echinostome Coinfection with Trematodes 284
12.3.2 Echinostome Coinfection with Tapeworms 286
12.3.3 Echinostome Coinfection with Nematodes 287
12.4 Echinostome Coinfection with Protozoans 287
12.4.1 Echinostome Coinfection with Plasmodium 287
12.4.2 Echinostome Coinfection with *Babesia* 289
12.4.3 Echinostome Coinfection with *Trypanosomes* 290
12.5 Concluding Remarks 290
References 291

Abstract Concomitant infection with helminths and other parasites is a frequent occurrence, due in part to the widespread distribution of helminths and chronic nature of infection. A multitude of studies have utilized echinostome parasites (Class: Trematoda) to examine the host response to coinfection, and this chapter discusses the use of echinostomes paired with (1) heterologous species of *Echinostoma*, (2) non-echinostome helminths, and (3) protozoan parasites in rodent models of coinfection. In general, the following trends were observed. Primary infection with echinostomes, which usually confers a high level of protection against homologous echinostome challenge, also tends to protect the infected host against heterologous echinostome challenge. Secondly, protection against echinostome challenge also appears to be induced by high-dose, patent, schistosome infection, as well as intestinal nematode infection. Finally, in contrast, primary chronic echinostome infection often enhances establishment of secondary schistosome infection, while likewise exacerbating sec-

G.S. Noland(✉) and T.K. Graczyk
Department of Pediatrics, Division of Infectious Diseases and Global Pediatrics Program, University of Minnesota, Minneapolis, MN, 55455, USA
e-mail: gsnoland@gmail.com

B. Fried and R. Toledo (eds.) *The Biology of Echinostomes*,

DOI: 10.1007/978-0-387-09577-6_12,

ondary protozoan infections. These interactions are dependent on the specific host used in the experiments, the specific parasites used and their dose, and the timing of the infections in relation to each other. Moreover, these studies suggest that echinostome infection results in a complex systemic host response that exhibits both specific and nonspecific effector mechanisms, though further study will be required to fully define the mechanistic determinants that influence host response to primary echinostome infection and to coinfection with other parasites.

12.1 Introduction

Infectious disease research has long recognized the importance of polyparasitism or coinfection – instances where a host is concomitantly infected with multiple pathogens – on disease susceptibility and outcome (Buck et al. 1978). Helminths are ubiquitous in tropical areas, and their distribution overlaps in epidemiologic time and space with that of many other tropical diseases. Additionally, chronic helminth infections can persist for years, providing ample opportunity for the host to be coinfected with a secondary infection. Recent studies have examined interactions between helminths and bacterial (Diniz et al. 2001; Resende et al. 2007) and viral pathogens (Berhe et al. 2007; Kamal et al. 2001), particularly HIV (Bentwich et al. 1999). Protozoan parasites, including *Plasmodium* and trypanosomes, also remain a primary source of morbidity and mortality in tropical areas. Many studies have therefore focused on helminth–protozoan interactions, as well as those between helminths and heterologous helminth species. Christensen and colleagues performed an extensive array of coinfection experiments throughout the 1980s that were summarized in a review written in 1987 (Christensen et al. 1987). Since then, a number of groups have continued to utilize various helminths to investigate host response to coinfection in rodent models of disease. As illustrated throughout this book, echinostomes serve as excellent models for biologic research. The present chapter thus aims to summarize those experiments that have employed these intestinal trematode parasites in coinfection research. Throughout, we will discuss the use of echinostomes paired with (1) heterologous species of *Echinostoma*, (2) nonechinostome helminths, and (3) protozoan parasites. Furthermore, we will attempt to synthesize results from these complex and often disparate interactions in order to gain insight into the nature of echinostome infection in the definitive laboratory host.

12.2 Echinostome Coinfection with Heterologous Echinostomes

Studies from Fried's laboratory have examined a number of important interactions in concurrent echinostome infections, particularly the effect of simultaneous infection with heterologous echinostome species on worm establishment, growth, and development. When golden hamsters (*Mesocricetus auratus*) were infected simultaneously with 20 ± 5 metacercariae of both *E. caproni* and *E. trivolvis* (referred to in the study as *E. liei* and *E. revolutum*, respectively), the mean number of *E. caproni* adult worms recovered was reduced compared to hamsters infected only with *E. caproni* (17.5% versus 54.4%) (Huffman et al. 1988). Additionally, *E. trivolvis* appeared to have an inhibitory effect on *E. caproni* establishment, as approximately three times as many (range: 2.3–4.0) *E. trivolvis* as *E. caproni* worms were recovered in animals given an equal inoculum of both parasites. Recovery of *E. trivolvis* worms was highly variable in both coinfected and singly infected rodents, and no significant differences were observed between hamsters infected only with *E. trivolvis* and those coinfected with *E. caproni*. Coinfection did result, however, in anterior relocalization of *E. trivolvis* adults from the ileum, where they are typically found in *E. trivolvis*-only infections, to the jejunum.

Similar experiments carried out with the same parasites in female ICR mice demonstrate that host differences play a major role in determining disease outcomes. In these experiments, *E. caproni* worm recovery was greater in coinfected mice given a dose of 25 metacercariae of both *E. trivolvis* and *E. caproni* compared to mice infected only with 25 *E. caproni* metacercariae (72.6% versus 37.8%) (Iorio et al. 1991). Recovery of *E. trivolvis* was also greater in coinfected mice compared to controls (14.2% versus 6.4%), but the difference was not statistically significant. Adult worms of both parasite species recovered from coinfected mice were similar in size at day 10 postinfection as those recovered from mice given an identical total single species inoculum. However, by day 14, worms from coinfected mice were significantly smaller than those from single species infections. No difference was found in these studies in the localization of either parasite in the gut between coinfected mice and those infected with either parasite alone.

Additional studies of simultaneous infection in outbred mice found that establishment of *E. caproni* was significantly reduced in mice given a simultaneous equivalent inoculum of *E. paraensei* compared to mice infected only with 25 *E. caproni* metacercariae (34% versus 61%) (Meece and Nollen 1996). Conversely, establishment of *E. paraensei* was greater in coinfected mice than in monoinfected controls (31% versus 15%). Such results reinforce the view that interaction between parasites during concomitant infection is species specific. Worms' growth was also affected by coinfection, as adult *E. paraensei* worms recovered from coinfected mice were initially larger at day 14 postinfection compared to those recovered from single species infection. However, at time points thereafter, *E. paraensei* worms were significantly smaller in comparison. Adult *E. caproni* worms from coinfected mice were larger or equal in size than those from monoinfected mice at all time points.

Primary infection with echinostomes generally results in protective immunity to subsequent homologous challenge (Sirag et al. 1980; Christensen et al. 1984). Investigators thus sought to determine whether this resistance was protective against secondary challenge with heterologous echinostome species. Hosier et al. (1988) found that a primary infection in outbred female ICR mice with 25 metacercariae of *E. trivolvis* (referred to as *E. revolutum*), which confers near complete protection against homologous *E. trivolvis* challenge (87% reduction in worm recovery), provided only minor protection (17%) against a secondary *E. caproni* infection (referred to as *E. liei*) given 21 days postprimary *E. trivolvis* infection. In contrast, primary *E. caproni* infection, which only conferred a moderate level of protection (42%) against homologous *E. caproni* challenge, resulted in a 79% decrease in *E. trivolvis* worm recovery when mice were challenged 23 days postprimary *E. caproni* infection. These challenge experiments were carried out at a time at which the primary parasite had been cleared either spontaneously (*E. trivolvis*) or through anthelmintic treatment (*E. caproni*) prior to secondary challenge. When *E. caproni* was allowed to persist in ICR mice, such that the secondary heterologous *E. trivolvis* challenge was given during patent *E. caproni* infection (21 days postinfection), a reduction in *E. trivolvis* worm establishment was also observed (55%).

Fujino et al. (1996a) found that a primary *E. trivolvis* infection of 40 metacercariae in male C3H mice conferred complete (100% reduction in worm recovery), or near complete (97% reduction), protection against a secondary heterologous challenge of 40 metacercariae of *E. caproni* when the challenge was made on day 10 or day 16, respectively, postprimary *E. trivolvis* infection. However, when *E. caproni* challenge was given at day 20 post-*E. trivolvis* infection, the level of protection fell to 48%. Mice with a 10-day-old primary *E. trivolvis* infection were also protected against challenge with 25, 5-day old *E. caproni* juvenile worms harvested from either mice or hamsters and intubated directly into the recipient gut (93 and 100% reduction in adult worm recovery, respectively). In primary infections only with *E. trivolvis*, C3H mice spontaneously expelled adult worms beginning around day 10 postinfection, with complete expulsion occurring by day 15 (Fujino et al. 1996a). This expulsion was closely correlated with goblet cell hyperplasia, which peaked between days 8 and 12 post-*E. trivolvis* infection, but remained elevated up to day 24 postinfection compared to uninfected controls. Therefore, it has been suggested that goblet cell mucins, which are involved in protection against primary *E. trivolvis* infection (Fujino et al. 1993), may also mediate resistance to secondary homologous and heterologous echinostome infection (Fujino et al. 1996a). However, experiments that directly evaluate this hypothesis have not been performed. Furthermore, the effector mechanisms mounted in response to echinostome infection are largely species specific (Toledo et al. 2006), and continued study is required to define the common, as well as unique, factors mediating resistance in each echinostome system.

12.3 Echinostome Coinfection with Other Helminths

12.3.1 Echinostome Coinfection with Trematodes

In addition to these studies exploring interactions between homologous and heterologous species of echinostomes, various laboratories have developed coinfection models using echinostomes paired with other nonechinostome helminths. Such studies have almost exclusively used *Schistosoma* spp. Early studies from Christensen's laboratory found that chronic, patent, 79- or 99-day-old *Schistosoma mansoni* infections in inbred female albino mice conferred complete protection against low-dose (six metacercariae) secondary *E. caproni* infection (often referred to as *E. revolutum* or *E. liei* in Christensen's studies), as measured by adult worm recovery following necropsy one week postechinostome challenge (Sirag et al. 1980). Likewise, *S. mansoni* infections of only 43 days resulted in significant, but incomplete (73% reduction in echinostome worm burden), resistance to echinostome challenge. The antagonistic response to secondary *E. caproni* infection was evident as early as 2 h postchallenge, with complete elimination of echinostome worms occurring within 24 h (Christensen et al. 1981b).

Such resistance appears to depend on the active production of schistosome eggs, as neither prepatent (Sirag et al. 1980) nor single-sex (90-day) (Christensen et al. 1981b) primary *S. mansoni* infections were able to induce significant protection against secondary echinostome challenge. However, some delay from the commencement of egg production to the acquisition of "resistant" phenotype seems to be required, as mice with newly patent (36 days) *S. mansoni* infections showed no difference in *E. caproni* establishment compared to schistosome-free mice (Christensen et al. 1981b). It was also shown that adult echinostomes persist in these schistosome coinfected mice, despite the fact that mice with similarly aged *S. mansoni* infection are resistant to *E. caproni* metacercarial challenge. These findings indicate that the factors contributing to resistance act primarily against the early stages of echinostome parasites – either through impairment of metacercarial excystation and/or direct killing of newly excysted larvae.

In contrast to these experiments, which were all performed with an inoculum of approximately 120 *S. mansoni* cercariae, subsequent experiments demonstrated that low-dose (≤10 worm pairs per mouse), prepatent (4-weeks old), primary *S. mansoni* infections had minimal effect on establishment of secondary *E. caproni* challenge in multiple strains of mice (Christensen et al. 1985). However, all coinfected mice showed a significant delay in the spontaneous resolution of echinostome infection, as measured by the cessation of fecal egg output. Such results illustrate that parasite–parasite interactions are also dose dependent.

Primary patent infection (100-day old) with high dose (120 cercariae) of the related parasite, *S. bovis*, also resulted in nearly complete resistance (97.7% reduction in worm burden) to secondary *E. caproni* challenge in outbred female albino mice (Christensen et al. 1981b). However, similar experiments from the same laboratory found that a 56-day-old primary *S. bovis* infection in inbred female albino mice resulted in markedly lower tissue egg counts and failed to confer resistance to secondary *E. caproni* challenge (Sirag et al. 1980).

While primary patent schistosome infection appears to generally induce resistance to heterologous echinostome challenge, echinostomes, when given as primary pathogen, often seem to enhance subsequent schistosome infection. It was first observed that patent 14- and 21-day-old *E. caproni* infections did not induce significant alteration to *S. mansoni* establishment (Sirag et al. 1980). However, a primary dose of 35 metacercariae given to inbred female albino mice at days 20, 24, or 33 postechinostome infection resulted in significant increase (66–91%) in the number of *S. mansoni* adults recovered, versus age-matched controls infected only with 120 *S. mansoni* cercariae (Christensen et al. 1981a).

A similar increase (70–75%) in adult schistosome worm recovery was observed in outbred female albino mice harboring a primary infection caused by 50 *E. caproni* metacercariae when challenged with *S. mansoni* at days 25 and 34 postechinostome infection (Christensen et al. 1988). Primary patent *E. caproni* infection appeared to have no enhancing effect, however, on establishment of *S. haematobium*, although recovery rates were low (approximately 2%) for both schistosome only and coinfected groups of mice.

A recent study by Maldonado et al. (2001) investigated experimental permutations of a naturally occurring coinfection system between *E. paraensei* and *S. mansoni* in the water rat *Nectomys squamipes*. In this host, 4-week-old patent primary experimental infections with 50 *E. paraensei* metacercariae reduced establishment of a wild strain of *S. mansoni* by 65–80% compared to rats infected only with *S. mansoni*. Conversely, there was an increase in the recovery of echinostome worms from coinfected rats, suggesting that coinfection with *S. mansoni* enables prolonged persistence of primary *E. paraensei* infections in this natural host. However, similar experiments using a laboratory strain of *S. mansoni* in rats, as well as in male albino mice, revealed no difference in schistosome worm recovery between rodents coinfected with *E. paraensei* and those infected only with *S. mansoni*. No consistent trends were observed with respect to *E. paraensei* recovery in rats or mice coinfected with the laboratory *S. mansoni* strain.

12.3.2 Echinostome Coinfection with Tapeworms

Christensen's group has also examined the interaction between echinostomes and tapeworms using *Hymenolepis diminuta* and *H. microstoma*. (Andreassen et al. 1990). When given simultaneously with a dose of 25 *E. caproni* metacercariae, no difference was found in number of *H. diminuta* worms recovered from outbred female NMRI mice versus tapeworm-only infected mice challenged with three cysticercoids. However, a significant reduction in *H. diminuta* establishment was observed when given to mice 7 (prepatent) or 25 (patent) days post-*E. caproni* infection with 25 metacercariae, or 7 days postinfection with six metacercariae. This result is somewhat surprising, as the resistance conferred by primary schistosome infection described appears to depend largely on the active secretion of eggs. The fact that primary echinostome infection provides protection against secondary tapeworm challenge even before the onset of egg production suggests that a distinct set of protective factors – possibly including the hypothesized secretion of mucins from goblet cells – may operate in the echinostome-tapeworm coinfection system. When a related tapeworm, *H. microstoma* was used in similar experiments with a primary 7-day-old *E. caproni* infection of 25 metacercariae, no reduction in *H. microstoma* establishment occurred compared to mice infected only with five *H. microstoma* cysticercoids (Andreassen et al. 1990). Unfortunately, experiments designed to determine the impact of patent *E. caproni* infection of *H. microstoma* were not reported.

Additional studies examining concurrent infection with three different parasite species found that simultaneous challenge with *H. diminuta* (three cysticercoids) and

H. microstoma (five cysticercoids) in mice with a primary 7-day-old *E. caproni* infection (25 metacercariae) reduced establishment of both tapeworm species compared to singly infected mice (Andreassen et al. 1990). However, when given simultaneously to mice in the absence of a pre-existing echinostome infection, a reduction in *H. diminuta*, but not *H. microstoma*, was observed. Previous studies have demonstrated that newly excysted and adult *H. diminuta* worms, but not those of *H. microstoma*, are susceptible to complement-mediated lysis (Bogh et al. 1986). Thus the reduction in *H. diminuta* may result from an *E. caproni*-induced increase in intestinal permeability and plasma influx into the lumen (Andreassen et al. 1990). However, why *H. microstoma* should be affected by triplicate infection with both *E. caproni* and *H. diminuta*, but neither in a dual infection, remains to be determined.

12.3.3 Echinostome Coinfection with Nematodes

Intestinal nematodes have also recently been employed in echinostome coinfection studies. Primary 8-day-old patent infections in outbred male C3H/HeN mice with approximately 500 *Nippostrongylus brasiliensis* L3 larvae induced complete resistance (0% recovery) to challenge with 40 metacercariae of either *E. trivolvis* or *E. caproni*, as determined by worm recovery 5 days postechinostome infection (Fujino et al. 1996b). Mice infected only with *N. brasiliensis* spontaneously and completely resolve nematode infection by day 11 postinfection in a process thought to depend largely on mucins produced by goblet cells (Fujino et al. 1996b; Uber et al. 1980). When secondary echinostome infection challenge was given at day 16 post nematode infection, a time at which mice no longer harbored adult worms and the goblet cell response had nearly returned to baseline levels, mice were only partially resistant to challenge with 40 metacercariae of *E. trivolvis* or *E. caproni* (44.3 and 22.7% reduction in worm recovery compared to echinostome-only infected controls). As discussed, protection against echinostome establishment is also thought to rely upon goblet cell mucin production (Fujino et al. 1996a), and it appears that activation of goblet cells by homologous and heterologous echinostomes or heterologous intestinal helminths is sufficient to induce expulsion of secondary echinostome challenge. However, the role for other factors in mediating resistance has yet to be conclusively excluded.

12.4 Echinostome Coinfection with Protozoans

12.4.1 Echinostome Coinfection with *Plasmodium*

The most frequently studied helminth coinfection system is probably that of helminths and protozoans. This is due to the widespread interest in helminth and malaria coinfection, which has been studied extensively in humans (Mwangi et al. 2006) and in rodents (Hartgers and Yazdanbakhsh 2006). While many of these studies have involved helminths other than echinostomes, a few groups have taken advantage of the echinostome rodent model system to elucidate interactions between these parasites. Here we will highlight these experiments as well as those studies examining echinostome coinfection with non-*Plasmodium* protozoans.

Christensen et al. (1988) found that a primary, 30-day-old *E. caproni* infection of 50 metacercariae in outbred female albino mice induced a significant increase in *P. yoelii* (nonlethal [NL] strain) parasitemia following infection with 10^3 blood stage parasites compared to worm-free *Plasmodium*-infected mice. However, this result was only observed in one experiment and could not be duplicated in several subsequent attempts. When they examined resistance to rechallenge with *P. yoelii*, they found that complete immunity to rechallenge was not impaired by concurrent *E. caproni* infection.

Noland and colleagues have similarly examined multiple aspects of *E. caproni* and *P. yoelii* coinfection in different strains of inbred mice. They found that the normal mild course of malaria induced by blood stage passage of 10^5–10^6 nonlethal *P. yoelii* (strain 17X NL) parasites in male BALB/c mice (Kopacz and Kumar 1999), was not altered by infection with 10 *E. caproni* metacercarial cysts given simultaneous to, or 5 or 10 days following, malaria infection (2008). However, a primary *E. caproni* infection given at times between 5 and 35 days prior to *Plasmodium* infection resulted in significantly enhanced malaria parasitemia (Noland et al. 2005). This enhancement led to fatalities in approximately 50% of mice, but was completely reversed when mice were treated with an anthelmintic 10 days prior to malaria challenge. Recent studies have shown that such exacerbation of malaria may be mediated by echinostome-induced immune regulation, as *E. caproni-P. yoelii* coinfected BALB/c mice were impaired in production of IFN-γ, a Th1 cytokine that is critical for control of *P. yoelii* (Noland et al. 2008). In contrast to BALB/c mice, C57BL/6 mice are susceptible to high-density malaria parasitemia (Kopacz and Kumar 1999). When *E. caproni* infection was initiated simultaneous with or 5–35 days prior to *P. yoelii* infection, no significant difference was observed in high peak parasitemia between coinfected and worm-free *Plasmodium*-infected mice (unpublished data). However, when *E. caproni* infection was initiated 5 or 10 days after *P. yoelii* infection, a significant reduction in peak parasitemia was observed (unpublished data). Additional unpublished studies found that enhancement of malaria in BALB/c mice during chronic *E. caproni* infection also occurred when the *Plasmodium* infection was initiated via sporozoite, as opposed to blood stage, challenge. Chronic *E. caproni* infection did not alter the course of a lethal strain of *P. yoelii* (17X L), as all malaria infected mice, whether alone, or in combination with 25-day-old *E. caproni* infection, died of hyperparasitemia within 10 days postmalaria infection (Noland et al. 2005).

Transmission of malaria parasites to mosquitoes was also enhanced by chronic, low-dose (ten metacercariae) *E. caproni* infection given 4 weeks prior to *P. yoelii* infection (Noland et al. 2007). *Anopheles stephensi* mosquitoes exposed to coinfected BALB/c mice 5 days post malaria infection had a significantly higher rate of infectivity (80.1%) and a twofold increase in the number of *P. yoelii* midgut oocysts (geometric mean = 19.2) compared to mosquitoes exposed to malaria-only infected mice (72.0%; 10.5, respectively). The number of *Plasmodium* sporozoites recovered from mosquito salivary glands was also greater in those exposed to coinfected mice. While the mechanisms underlying such enhanced transmission are not yet clear, we hypothesize that this results from an echinostome-induced impairment of host response to transmissible stages of *Plasmodium*. Furthermore, such results suggest that by modulating malaria transmission potential, helminth coinfection not only influences disease outcome in the individual host, but also in the greater host community.

These two laboratories have also examined the effect of *Plasmodium* infection on echinostome establishment, challenge, and fecundity. A primary *P. yoelii* infection in outbred female albino mice with 10^3 blood stage parasites 10–12 days before a secondary *E. caproni* challenge increased the proportion of helminth infected mice and the percentage of overall adult worm recovery compared to those infected only with *E. caproni* (Christensen et al. 1988). A primary 2–5 week-old infection of 10–15 *E. caproni* metacercariae in outbred female albino mice induces high-level resistance (89–100%) to secondary homologous challenge with *E. caproni* (Christensen et al. 1984, 1988). However, when a *P. yoelii* infection was given 4 days after the primary *E. caproni* infection, resistance was completely impaired, as a similar number of worms were recovered from coinfected mice as those infected only with a single challenge dose of *E. caproni* (Christensen et al. 1988). Noland et al. (2005) found that superimposed *P. yoelii* infection had no effect on *E. caproni* worm persistence or fecal egg output. Subsequent studies, though, revealed that concurrent *P. yoelii* infection markedly suppressed the IL-4 response normally generated during infection only with *E. caproni* (Noland et al. 2008)

12.4.2 Echinostome Coinfection with Babesia

Examination of additional protozoans in echinostome coinfections has only included other hemoprotozoans. One such organism is *Babesia microti*. Even though *Babesia* is closely related to *Plasmodium*, the effects between the two on echinostome coinfection are markedly different. Infection with 10^6 *B. microti* 1 week prior to, simultaneous with, or 1 week following *E. caproni* infection did not alter the expulsion of a primary low-dose (six metacercariae) *E. caproni* infection (Christensen et al. 1988). *B. microti* also had no effect on the resistance to secondary homologous challenge with *E. caproni*, when given simultaneous to a primary *E. caproni* infection of 15 metacercariae (Christensen et al. 1988). No explanation has been proposed as to why these similar parasites should induce such distinct susceptibilities. Interestingly, when examining the effect of primary echinostome infection on superimposed *B. microti* pathogenesis, results were consistent with those from *P. yoelii* infections. A high dose of *E. caproni* (50 metacercariae) was found to consistently induce an increase in *B. microti* parasitemia, and also not compromise immunity to homologous *B. microti* challenge (Christensen et al. 1988).

12.4.3 Echinostome Coinfection with Trypanosomes

While differences in outcome were seen between *Plasmodium*-echinostome and *Babesia*-echinostome model systems, the interactions between echinostomes and the hemoflagelate *Trypanosoma brucei* closely parallel those of *Plasmodium*, even though *T. brucei* is more distantly related to *Plasmodium* than *Babesia*. A primary *T. brucei* infection in outbred female albino mice with 10^4 parasites 18–20 days before challenge with 12 *E. caproni* metacercariae increased the proportion of helminth-infected mice and the percentage of overall adult worm recovery compared to those infected only with *E. caproni* (Christensen et al. 1988). More detailed studies at a lower echinostome infection dose (six metacercariae) demonstrated that *E. caproni* worms persisted for up to 6 weeks in mice given 10^4*T. brucei* parasites 1–3 weeks prior to echinostome infection, whereas mice infected only with *E. caproni* spontaneously and completely resolved infection by week 5 postechinostome infection (Christensen et al. 1984). Mice infected with *T. brucei* 2 or 3 weeks post-*E. caproni* infection were ultimately able to resolve echinostome infection, but experienced a significant delay in worm expulsion. As previously mentioned, a primary 2–5 week-old infection of 10–15 *E. caproni* metacercariae in outbred female albino mice induces high-level resistance (89–100%) to secondary homologous challenge with *E. caproni* (Christensen et al. 1984, 1988). When a *T. brucei* infection was given 1 week prior to, or 1 and 3 weeks following, the primary *E. caproni* infection, resistance was completely impaired, as there was no significant difference in worm recovery between coinfected mice and those infected only with a single challenge dose of *E. caproni* (Christensen et al. 1984, 1988).

12.5 Concluding Remarks

Results from these detailed coinfection studies using echinostomes in combination with other parasites reveal a spectrum of biological and immunological interactions. As with any host-pathogen model, these interactions appear dependent on the specific host used in the experiments, the specific parasite used, the dose of the infections, and the timing of the infections in relation to each other. Nonetheless, several trends seem to emerge from these studies. Firstly, primary infection with echinostomes, which usually confers a high level of protection against homologous echinostome challenge, also tends to protect the infected host against heterologous echinostome challenge. Secondly, protection against echinostome challenge also appears to be induced by high-dose, patent, schistosome infection, as well as intestinal nematode infection. Thirdly, in contrast, primary chronic echinostome infection often enhances establishment of secondary schistosome infection and exacerbates secondary protozoan infection. These generalizations are not absolute, but they do provide a framework for interpreting interactions during coinfection and developing future hypotheses.

Throughout this chapter, we have considered how the host response to echinostome infection, which has been covered in detail elsewhere in this book (see Chap. 8), effects, and is affected by, concurrent infection. One of the benefits of coinfection research is that it illuminates host response to single infection: we are able to view that primary host response from a different perspective by observing how it is altered in the context of concomitant infection. For echinostomes, in particular, host resistance to homologous echinostome challenge is predicted to rely on nonspecific mechanisms such as goblet cell hyperplasia and subsequent mucin production (Fujino et al. 1996a). It makes sense, then, that this response seems to have similar activity against subsequent infection with an intestinal parasite that inhabits the same niche–whether that parasite is a heterologous echinostome or a heterologous helminth. However, such mechanisms are unable to act against parasites colonizing disparate locations. As echinostome infection often has a dramatic effect on host susceptibility to both blood- and tissue-borne parasites, interactions induced by echinostome coinfection must also involve systemic alterations. Indeed, many previous studies have shown that echinostome infection induces a significant antibody response (Toledo et al. 2006). Likewise, murine hosts mount a significant cytokine response to infection (Brunet et al. 2000; Noland et al. 2008) that we hypothesize is involved in exacerbation of malaria. Furthermore, unpublished data from our laboratory demonstrate that echinostome infection adversely affects antibody response to a DNA-based malaria vaccine. Taken together, these results indicate that echinostome infection results in a complex systemic host response that exhibits both specific and nonspecific effector mechanisms. Continued pursuit of these exciting studies will be important in elucidating the various interactions observed during primary echinostome infection and during coinfection with other parasites.

References

Andreassen, J., Odaibo, A. B., and Christensen, N. O. 1990. Concurrent infections of the trematode *Echinostoma caproni* and the tapeworms *Hymenolepis diminuta* and *Hymenolepis microstoma* in mice. Journal of Parasitology 76: 573–575.

Bentwich, Z., Kalinkovich, A., Weisman, Z., Borkow, G., Beyers, N., and Beyers, A. D. 1999. Can eradication of helminthic infections change the face of AIDS and tuberculosis? Immunology Today 20: 485–487.

Berhe, N., Myrvang, B., and Gundersen, S. G. 2007. Intensity of *Schistosoma mansoni*, hepatitis B, age, and sex predict levels of hepatic periportal thickening/fibrosis (PPT/F): a large-scale community-based study in Ethiopia. American Journal of Tropical Medicine and Hygiene 77: 1079–1086.

Bogh, H. O., Christensen, J. P., and Andreassen, J. 1986. Complement-mediated lysis in vitro of newly excysted tapeworms: *Hymenolepis diminuta*, *Hymenolepis microstoma*, *Hymenolepis nana* and *Hymenolepis citelli*. International Journal for Parasitology 16: 157–161.

Brunet, L. R., Joseph, S., Dunne, D. W., and Fried, B. 2000. Immune responses during the acute stages of infection with the intestinal trematode *Echinostoma caproni*. Parasitology 120: 565–571.

Buck, A. A., Anderson, R. I., and MacRae, A. A. 1978. Epidemiology of poly-parasitism. IV. Combined effects on the state of health. Tropenmedizin und Parasitologie 29: 253–268.

Christensen, N. O., Nydal, R., Frandsen, F., and Nansen, P. 1981a. Homologous immunotolerance and decreased resistance to *Schistosoma mansoni* in *Echinostoma revolutum*-infected mice. Journal of Parasitology 67: 164–166.

Christensen, N. O., Nydal, R., Frandsen, F., Sirag, S. B., and Nansen, P. 1981b. Further studies on resistance to *Fasciola hepatica* and *Echinostoma revolutum* in mice infected with *Schistosoma* sp. Zeitschrift für Parasitenkunde 65: 293–298.

Christensen, N. O., Fagbemi, B. O., and Nansen, P. 1984. *Trypanosoma brucei*-induced blockage of expulsion of *Echinostoma revolutum* and of homologous *E. revolutum* resistance in mice. Journal of Parasitology 70: 558–561.

Christensen, N. O., Knudsen, J., Fagbemi, B., and Nansen, P. 1985. Impairment of primary expulsion of *Echinostoma revolutum* in mice concurrently infected with *Schistosoma mansoni*. Journal of Helminthology 59: 333–335.

Christensen, N. O., Nansen, P., Fagbemi, B. O., and Monrad, J. 1987. Heterologous antagonistic and synergistic interactions between helminths and between helminths and protozoans in concurrent experimental infection of mammalian hosts. Parasitology Research 73: 387–410.

Christensen, N. O., Furu, P., Kurtzhals, J., and Odaibo, A. 1988. Heterologous synergistic interactions in concurrent experimental infection in the mouse with *Schistosoma mansoni*, *Echinostoma revolutum*, *Plasmodium yoelii*, *Babesia microti*, and *Trypanosoma brucei*. Parasitology Research 74: 544–551.

Diniz, L. M., Zandonade, E., Dietze, R., Pereira, F. E., and Ribeiro-Rodrigues, R. 2001. Short report: do intestinal nematodes increase the risk for multibacillary leprosy? American Journal of Tropical Medicine and Hygiene 65: 852–854.

Fujino, T., Fried, B., and Tada, I. 1993. The expulsion of *Echinostoma trivolvis*: worm kinetics and intestinal cytopathology in conventional and congenitally athymic BALB/c mice. Parasitology 106: 297–304.

Fujino, T., Fried, B., Ichikawa, H., and Tada, I. 1996a. Rapid expulsion of the intestinal trematodes *Echinostoma trivolvis* and *E. caproni* from C3H mice by trapping with increased goblet cell mucins. International Journal for Parasitology 26: 319–324.

Fujino, T., Yamada, M., Ichikawa, H., Fried, B., Arizono, N., and Tada, I. 1996b. Rapid expulsion of the intestinal trematodes *Echinostoma trivolvis* and *E. caproni* from C3H/HeN mice after infection with *Nippostrongylus brasiliensis*. Parasitology Research 82: 577–579.

Hartgers, F. C., and Yazdanbakhsh, M. 2006. Co-infection of helminths and malaria: modulation of the immune responses to malaria. Parasite Immunology 28: 497–506.

Hosier, D. W., Fried, B., and Szewczak, J. P. 1988. Homologous and heterologous resistance of *Echinostoma revolutum* and *E. liei* in ICR mice. Journal of Parasitology 74: 89–92.

Huffman, J. E., Alcaide, A., and Fried, B. 1988. Single and concurrent infections of the golden hamster, *Mesocricetus auratus*, with *Echinostoma revolutum* and *E. liei* (Trematoda: Digenea). Journal of Parasitology 74: 604–608.

Iorio, S. L., Fried, B., and Hosier, D. W. 1991. Concurrent infections of *Echinostoma caproni* and *Echinostoma trivolvis* in ICR mice. International Journal for Parasitology 21: 715–717.

Kamal, S. M., Bianchi, L., Al Tawil, A., Koziel, M., El Sayed Khalifa, K., Peter, T., and Rasenack, J. W. 2001. Specific cellular immune response and cytokine patterns in patients coinfected with hepatitis C virus and *Schistosoma mansoni*. Journal of Infectious Diseases 184: 972–982.

Kopacz, J., and Kumar, N. 1999. Murine gamma delta T lymphocytes elicited during *Plasmodium yoelii* infection respond to *Plasmodium* heat shock proteins. Infection and Immunity 67: 57–63.

Maldonado, A., Jr., Coura, R., da SIlva, G., Lanfredi, R. M., and Rey, L. 2001. Changes on *Schistosoma mansoni* (Digenea: Schistosomatidae) worm load in *Nectomys squamipes* (Rodentia: Sigmodontinae) concurrently infected with *Echinostoma paraensei* (Digenea: Echinostomatidae). Memórias do Instituto Oswaldo Cruz 96: S193–S198.

Meece, J. K., and Nollen, P. M. 1996. A comparison of the adult and miracidial stages of *Echinostoma paraensei* and *E. caproni*. International Journal for Parasitology 26: 37–43.

Mwangi, T. W., Bethony, J. M., and Brooker, S. 2006. Malaria and helminth interactions in humans: an epidemiological viewpoint. Annals of Tropical Medicine and Parasitology 100: 551–570.

Noland, G. S., Graczyk, T. K., Fried, B., Fitzgerald, E. J., and Kumar, N. 2005. Exacerbation of *Plasmodium yoelii* malaria in *Echinostoma caproni* infected mice and abatement through anthelmintic treatment. Journal of Parasitology 91: 944–948.

Noland, G. S., Graczyk, T. K., Fried, B., and Kumar, N. 2007. Enhanced malaria parasite transmission from helminth co-infected mice. American Journal of Tropical Medicine and Hygiene 76: 1052–1056.

Noland, G. S., Urban, J. F., Jr., Fried, B., and Kumar, N. 2008. Counter-regulatory anti-parasite cytokine responses during concurrent *Plasmodium yoelii* and intestinal helminth infections in mice. Experimental Parasitology 119: 272–278.

Resende Co, T., Hirsch, C. S., Toossi, Z., Dietze, R., and Ribeiro-Rodrigues, R. 2007. Intestinal helminth co-infection has a negative impact on both anti-*Mycobacterium tuberculosis* immunity and clinical response to tuberculosis therapy. Clinical and Experimental Immunology 147: 45–52.

Sirag, S. B., Christensen, N. O., Frandsen, F., Monrad, J., and Nansen, P. 1980. Homologous and heterologous resistance in *Echinostoma revolutum* infections in mice. Parasitology 80: 479–486.

Toledo, R., Esteban, J. G., and Fried, B. 2006. Immunology and pathology of intestinal trematodes in their definitive hosts. Advances in Parasitology 63: 285–365.

Uber, C. L., Roth, R. L., and Levy, D. A. 1980. Expulsion of *Nippostrongylus brasiliensis* by mice deficient in mast cells. Nature 287: 226–228.

Chapter 13
Use of *Echinostoma* spp. in Studies on Chemotherapy and Metabolic Profiling

Jasmina Saric, Jia V. Li, Yulan Wang, Elaine Holmes, Jürg Utzinger, and Jennifer Keiser

Contents

13.1	Introduction	296
13.2	Use of *Echinostoma spp.* in Trematocidal Drug Discovery and Development	297
	13.2.1 Evaluation of Echinostomicidal Activity *In Vivo*	297
	13.2.2 Drug Sensitivity Assay with *Echinostoma caproni In Vitro*	300
	13.2.3 Scanning Electron Microscopic (SEM) Studies	300
	13.2.4 Spectroscopy	302
13.3	Metabonomic Investigation on *Echinostoma caproni* in the Mouse Model	302
	13.3.1 What Is Metabonomics?	302
	13.3.2 Experimental Protocol	305

J. Keiser (✉)
Department of Medical Parasitology and Infection Biology, Swiss Tropical Institute, P.O. Box, CH-4002 Basel, Switzerland
e-mail: Jennifer.keiser@unibas.ch

J. Saric and J. V. Li
Department of Public Health and Epidemiology, Swiss Tropical Institute, P.O. Box, CH-4002 Basel, Switzerland; Department of Biomolecular Medicine, Division of Surgery, Oncology, Reproductive Biology and Anaesthetics (SORA), Faculty of Medicine, Imperial College London, Sir Alexander Fleming Building, South Kensington, London SW7 2AZ, UK

Y. Wang
Department of Biomolecular Medicine, Division of Surgery, Oncology, Reproductive Biology and Anaesthetics (SORA), Faculty of Medicine, Imperial College London, Sir Alexander Fleming Building, South Kensington, London SW7 2AZ, UK

Present address: State Key Laboratory of Magnetic Resonance and Atomic and Molecular Physics, Wuhan Center for Magnetic Resonance, Wuhan Institute of Physics and Mathematics, Chinese Academy of Sciences, Wuhan 430071, People's Republic of China

E. Holmes
Department of Biomolecular Medicine, Division of Surgery, Oncology, Reproductive Biology and Anaesthetics (SORA), Faculty of Medicine, Imperial College London, Sir Alexander Fleming Building, South Kensington, London SW7 2AZ, UK

J. Utzinger
Department of Public Health and Epidemiology, Swiss Tropical Institute, P.O. Box, CH-4002 Basel, Switzerland

B. Fried and R. Toledo (eds.) *The Biology of Echinostomes*,
DOI: 10.1007/978-0-387-09577-6_13,

13.3.3 Biomarker Identification and Diagnostic Potential of Biofluids in an *Echinostoma caproni* Infection 306
13.3.4 Biomarker Identification in Tissue via MAS^{1}H NMR Spectroscopy 312
13.3.5 Changes in the Intestinal Profiles 314
13.3.6 Integrative Metabonomics: An Approach to Understand Host–Parasite–Gut Microflora Interactions 317
13.4 Concluding Remarks 319
References 320

Abstract In this chapter we briefly summarize the use of echinostomes in chemotherapeutic studies and provide a comprehensive overview of metabolic profiling using an *Echinostoma caproni*-mouse model. *In vitro* and *in vivo* assays with *Echinostoma* spp. are suitable means for screening of anthelminthic drugs. Indeed, these models have been utilized for more than three decades and play an important role for discovery and development of novel drugs and treatment options against major trematode infections. Rapid diagnosis of infection and monitoring of disease progression and resolution are of pivotal importance for personalized health care and disease control at the population level. Spectroscopic profiling of the host metabolism, in combination with multivariate statistical methods, is a powerful approach for biomarker discovery that give rise for diagnosis and prognosis of infection and disease stages, and for subtle monitoring of control interventions. The comparison of biofluids and tissue samples obtained from mice experimentally infected with *E. caproni* and from noninfected control animals revealed a high number of biomarkers in different kinds of biofluids (e.g., blood plasma, fecal water, and urine). Several tissues also showed significant metabolic changes after a chronic infection with *E. caproni*. While urine and blood plasma are obvious and convenient choices for diagnostic purposes, the study on tissue samples complemented the information extracted from biofluids and aided obtaining a deeper understanding of biochemical mechanisms of disease.

13.1 Introduction

This chapter reviews the use of echinostomes in chemotherapeutic studies and in metabolic profiling. Chemotherapy is one of the main pillars for the control of food-borne trematode infections, including echinostomiasis. However, the current chemotherapeutic arsenal is limited to two drugs; namely (1) triclabendazole for the treatment of fascioliasis, and (2) praziquantel against clonorchiasis, opisthorchiasis, paragonimiasis, and intestinal fluke infections (Keiser and Utzinger 2004). Disposing of only two drugs for diseases that affect at least 40 million individuals, with some 750 million people at risk (more than 10% of the world's population) is potentially dangerous. Moreover, food-borne trematodiasis is emerging (Chai et al. 2005; Keiser and Utzinger 2005), and there is considerable concern regarding the development of drug-resistant parasite strains as unexpectedly low cure rates have been observed in clonorchiasis patients following praziquantel administration (Tinga et al. 1999). A vaccine against food-borne trematodiasis is not expected to arrive for at least another decade or so. Changing human behavior as a means to reduce transmission of the diseases is a formidable challenge due to cultural habits (e.g., consumption of raw fish and other aquatic products) (Haswell-Elkins and Levri 2003). Hence, development of novel trematocidal drugs is of great importance in the period while praziquantel and triclabendazole remain effective. An *Echinostoma*-mouse model holds promise in trematocidal drug discovery and development, and metabolic profiling can serve as a platform for biomarker discovery with diagnostic, prognostic, and drug/vaccine target potential.

In the next part we summarize studies that have utilized *Echinostoma* spp. as a drug discovery tool. Emphasis is placed on *in vivo*, *in vitro*, and scanning electron microscopic (SEM) investigations. Spectroscopy is another interesting tool that has been applied in drug discovery and development (Borges and Figueroa-Villar 2001). However, while spectroscopy has been widely used in drug research and development (R&D) for antiparasitic drugs (Akoachere et al. 2005) to our knowledge, it has not been specifically applied to echinostomiasis. Advances in spectroscopic technology lend themselves to profiling the metabolic response of the host organism to parasite invasion. Hence, the third part of the chapter is focused on illuminating the role of the spectroscopic strategies in profiling host–parasite and host–parasite–drug interactions. The main focus of its application to *E. caproni* has been to establish a comprehensive metabolic response of the mouse to an *E. caproni* infection (Saric et al. 2008a), and to compare the biochemical profiles in the biofluids and tissues of infected with noninfected control mice.

13.2 Use of *Echinostoma spp.* in Trematocidal Drug Discovery and Development

13.2.1 Evaluation of Echinostomicidal Activity In Vivo

Rodents infected with *Echinostoma* spp. are suitable models for anthelminthic studies, justified as follows. First, it is relatively straightforward to maintain the life cycle in the laboratory. In the case of *E. caproni*, for example, mice are used as the definitive host and *Biomphalaria glabrata* snails as first and second intermediate host (Fried and Huffman 1996). In contrast, studies utilizing *Clonorchis sinensis*, one of the most important liver flukes (Lun et al. 2005), require rats as the definitive host, a snail and a fish as first and second intermediate hosts (Yoshimura, 1965), and hence the life cycle is considerably more challenging and commonly not maintained in the laboratory. Second, *E. caproni* has a short development period of only 10–14 days in rodents (Fried and Huffman 1996). This model therefore produces rapid results and is more costeffective than host–parasite models with a longer development and a more complex life cycle. Moreover, the *Echinostoma*-mouse model is not only useful for anthelminthic studies on echinostomes, but can also serve as a prescreen to test for *in vivo* trematocidal activity of compounds. However, it should be kept in mind that the feeding behavior (echinostomes do not feed on blood) and the location of

Echinostoma spp. in the definitive host differ from the major food-borne trematodes, the liver flukes (e.g., *C. sinensis*, *Fasciola hepatica*, and *Opisthorchis viverrini*) and the lung flukes (*Paragonimus* spp.). So for example, depending on the pharmacologic characteristics (e.g., absorption and distribution) of the drug, the *in vivo* drug efficacy between the individual fluke infections might vary.

Table 13.1 summarizes *in vivo* studies that employed *Echinostoma* spp. for drug screening studies. The first investigation we could identify dates back to 1971 (Krotov and Gusel'nikova 1971). The majority of studies employed *E. caproni* in the mouse, and also *E. ilocanum* in rats and gerbils (Cross and Basaca-Sevilla 1986), and *E. miyagawai* in mice (Krotov and Gusel'nikova 1971) were utilized. Drug efficacy was analyzed as early as 24 h post-treatment (Leger and Notteghem 1975). High activities against *E. caproni* and *E. miyagawai* were observed with single oral doses of chloxyl, phenasan, niclosamide, niclofolan, and hexachlorofen (Krotov and Gusel'nikova 1971; Leger and Notteghem 1975). Some of these drugs were commonly used in the first half of the twentieth century to treat infections with different food-borne trematodes in humans; however, patients often suffered from severe adverse events (Lämmler 1968; Rim 1984). Four studies analyzed the echinostomicidal effect of different benzimidazoles. Several of the benzimidazoles (e.g., thiabendazole) showed no activity even after long treatment courses or when administered at high doses (Maurer et al. 1996; Notteghem et al. 1980). On the other hand, complete worm burden reductions were observed in two studies with multiple treatment courses of mebendazole (Notteghem et al. 1980; Cross and Basaca-Sevilla 1986).

Praziquantel, tribendimidine, artemisinin and two of its derivatives (artemether and artesunate), and the synthetic trioxolane OZ78 were tested recently in the *E. caproni*-mouse model (Keiser et al. 2006a–2006c). A single 50-mg kg^{-1} oral dose of praziquantel resulted in 100% worm burden reduction. High doses of artesunate (700 mg kg^{-1}), OZ78 (1,000 mg kg^{-1}), artemether (1,100 mg kg^{-1}), and artemisinin (1,500 mg kg^{-1}) were required to yield cure rates of 99–100%. The need for high doses of artemisinin and two of its main derivatives to achieve high efficacy in the *E. caproni*-mouse model might be explained with *E. caproni* being a nonhematophagous feeder. Hence, this intestinal fluke does not degrade hemoglobin to generate free heme, which is a possible activator/receptor for the artemisinins (Keiser and Utzinger 2007a). Indeed, previous studies found that considerably lower doses of the artemisinins and synthetic trioxolanes administered to rats infected with *F. hepatica*, or mice infected with the blood fluke *Schistosoma mansoni* resulted in high worm burden reductions (Keiser et al. 2006d; Utzinger et al. 2007; Xiao et al. 2007). Finally, tribendimidine is considered a promising broad spectrum anthelminthic drug, which has recently been approved by Chinese food and drug regulatory authorities (Xiao et al. 2005). Complete worm burden reductions were achieved in the *E. caproni*-mouse model following the administration of a single 125-mg kg^{-1} oral dose or higher (Keiser et al. 2006a).

Table 13.1 Use of *Echinostoma* spp. for chemotherapeutic studies

Year (Reference)	Model	Treatment	Dose (mg kg^{-1})	Worm burden reduction (%)
1971 (Krotov and Gusel'nikova 1971)	*E. miyagawai* in mice	Male fern extract	1 × 400	64.5
		Acriquine	1 × 300	86
		Dichlorophen	1 × 200	80
		Trichlorophen	1 × 200	84
		Chloxyl	1 × 100	100
		Bithionol	1 × 100	100
		Phenasal	1 × 100	100
		Piperasine salt of phenasal	1 × 100	100
1975 (Leger and Notteghem 1975)	*E. caproni* in mice	Brotianide	1 × 75	100
		Niclofolan	1 × 10	100
		Nitroxynil	1 × 25	100
		Oxyclozanide	1 × 50	100
		Rafoxanide	1 × 50	100
		Tribromsalan	1 × 100	100
		Niclosamide	1 × 200	100
		Hexachlorophene	1 × 200	100
		Bithionol	1 × 300	80
1979 (Notteghem et al. 1979)	*E. caproni* in mice	Flubendazole	1 × 200	0
			5 × 200	100
1980 (Notteghem et al. 1980)	*E. caproni* in mice	Thiabendazole	5 × 200	0
		Oxybendazole	5 × 200	0
		Cambendazole	5 × 200	5
		Fenbendazole	5 × 200	20
		Parbendazole	5 × 200	50
		Mebendazole	5 × 200	100
1986 (Cross and Basaca-Sevilla 1986)	*E. ilocanum* in rats and gerbils	Mebendazole	50 mg per animal	100
1996 (Maurer et al. 1996)	*E. caproni* in ICR mice	Clorsulon	1 × 100	100
		Rafoxanide	1 × 50	100
		Mebendazole	1 × 1,000	No effect
		Arprinocid	1 × 100	No effect
1998 (Schmidt 1998)	*E. caproni* in BALB/c mice	Albendazole	1 × 100	Not reported
2006 (Keiser et al. 2006a)	*E. caproni* in NMRI mice	Tribendimidine	1 × 125	100
2006 (Keiser et al. 2006c)	*E. caproni* in NMRI mice	OZ78	1 × 1,000	100
2006 (Keiser et al. 2006b)	*E. caproni* in NMRI mice	Artemether	1 × 1,100	100
	E. caproni in NMRI mice	Artesunate	1 × 700	100
	E. caproni in NMRI mice	Artemisinin	1 × 1,500	99
	E. caproni in NMRI mice	Arteether	1 × 1,300	75
	E. caproni in NMRI mice	Praziquantel	1 × 50	100

13.2.2 Drug Sensitivity Assay with Echinostoma caproni In Vitro

Several studies have examined the echinostomicidal properties of different drugs *in vitro* (Leger et al. 1973; Leger and Notteghem 1975; Keiser et al. 2006b). Since echinostomes have a short development period *in vivo*, these trematodes can be obtained within a couple of days for subsequent *in vitro* studies. Another advantage of this model is that mice tolerate high infection intensities with echinostomes, leading to a large quantity of worms (up to 100) that can be recovered from the intestine of mice (Huffman and Fried 1990). In contrast, only a low number of flukes can be obtained from other food-borne trematode-rodent models, e.g., a *F. hepatica*-rat model delivers, on average, ~5–8 flukes (Keiser et al. 2006a). For *in vitro* chemotherapeutic studies, echinostomes were recovered from mice 11–14 days postinfection and incubated in the presence of the compound to be analyzed in NCTC 135 culture medium, or Hanks' balanced salt solution (HBSS), supplemented with antibiotics. Worms were observed for up to 5 days and mortality recorded (Leger et al. 1973; Leger and Notteghem 1975; Keiser et al. 2006d). Drug effects might also be assessed by motility disturbances (e.g., activity or paralysis), or morphological changes such as relaxation, shrinkage, curling, tegumental disruption, or worm disintegration.

Excysted juvenile *Echinostoma* worms (Fried et al. 1997) might also prove useful for *in vitro* chemotherapeutic studies. Assays with juvenile *E. caproni* obtained from metacercariae, treated in a trypsin-bile salts excystation medium at 39°C (Fried et al. 1997), for example, do not require the definitive host, hence are ethically sound, rapid to obtain, and inexpensive.

13.2.3 Scanning Electron Microscopic (SEM) Studies

The tegument of trematodes is one of the most susceptible sites of damage following chemotherapy, and drug effects are readily examined by the evaluation of treated worm specimen by SEM. Indeed, SEM analysis deepens our knowledge of the sites and mechanisms of action of anthelminthics (Halton 2004). Structural changes in the tegument following drug administration can be analyzed by means of SEM under *in vitro* and *in vivo* conditions. Tegumental responses are often analyzed with increasing incubation periods *in vitro* or progressing time post-treatment *in vivo*.

While surface structures of normally developed control specimens of *E. caproni* have been studied in detail using SEM (Fried et al. 1990), thus far only one study has evaluated the effect of drugs on this intestinal fluke by the same tool (Keiser et al. 2006a). Figure 13.1 depicts the oral sucker (OS) and the ventral sucker (VS) of an adult *E. caproni* specimen recovered from an untreated control mouse. For comparison, Fig. 13.2 shows an *E. caproni* trematode recovered from a mouse treated with a single 150-mg kg^{-1} oral dose of tribendimidine (Fig. 13.2a,b), and 700 mg kg^{-1} artesunate (Fig. 13.2c,d). Typical features of drug-induced damage included a reduced opening of the oral sucker (Fig. 13.2a,b,d), focal damage of the tegument as eruption (Fig. 13.2c), furrowing, or sloughing (no image shown).

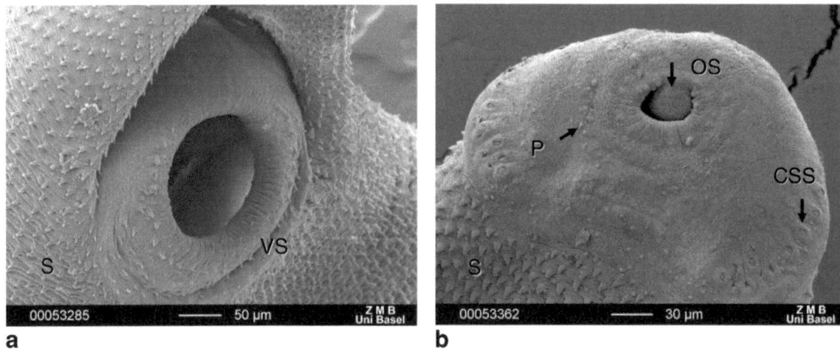

Fig. 13.1 SEM observation of an *E. caproni* trematode recovered from a mouse 14 days after an experimental infection. (**a**) Ventral sucker (VS), and tegumental spines (S). (**b**) Anterior part with oral sucker (OS), papillae (P), collar spine sockets (CSS), and spines (S)

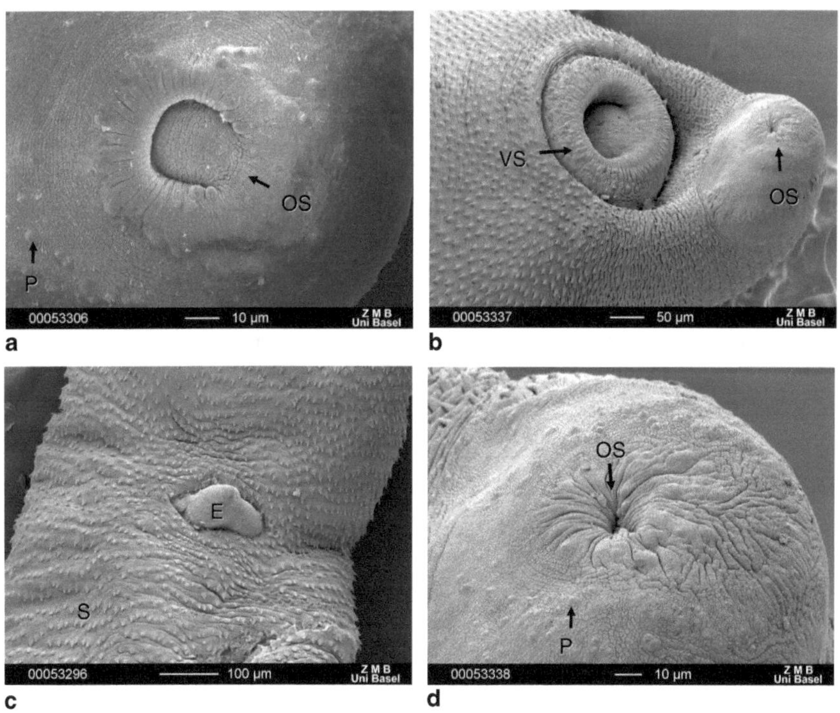

Fig. 13.2 SEM observation of *E. caproni* trematodes recovered from mice following treatment with tribendimidine (**a**, **b**), and artesunate (**c**, **d**). (**a**) Reduced opening of oral sucker (OS) 4 h after administration of a single 150-mg kg^{-1} oral dose of tribendimidine. (P) papillae. (**b**) Reduced opening of oral sucker (OS) 8 h after administration of a single 150-mg kg^{-1} oral dose of tribendimidine. (VS) ventral sucker. (**c**) Eruption (E), and tegumental spines (S) on the ventral midbody surface 4 h after administration of a single 700-mg kg^{-1} oral dose of artesunate. (**d**) Reduced opening of oral sucker (OS) 8 h after administration of a single 700-mg kg^{-1} oral dose of artesunate. (P) papillae

13.2.4 Spectroscopy

Both nuclear magnetic resonance (NMR) spectroscopy and mass spectrometry (MS) have been used to elucidate the mechanism of action of antiparasitic drugs. The general premetabonomic applications of NMR and MS include structural identification of possible target proteins and pathway steps for new drug developmental angles, and the monitoring of drug degradation (Shukla-Dave et al. 2000; Azzouz et al. 2007; de Aquino et al. 2008). For example, it was demonstrated that the antibiotics pentalenolactone and koningic acid inhibit the glycosomal enzyme of *Trypanosoma brucei*. NMR analysis was able to indicate that these compounds bind to the cysteine residue of the active site of the enzyme (Willson et al. 1994). Other examples include the investigation of the inhibition of the merozoite surface protein in *Plasmodium falciparum* by suramin and its analogs (Fleck et al. 2003), the interaction of antiparasitic guanyl hydrazones with the cell membrane of *Trypanosoma cruzi* (Borges and Figueroa-Villar 2001), and the evaluation of methylene blue as an antimalarial drug candidate against drug-resistant parasite strains (Akoachere et al. 2005).

Further extension of spectroscopy has allowed direct application to profiling metabolic changes in the tissues and biofluids of host organisms infected with parasites. The advent of postgenomic disciplines such as transcriptomics, proteomics, metabonomics, and lipidomics has opened up a wealth of new possibilities in characterizing the response of an organism to physiological and pathological challenges (Nicholson et al. 2002), and has facilitated the increasingly popular "systems biology" approach. A clear role for genomics, proteomics (see Chap. 9), and metabonomics has been established in parasitology with respect to diagnostics, characterizing host–parasite interactions, and in monitoring and evaluating therapeutic responses to intervention.

13.3 Metabonomic Investigation on *Echinostoma caproni* in the Mouse Model

13.3.1 What Is Metabonomics?

Metabonomics, also known as metabolomics and metabolic profiling (Nicholson et al. 1999, 2002; Fiehn 2002), involves the generation of multivariate profiles or fingerprints of biological samples using multivariate statistics, applied to high-resolution analytical methods such as NMR spectroscopy, MS, and capillary electrophoretic chromatography. The approach has been used for the characterization of metabolic phenotypes of healthy, diseased, and treated subjects (for a recent review see Schlotterbeck et al. 2006), and the diagnosis and monitoring of a wide variety of clinical and physiological conditions, ranging from cardiovascular disease (Ordovas and Mooser 2006) to nutritional intervention (Wang et al. 2007). Although gas chromatography (GC)-MS, capillary electrophoresis chromatography (CE), and infrared (IR) spectroscopy have all been used successfully as metabonomics tools, applications in parasitology have been limited largely for NMR-based studies with liquid chromatography (LC)-MS application beginning to emerge. Studies by Vasta and colleagues have used high-performance thin-layer chromatographic (HPLC) analyses to examine metabolic profiling in mice infected with *E. caproni*. Significant changes were discovered in the neutral lipid profiles in both feces and urine of mice infected with *E. caproni* relative to the uninfected controls (Vasta et al. 2008).

In the current chapter we will focus on NMR-based metabolite profiling with a brief discussion of the current capability and potential of LC-MS for diagnosing and characterizing parasite infections with particular consideration to *E. caproni*.

13.3.1.1 NMR Spectroscopy

NMR has been an important tool in analytical and organic chemistry for several decades. Early studies on biofluids (typically urine and blood) were performed using visual means of interpreting molecular changes in the spectra. However, in 1989 a breakthrough occurred in spectral analysis involving the application of computer-based methods for data reduction and assessing similarity/dissimilarity between various spectra (Nicholson and Wilson 1989), and NMR spectroscopy (especially ^{1}H NMR spectroscopy) became an efficient and convenient analytical tool used in a diversity of medical and biological fields, and currently represents an important tool in systems biology (Rochfort 2005).

NMR spectroscopy measures the interaction of nuclei with an applied electronic magnetic field. The frequency or position of the chemical shift scale will be determined by the electron cloud surrounding a particular nucleus which provides a shielding effect from the magnetic field. Hence, the exact resonance of a nucleus is related to its chemical environment and therefore gives information regarding chemical structure of a molecule. In addition, neighboring nuclei interact *via* bonding electrons, and hence carry information regarding the neighboring nuclei. Coupled nuclei show multiplicity of peaks, which is determined by the number of protons on the adjacent atoms and is given by $n+1$ peaks, where n is the number of neighboring spin-coupled nuclei. Furthermore, the intensity of a resonance peak in NMR spectra is proportional to the concentration of nuclei present in a particular chemical environment. Therefore, NMR spectroscopy can provide quantitative information as well as structural information on the metabolites or compounds in a biological sample. For methodological details the reader is referred to Claridge (1999).

13.3.1.2 Mass Spectrometry

The initial application of MS to metabolic profiling originated from cell culture systems and plant studies (Fiehn 2002). Nowadays MS is increasingly utilized in the metabolic profiling of biofluids, because of its inherently high sensitivity and its ability to target different classes of molecules in the metabolic spectrum, depending on the chosen type of MS platform and the characteristic of the sample. Thus MS is a highly complementary technique to NMR. MS is ideally coupled with a previous, separation of the mixture *via* LC or solid-phase extraction. LC-MS is the most popular method of metabolic profiling after NMR spectroscopy (Lenz et al. 2005; Nicholson 2006), but also GS-MS and MS-MS find increasing application. In practice, the combination of multiple analytical platforms will provide the most precise information, because every single technique contributes a unique set of data shaping the overall metabolic picture (Lindon et al. 2000; Ohdoi et al. 2003).

For MS analysis of biological matrices, after appropriate chromatographic separation, the fraction is introduced into the ionization source, where the molecules gain (positive ionization mode) or lose (negative ionization mode) one ion, and are further separated by MS according to the mass-to-charge ratio (m/z) of the ions. Where hard ionization sources are used, the molecular fragments can provide structural information as well as an accurate relative molecular mass, which can be useful in molecular assignment. The data generated consist of a chromatographic retention time and a mass-to-charge (m/z) ratio, and thus there is significantly less overlap with metabolite signals generated by LC-MS than one-dimensional (1D) NMR analysis since the data are already dispersed in two dimensions. However, analytical drift in either the chromatographic conditions or the m/z value can be problematic, and several algorithms have been produced to address the peak or signal alignment. An additional factor which requires consideration in MS analysis is the fact that differential ion suppression effects across samples can produce artifacts and render quantification of metabolites difficult.

13.3.1.3 Chemometric Tools

Spectra from biological samples, such as blood, urine, fecal water, and tissues, generally contain several thousands of resonances, which are partly overlapping and thus can obscure important information. In order to extract the maximum information from the large amount of raw, complex data, multivariate statistical methods are used for data reduction and to visualize the data in simple graphs.

Prior to multivariate analysis a series of preprocessing steps are usually applied to the spectral data to enable more accurate interpretation. These include procedures such as peak alignment and registration, smoothing, centering of baselines, and normalization of each spectrum to a constant sum such that concentration differences between samples can be removed.

There are two basic types of multivariate statistical methods: (1) unsupervised, and (2) supervised approaches. Unsupervised methods are generally used for reducing the complexity and the volume of the data, and use no prior knowledge of class or measure of response. Following data reduction approaches according to defined algorithms, the data are displayed as co-ordinates in a 2- or 3-dimensional (2D or 3D) plot, which can be examined to obtain information on outliers, trends, and similarities within the data. The most widely applied unsupervised method is principal component analysis (PCA) (Fig. 13.3). In contrast with unsupervised methods, the main

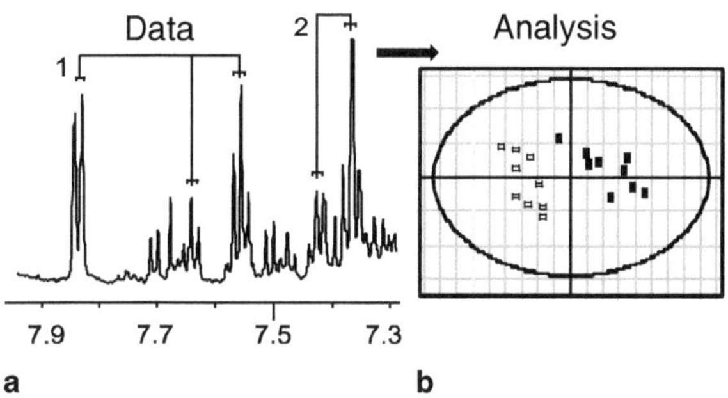

Fig. 13.3 2D and 3D time trajectory of urine samples from *E. caproni*-infected mice obtained from the mean PC1 and PC2 values over a 33-day time period. Biological samples were collected at seven time points postinfection (e.g., days 1, 5, 8, 12, 19, 26, and 33). The error bars reflect two standard deviations of the mean

feature of supervised methods is to use knowledge of class membership to optimize the difference between two or more classes. Thus in this context supervised methods are generally used to predict the biological "state" of an organism, e.g., infected or noninfected. For supervised methods, cross-validation is necessary by using a part of the data as "training set" in order to construct the model and the remaining part of the data as a "test set" to validate the model. Some of the main supervised methods are soft independent modeling of classification analogy (SIMCA), partial least squares (PLS) analysis (Trygg et al. 2007) and, more recently, also Bayesian methods (Lindon et al. 2001; Bang et al. 2008). At its simplest these linear projection methods allow rearrangement of the original variables (in this case spectral data points) to form a new component which describes the maximum direction or dimension of variation in the data. Subsequent components are calculated to be orthogonal and describe progressively less variation. Thus a plot of the first two or three components allows visualization of similarity between objects (in this case biological samples) based on their spectral composition.

Further improvement in data modeling can also be obtained by implementation of a mathematical filter, such as orthogonal signal correction, to remove systematic variation in the data set that is unrelated to the property of interest. For example, in determining whether an animal is infected, systematic variation introduced by age or sex differences may compromise the model. Orthogonal PLS discriminant analysis (O-PLS-DA) (Trygg and Wold 2002) has been used successfully to remove the influence of such potential sources of variation.

13.3.2 Experimental Protocol

Here we use NMR spectroscopy to characterize the metabolic features of an *E. caproni*-mouse model and to illustrate our metabolic profiling strategy. A standard metabonomic protocol for exploring the effects of parasite infection is sum-

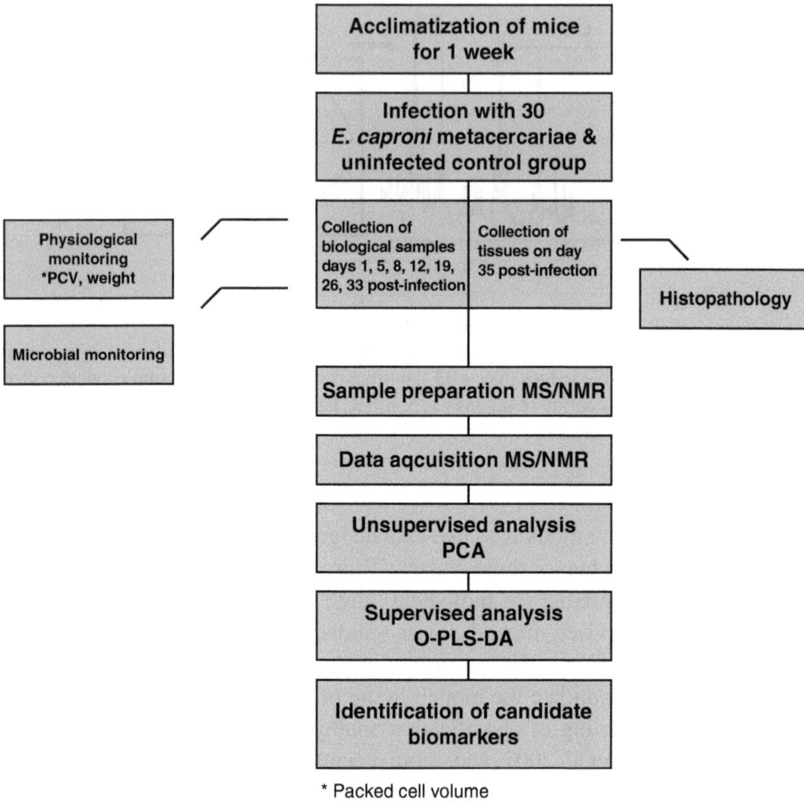

Fig. 13.4 Schematic draft of the experimental protocol used to acquire biological samples from mice infected with *E. caproni* for metabolic profiling

marized in Fig. 13.4. Briefly, urine, plasma, and stool from *E. caproni*-infected, and age- and sex-matched uninfected mice were collected at weekly intervals over a 33-day time course, and tissues were removed upon dissection, i.e. liver, kidney, spleen, ileum, jejunum, and colon. The biofluids were measured using conventional ^{1}H NMR spectroscopy, whilst tissue spectra were acquired by magic angle spinning (MAS) ^{1}H NMR. All samples were stored frozen at −80°C pending analysis by ^{1}H NMR spectroscopy or MS. A suite of multivariate statistical methods beginning with unsupervised methods, such as PCA, and progressing to supervised approaches, such as PLS-DA, were used to analyze the spectra.

13.3.3 Biomarker Identification and Diagnostic Potential of Biofluids in an Echinostoma caproni *Infection*

Characteristic spectral profiles of several mammalian biofluids are already well established, including those of urine, plasma, and fecal water extracts (Foxall et al. 1993a,b; Nicholson et al. 1995; Keun et al. 2002; Ebbels et al. 2004; Beckonert et al. 2007). Urine and plasma have already been assessed for their diagnostic potential in various single parasite-rodent models including *Schistosoma mansoni* (Wang et al. 2004), *Schistosoma japonicum* (Wang et al. 2006), *Trypanosoma brucei brucei* (Wang et al. 2008), *Trichinella spiralis* (Martin et al. 2006), and *Plasmodium berghei* (Li et al. 2008). More recently, fecal water composition in relation to physiological variation and characterization has been assessed in more detail (Marchesi et al. 2007; Saric et al. 2008b). These biofluids are relatively straightforward to obtain in a minimally invasive manner and are present in reasonably large quantities. Hence, they can serve as suitable biological matrices for disease diagnostics. Furthermore, urine, plasma and stool are to some extent representative of the state of the organism as a whole, as they reflect to a high degree malfunctions and impaired homoeostasis in metabolic composition. Typical metabolites detected in each of these three types of biological samples are listed in Table 13.2.

In the *E. caproni*-mouse model, all three biological samples were found to alter with infection and progressive changes in the spectral profiles were monitored over time. Alterations in the metabolic composition of the samples reflect the response of the host organism as it strives to compensate for the invasion of the parasite and maintain homeostatic control.

Table 13.2 List of metabolites found in urine (ur), plasma (pl), and stool (st) of 7–8 week-old mice (NMRI strain, female mice)

Metabolite	Coupling pattern (Hz)	δ (ppm)	Biofluid
2-Hydroxyisobutyrate	$2\times CH_3$	1.36(s)	ur
2-Hydroxyisovalerate	α-CH, β-CH, γ-CH_3, γ′-CH_3	3.85(d), 2.02(m), 0.79(d), 0.84(d)	st
2-Ketoisocaproate	α-CH, β-CH, γ-CH_3, γ′-CH_3	3.85(d), 2.02(m), 0.79(d), 0.84(d)	ur, st
2-Ketoisovalerate	CH, $2\times CH_3$	3.02(m), 1.13(d)	pl, st
2-Oxoglutarate	β-CH_2, γ-CH_2	3.02(t), 2.50(t)	ur
3-Aminopropionate (β-alanine)	NCH_2, CH_2COOH	3.19(t), 2.56(t)	st
3-hydorxyphenyl propionic acid	half α-CH_2 half α-CH_2, β-CH, γ-CH_3	2.32(m), 2.42(m), 4.16(m), 1.21(d)	pl
3-hydorxybutyrate	α-CH_2, β-CH_2, 2-CH	2.47(t), 2.85(t), 6.80(m)	st
3-Methyl-2-oxovalerate	β-CH, half γ-CH_2, half γ-CH2, γ-CH_3, β-CH_3	2.93(m), 1.70(m), 1.46(m), 0.90(t), 1.10(d)	st
5-Aminovalerate	5-CH_2, 2-CH_3, 3,4-CH_2	3.02(t), 2.24(t), 1.65(m)	st
acetate	CH_3	1.90(s)	pl, ur, st
acetoacetate	α-CH_2, γ-CH_3	2.29(s), 3.45(s)	pl
alanine	α-CH, β-CH_3	3.81(q), 1.48(d)	pl, ur, st
allantoin	CH	5.40(s)	pl, ur
arginine	α-CH, β-CH_2, γ-CH_2, δ-CH_2	3.76(t), 1.89(m), 1.59(m), 3.17(t)	st
asparagine	α-CH, half β-CH_2, half β-CH_2	4.01(m), 2.70(m), 2.81(m)	st
aspartate	α-CH, half β-CH_2, half β-CH_2	3.92(m), 2.70(m), 2.81(m)	st

(continued)

Table 13.2 (continued)

Metabolite	Coupling pattern (Hz)	δ (ppm)	Biofluid
bile acids	CH_3	0.70(m)	st
butyrate	α-CH_2, β-CH_2, γ-CH_3	2.16(t), 1.56(m), 0.90(t)	st
choline	3×CH_3, α-CH_2, β-CH_2	3.21(s), 4.07(m), 3.52(m)	pl
citrate	1-CH_2, 3-CH_2	2.69(d), 2.54(d)	pl, ur, st
creatine	CH_3, CH_2	3.04(s), 3.93(s)	pl, ur, st
creatinine	CH_3, CH_2	3.05(s), 3.93(s)	ur
dihydorxythymine	CH_2, CH, CH_3	3.17(m), 2.47(m), 1.07(d)	pl
dimethylamine	2×CH_3	2.71(s)	ur
dimethylglycine	2×CH_3, CH_2	2.89(s), 3.71(s)	ur
ethanolamine	NH–CH_2, HO–CH_2	3.15(t), 3.78(t)	st
formate	CH	8.45(s)	pl, ur, st
fumarate	CH	6.53(s)	st, ur
glutamate	α-CH, β-CH_2, γ-CH_2	3.78(m), 2.06(m), 2.36(m)	st
glutamine	α-CH, β-CH_2, γ-CH_2	3.78(m), 2.15(m), 2.46(m)	st
glycerol	half α, γ-CH_2, half α γ-CH_2, β-CH	3.56(dd), 3.64(dd), 3.87(m)	st
glycerophosphocholine	3×CH_3, half α-CH_2, half α-CH_2, half β-CH_2 half β-CH_2, γ-CH_2	3.23(s), 4.32(t), 3.6(dd), 3.68(t), 3.89(m), 3.72(dd)	pl
glycine	CH_2	3.55(s)	st
glycolate	CH_2	3.94(s)	ur
gnanidinoacetate	CH_2	3.80(s)	ur
hippurate	CH_2, 2,6-CH, 3,5-CH, 4-CH	3.97(d), 7.84(d), 7.55(t), 7.64(t)	ur
hypoxanthine	3-CH, 7-CH	8.10(s), 8.11(s)	st
indoxylsulfate	5-CH, 6-CH, 4-CH, 7-CH	7.20(t), 7.27(t), 7.51(d), 7.7(d)	ur
isoleucine	α-CH, β-CH, half γ-CH_2, half γ-CH_2, δ-CH_3, β-CH_3	3.68(d), 1.93(m), 1.25(m), 1.47(m), 0.99(d), 1.02(d)	pl, st
lactate	CH, CH_3	4.12(q), 1.33(d)	pl, ur, st
leucine	α-CH, β-CH_2, γ-CH, δ-CH_3, δ-CH_3	3.72(t), 1.96(m), 1.63(m), 1.69(m), 0.91(d), 0.94(d)	pl, st
lysine	α-CH, β-CH_2, γ-CH_2, δ-CH_2, ε-CH_2	3.77(t), 1.92(m), 1.73(m), 1.47(m), 3.05(t)	st
mannitol	α-CH_2, β-CH. Y′-CH	3.78(m), 3.88(dd), 3.68(dd)	ur
methanol	CH_3	3.36(s)	pl
methionine	α-CH, β-CH_2 γ-CH_2, CH_3	3.87(m), 2.10(m), 2.65(dd), 2.15(s)	pl, st
methylcrotonate	CH, CH_3, CH_3	5.65(s), 1.70(s), 1.71(s)	ur
methylamine	CH_3	2.61(s)	ur
methylguanidine	CH_3	2.83(s)	ur
myo-inositol	1,3-CH, 2-CH, 5-CH, 4,6-CH	3.53(dd), 4.06(t), 3.28(t), 3.63(t)	st
n-Methyl-nicotinamide	CH_3, 6-CH, 2-CH, 5-CH, 4-CH	4.48(s), 8.97(d), 9.28(s), 8.19(t), 8.90(d)	ur
p-Cresolglucuronide	2-CH, 6-CH, 3-CH, 5-CH	7.06(d), 7.23(d), 2.30(s), 5.07(d)	ur
phenylacetic acid	CH2, 2,6-CH, 3,5-CH	3.52(s), 7.29(t), 7.36(t)	st

(continued)

Table 13.2 (continued)

Metabolite	Coupling pattern (Hz)	δ (ppm)	Biofluid
phenylacetylglycine	2,6-CH, 3,5-CH, 4-CH, $ArCH_2$, NCH_2	7.43(m), 7.36(s), 7.37(m), 3.75(d), 3.68(s)	ur
phenylalanine	2,6-CH, 3,5-CH, 4-CH, CH_2, CH	7.44(m), 7.39(m), 7.33(m), 3.17(dd), 3.30(dd), 3.99(dd)	st
proline	α-CH, half β-CH_2, half β-CH_2, γ-CH_2, δ-CH_2	4.15(dd), 2.05(m), 2.38(m), 2.00(m), 3.39(m)	st
propionate	CH_2, CH_3	2.19(q), 1.06(t)	st
pyridoxamine-5-phosphate	OCH_2, CH_{2N}, CH_3	7.67(s), 4.34(s), 2.48(s)	ur
scyllo-Inositol	6×CH	3.35(s)	pl
succinate	2×CH_2	2.40(s)	ur, st
taurine	CH_2N, CH_2S	3.27(t), 3.43(t)	ur
threonine	α-CH, β-CH, γ-CH_3	3.60(d), 4.26(m), 1.33(d)	st
trimethylamine	3×CH_3	2.88(s)	ur
trimethylamine-*N*-oxide	3×CH_3	3.27(s)	ur
tryptophan	4-CH, 7-CH, 2-CH, 5-CH, 6-CH, α-CH, half β-CH_2, half β-CH_2	7.79(d), 7.56(d), 7.34(s), 7.29(t), 7.21(t), 4.06(dd), 3.49(dd), 3.31(dd)	st
tyrosine	2,6-CH, 3,5-CH, CH_2, α-CH	7.23(d), 6.91(d), 2.93(t), 3.25(t)	st
uracil	5-CH, 6-CH	5.81(d), 7.59(d)	st
ureidopropanoate	α-CH_2, β-CH_2	2.38(t), 3.30(t)	ur
urocanate	α-CH, β-CH, 5-CH, 2-CH	6.40(d), 7.13(d), 7.41(s), 7.89(s)	ur, st
valine	α-CH, β-CH, γ-CH_3, γ′-CH_3	3.62(d), 2.28(m), 0.98(d), 1.03(d)	pl, st
α-Glucose	1-CH, 2-CH, 3-CH, 4-CH, 5-CH, half 6-CH_2, half 6-CH_2	5.24(d), 3.56 (dd), 3.70(t), 3.40(t), 3.83(m), 3.72(dd), 3.85(m)	pl, ur, st
β-Glucose	1-CH, 2-CH, 3-CH, 4-CH, 5-CH, half 6-CH_2, half 6-CH_2	4.65(d), 3.25 (dd), 3.47(t), 3.40(t), 3.47(ddd), 3.78(dd), 3.90(dd)	pl, ur, st
lipids components	CH_3	0.84(t)	pl
	$(CH_2)_n$	1.25(m)	pl
	β-CH_2CH_2CO	1.57(m)	pl
	$CH_2C{=}C$	1.97(m), 2.00(m)	pl
	CH_2CO	2.23(m)	pl
	$C{=}CCH_2C{=}C$	2.69(m), 2.71(m), 2.72(m)	pl
	CH=CH	5.23(m), 5.26(m), 5.29(m)	pl

Key: d, doublet; dd, doublet of doublets; ddd, doublet of doublet of doublet; m, multiplet; q, quadruplet; s, singlet; t, triplet.

13.3.3.1 Plasma

The strongest differentiation of the *E. caproni*-infected group of mice based on the H NMR plasma profile occurred between days 12 and 26 postinfection, and the most striking differences related to increases in the concentrations of lipoproteins, triacylglycerols, and saturated and unsaturated fatty acids after day 12 postinfection. This observation is consistent with the well-documented degradation of biological membranes by this intestinal fluke (Horutz and Fried 1995). Other infection-related changes included a depletion of the branched chain amino acids (BCAAs; e.g., leucine, isoleucine, and valine) and choline/glycerophosphocholine.

13.3.3.2 Urine

Urine has been emphasized to represent an ideal diagnostic template (Merchant and Klein 2007; Deras et al. 2008), as it collects the waste products of the whole organism and can reflect the metabolic status of an individual. Some urinary biomarkers can be directly related to kidney failure, organ damage, nutrient uptake, microbiotal composition, etc. (Clayton et al. 2003; Wang et al. 2007). Urine is relatively easy to obtain and probably the best metabolically characterized biological template of all (Foxall et al. 1993a; Keun et al. 2002; Ebbels et al. 2004; Beckonert et al. 2007).

E. caproni-infected mice showed decreased levels of 2-ketoisocaproate in urine, which could be due to the depletion of leucine in plasma, which acts as precursor in a transamination reaction (Mitch and Chan 1979). Significantly lower level of urinary taurine was a consistent feature in infected mice and is likely to derive from the high level of lipid degradation in the gut, caused by *E. caproni*, thus creating a higher demand for lipid digestion. As taurine is involved in bile acid conjugation and fat emulsification, it is likely that the urinary depletion of the compound is due to the higher demand. Taurine is conjugated with chenodeoxycholic acid and cholic acid in the liver, where primary bile salts are formed and excreted in urine. Deconjugation of taurine happens *via* gut bacterial species, after the lipids formed a micellar form together with the bile salts, in order to cross the intestinal wall. Hepatic taurine is then absorbed by the blood circulation (Bender 2002).

It is conceivable that an infection with intestinal flukes results in changes in the bacterial composition due to the close proximity and direct influence of *E. caproni*. Indeed, the most prevalent alterations in the urine profile were the relative concentrations of gut microbial synthesized compounds, such as decreased 2-ketoisocaproate, acetate, hippurate, and increased phenylacetylglycine, *p*-cresol glucuronide, and trimethylamine. The specific change in *p*-cresol glucuronide is likely to reflect an alteration in either the presence or activity of *Clostridium* subspecies, as *p*-cresol is a known metabolite of both *C. difficile* and *C. scatologenes* (Selmer and Andrei 2001; Yu et al. 2006). *p*-cresol is taken up by the blood stream where it binds to serum proteins and is later excreted, after conjugation with glucuronic acid in the liver and kidney (Lesaffer et al. 2003). Other bacteria in the gut degrade dietary choline and carnitine to trimethylamine (TMA), which is then

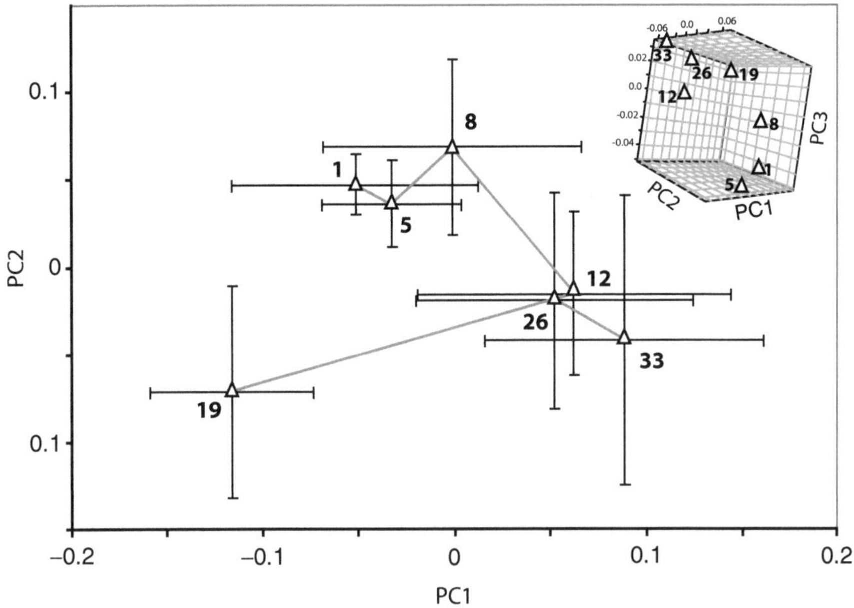

Fig. 13.5 Schematic of sample acquisition and data handling; (**a**) acquisition of metabolite profiles followed by (**b**) mathematical modeling of the data to establish pre- (*filled square*) and postinfection (*open square*) phenotypes for each parasite. Key: 1, hippurate; 2, phenylacetylglycine

oxidized further in the liver to trimethylamine-*N*-oxide (TMAO) (Smith et al. 1994; Seibel and Walsh 2002). Again, an *E. caproni*-induced change of the microbiotal community is suggested by the disturbance of the TMA and TMAO equilibrium in urine and the subsequent depletion of choline/glycerophosphocholine in the plasma.

The results generated by multivariate data analysis can be ordered chronologically to display the metabolic response in a host–parasite model over the duration of the time course of infection. This can be a useful tool in characterizing the stage of an infection and may provide information in determining a therapeutic strategy for disease management. An example of such a response or "trajectory" is depicted in Fig. 13.5, which illustrates the evolving response in the urine profile. Key features include decreased levels of hippurate, 2-ketoisocaproate, and taurine, and subsequent increases in trimethylamine, *p*-cresol glucuronide, TMAO, and phenylacetylglycine.

13.3.3.3 Stool

As with the urine profiles, the fecal water samples from mice infected with *E. caproni* were characterized by alteration in gut microbial metabolites. Spectra from fecal extracts contained bile acids, the short chain fatty acids (SCFAs) butyrate, acetate, and propionate, and the amino acids including lysine, arginine, tryptophan and glutamine, among others which were also present in plasma and/or urine (Table 13.2).

Nearly all changes observed in the fecal samples due to an experimental infection with *E. caproni* can be attributed to a changed gut microbial composition. The SCFAs (e.g., propionate, butyrate, and acetate) are produced by colonic bacteria *via* fermentation and undergo a decrease in the fecal samples of infected animals compared to the noninfected control group. Acetate is the main SCFA in the colon, and is, together with propionate, taken up into the peripheral blood circulation where they have antagonistic functions in cholesterol synthesis, e.g., acetate increases, while propionate is known to decrease synthesis. Butyrate is the main source of energy for the colonic mucosa (Wong et al. 2006). The changed gut microbiota is also reflected in the increased concentration of 5-aminovalerate in infected animals, which is normally degraded by a *Clostridium* subspecies (e.g., *C. aminovalericum*) to propionate and acetate (Barker et al. 1987; Ramsay and Pullammanappallil 2001). The increase of this metabolite could indicate a depletion of this bacterial substrain. Increased levels of the BCAAs in stool are consistent with this effect but could also be due to malabsorption caused by the parasite.

13.3.3.4 Utility of Biofluids as a Matrix for Diagnostics

Based on our findings with the *E. caproni*-mouse model, the combination of all three assessed biofluids would deliver the most comprehensive metabolic fingerprint and would most accurately reflect the overall picture of the infection impact, (Saric et al. 2008a). However, in reality it is more conceivable to select the most suitable template and to further optimize it, since collection of all three biofluids in human populations may not always be practical. Comparing the three biofluids in terms of their biomarker potential, plasma would be the diagnostic matrix of choice in terms of biomarker quantity. Urine also has good potential, with ten identified biomarkers, and this biofluid is particularly convenient for sample collection. Fecal samples would be the least suitable template, as the inter- and intraindividual variation in healthy individuals is already considerable, which might be due to the extremely variable dynamics over time and diversity of bacterial strains in general.

13.3.4 Biomarker Identification in Tissue via MAS ^1H NMR Spectroscopy

MAS NMR spectroscopy allows solid and semisolid tissues to be profiled (Andrew et al. 1959). The utility of this technology lies mainly in the evaluation of disease mechanisms in animal models, since it is a necessarily invasive approach. However, under some circumstances, e.g., differentiation of benign and malignant tumors, it is a potentially valuable clinical tool. Since candidate biomarkers identified in biofluids such as urine or plasma can originate from different tissues or organs, direct analysis of these tissues can provide a better understanding of the etiology of the biomarker. Each tissue has a characteristic and unique metabolic profile reflecting the different cellular structures and functions specific to that tissue.

13.3.4.1 Kidney

The heterogeneous structure of the kidney is reflected in the region specific metabolic profiles. Both the cortex and the medulla contain various amino acids, e.g., glutamate/glutamine, aspartate, lysine, threonine, the aromatic amino acids phenylalanine, tyrosine, and the BCAAs (Fig. 13.5d). However, characteristic for the renal medulla is the presence of higher concentrations of various osmolytes, like taurine, betaine, *myo*- and *scyllo*-inositol, together with glucose, choline, glycerophosphocholine (GPC), phosphocholine, and creatine. Analysis of *E. caproni*-infected and noninfected control animals showed no difference in the biochemical composition of the renal medulla, while the cortex manifested in increase of alanine, glutamate, aspartate, and *scyllo*-inositol in infected animals. GPC was found to be in lower concentrations in the renal cortex of infected mice, which could reflect the decrease of the metabolite in the plasma, and could also indicate a changed ionic cellular environment, e.g., NaCl, as GPC uptake into cells is known to be increased in case of high tonicity outside the cell membrane (Zablocki et al. 1991; Kwon et al. 1996). Other osmotically active substances like betaine and inositols, e.g., *scyllo*-inositol, are found in kidney, and also in other organs such as the brain or liver are known to change inversely with the GPC concentration. In fact, infection with *E. caproni* appeared to be positively associated with *scyllo*-inositol concentrations in the cortical part of the kidney.

Amino acid reabsorption occurs predominantly in the proximal tubules of the kidney. The dicarboxylic amino acids glutamate and aspartate have separate carrier molecules, while alanine belongs to the neutral amino acids, which are carried by specific transporters (Wright et al. 1993; Thwaites and Anderson 2007).

It is probable that the resulting immune response, initiated by the infection, plays a central role in the renal amino acid changes, namely, cytokines and antigen-antibody complexes, which circulate through the kidney. In the literature, the immunological consequence of an infection with *E. caproni* seems to be inconsistent, based on the varying ratio of T-helper cell subgroups (e.g., T_h1/T_h2), which is partially responsible for the establishment of an infection, and induces different groups of cytokines and immunoglobulin (Brunet et al. 2000). However, both antibody-antigen complexes and cytokines are known to be involved in pathological processes in the kidney and are likely to induce minor biochemical changes.

Secretory products of *E. caproni*, could hypothetically act as renal stimulators as well, but due to low concentration in plasma the compounds were below detection limit of NMR.

13.3.4.2 Liver

Phosphocholine and betaine were present in significantly higher concentrations in the liver of *E. caproni*-infected mice. However, choline and GPC levels, which are in the same pathway, remained unaffected. We speculate that the increased amount of free amino acids and triacylglycerols in plasma is taken up into the hepatic cell in the form of phosphatidylcholine, degraded to GPC and subsequently to choline with the same rate of turnover. In a further degradation step choline is converted to betaine (irreversible) and phosphocholine, which are present in relatively higher levels in the liver of infected mice, compared to the noninfected control animals (Dawson 1955; Zeisel et al. 2003).

13.3.4.3 Spleen

In accordance with the studies of Brunet et al. (2000), where the mesenteric lymph nodes showed immunologic activity as response to an infection with *E. caproni*, but not the spleen, no infection-related changes in metabolic composition were observed in the spleen.

13.3.5 Changes in the Intestinal Profiles

The metabolic monitoring of the gut is of particular interest after an infection with an intestinal fluke for obvious reasons. On the one hand, there is well-documented structural damage following an infection with *E. caproni*, whereby the most obvious histopathological changes in the host are derived from the direct action of the fluke, which remains after excystation in ICR mice, for 8 weeks in the jejunum and ileum of the small intestinal part, and then shifts to duodenum and jejunum, which is more anterior (Huffman and Fried 1990).

On the other hand, analysis of the biofluids indicated a strong infection-induced effect on the host microbiota. Indeed, it is unthinkable that a parasitic worm, living in such proximity to the microbial community, would not influence the dynamics of the microorganisms in one way or another, which in turn bears potential consequences for the intestinal tissue in particular, and the whole organism in general.

Jejenum and ileum are functionally and histologically quite similar, with the main difference being that the ileum contains the highest density of Peyer's patches, which are immunologically active sites. The jejunum is larger in diameter, and hence its overall absorptive capacity is higher than that of the ileum. Digestion is completed in the small intestine, and the products and fluids are absorbed. In contrast, the colon (large intestine) bears the highest density of bacteria and is the major site for fermentation of dietary carbohydrates to SCFAs (e.g., butyrate, acetate, and propionate). Acetate is known to increase the colonic blood flow and is the main SCFA in the colon. It enters the blood circulation and can be taken up by muscle cells (e.g., longitudinal muscle of the small intestine) in order to generate energy, while propionate is taken up by the liver cells as substrate for gluconeogenesis. Butyrate is the main energy source for colonocytes (Wong et al. 2006). Comparing the spectral profiles of the two parts of the small intestine (jejunum and ileum) with the large intestine, the SCFAs account for the main difference. Butyrate and propionate were only found in colon spectra, while acetate was a feature of all three regions of the gut, but was found in much higher relative concentrations in colon than in ileum/jejunum. The characteristic spectral composition for noninfected tissues for ileum is illustrated in Figs. 13.6 and 13.7.

All three parts of the gut showed significantly different metabolite profiles between infected and noninfected animals with the greatest effect on ileum where a disturbance of the intestinal transport (e.g., a disturbed Na^+ electrochemical

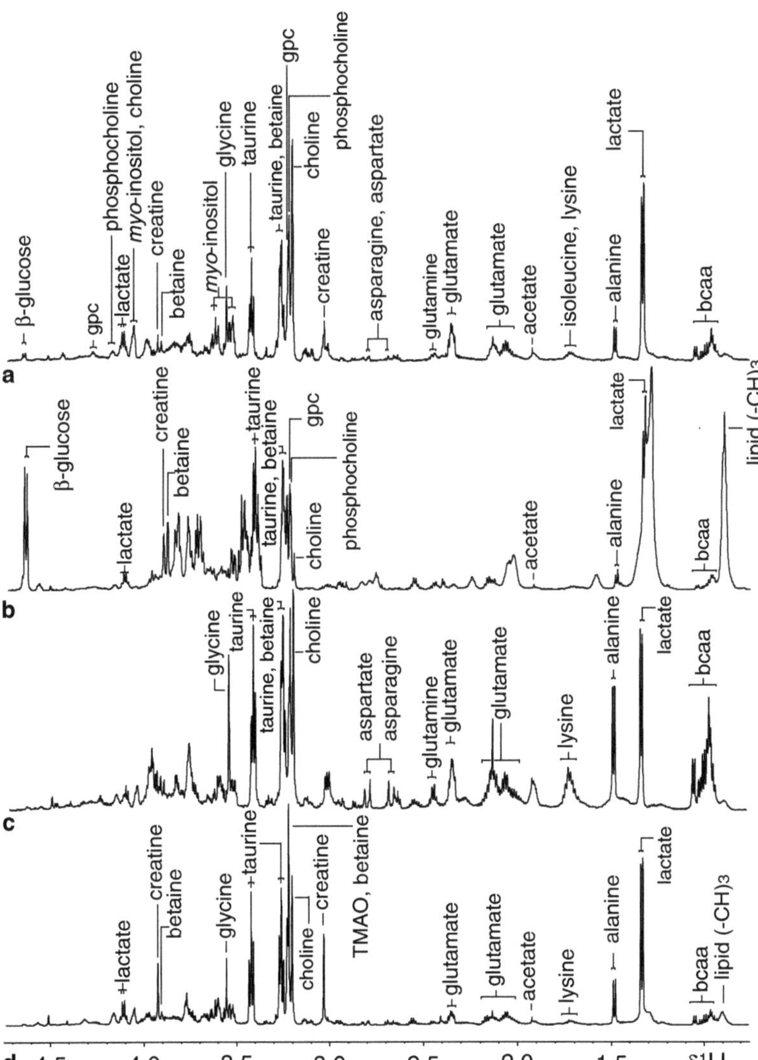

Fig. 13.6 Typical 600-MHz ¹H-1D spectra of four selected tissues of an uninfected control mouse, aged 11 weeks; (**a**) kidney cortex, (**b**) liver, (**c**) spleen, and (**d**) ileum. Key: bcaa, branched chain amino acids; gpc, glycerophosphocholine

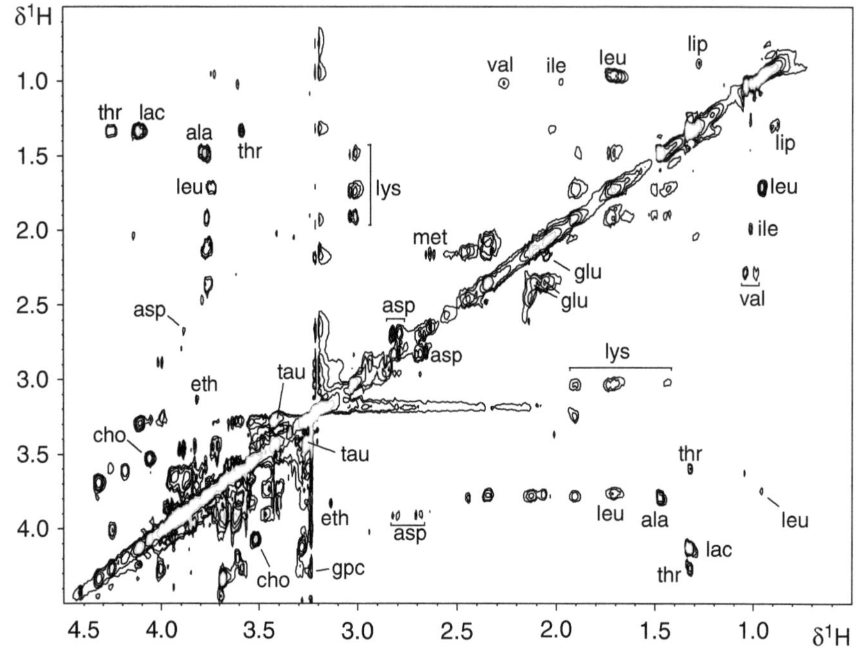

Fig. 13.7 MAS 600 MHz ^{1}H-^{1}H 2D TOCSY spectra of ileum from an uninfected control mouse. Keys: ala, alanine; asp, asparagine; cho, choline; eth, ethanolamine; glu, glutamate; gpc, glycerophosphocholine; lac, lactate; leu, leucine; lip, lipid; lys, lysine; ile, isoleucine; met, methionine; tau, taurine; thr, threonine; val, valine

gradient) was evident. Increased tissue concentration of glucose was found in the ileum, which is consistent with the documented increase in serum mucosal flux induced by proinflammatory cytokines such as IL-1α, IL-6, and IL-8, which leads to an increased uptake of glucose into the ileal cells *via* the sodium-glucose linked cotransporter (SGLT1) (Wright et al. 1993; Thwaites and Anderson 2007). The high uptake rate of glucose resulted in lower concentrations of glycine in infected animals, due to a glucose competitive mechanism (Cook 1971). Further metabolic change included a decrease in tissue concentration of ascorbate, alanine, glutamine, aspartate, isoleucine, and valine in the ileum of infected animals 36 days postinfection. This may be associated with inhibition of transporters, including SLC1A1 (excitatory amino acid carrier), B^{0}AT1/SLC6A19 (neutral amino acid transport system), and SVCT1 (ascorbic acid transporter), which are known to regulate the uptake of anionic amino acids (aspartate), neutral amino acids, and the ascorbate, respectively (Thwaites and Anderson 2007). Additionally, the *E. caproni*-induced hypertrophy leads to an increased consumption of glutamine in the intestinal, overproliferating cells and thus contributes to the depletion of glutamine stores in the ileum (Labow and Souba 2000; Balfour et al. 2001).

Concentrations of the two aromatic amino acids tyrosine and phenylalanine were increased in jejunum, but even though they share the same mechanism of reabsorption as the other neutral amino acids, which were depleted in ileum, their increase appears to contradict to a higher demand in the jejunal tissue, rather than a disturbed electrochemical gradient.

The observed decrease on *myo*-inositol signals in the metabolic profiles results from mechanic disruption of the intestinal mucosa cells. *Myo*-inositol, taurine, and betaine are small organic solutes, which are kept in the cell to maintain osmotic homoeostasis. This is obviously important in intestinal epithelia cells due to the digestive activities, which can potentially induce a hypertonic condition (Yancey et al. 1982; Miyakawa et al. 1999). Increased levels of trimethylamine were observed in the colon of infected animals, which, together with increased urinary excretion, would appear to indicate a general increase in the production of this metabolite (Smith et al. 1994; Seibel and Walsh 2002).

13.3.6 Integrative Metabonomics: An Approach to Understand Host–Parasite–Gut Microflora Interactions

The direct impact of a pathological agent on an individual tissue can be ascertained by histology or by measurement, for example, of inflammatory markers. However, pathologies rarely target single organs, and in order to achieve an accurate mechanistic understanding of how a disease progresses, it is usually necessary to evaluate the response of an organism across several physiological compartments. The application of metabolic profiling allows the extraction of latent information derived from numerous biofluid and tissue matrices and by integration of this knowledge can deliver a holistic framework of responses at a global systems level.

13.3.6.1 *Echinostoma caproni*-Induced Consequences on Osmotic Micro Environment and Cross Cellular Transport

Various osmotically active substances showed changes in relative concentrations during the course of an *E. caproni* infection, such as betaine which increased in the liver, *scyllo*-inositol (increased), and GPC (decreased) in the kidney, and *myo*-inositol in jejunum which decreased. All these metabolites serve as osmotic regulators in the tissue cells, whereas GPC is the only osmolyte in which uptake into the cell is induced by high NaCl levels. Thus the lower GPC levels could be either due to the changed tonicity (NaCl-to-water ratio) or directly relating to the depletion of plasma concentrations. However, its decrease seems to induce an increased intracellular concentration of *scyllo*-inositol in the renal tubes, which is known to change concentration inverse to GPC. The decrease of *myo*-inositol in the ileum is probably due to mechanical disruption of the intestinal mucosa cells, rather than due to osmotic necessity, as this would imply a higher uptake of an osmolyte to increase the cell influx of liquid and subsequently nutrients.

In the gut, it appears that the Na^+-dependent transport is disturbed, which may be due to disruption of the NaCl concentration with the resulting changes in membrane potential and gradient, which regulate carrier-dependent as well as independent transport. The malabsorption of the anionic amino acids aspartate, ascorbate and the neutral amino acids alanine, isoleucine, and valine perhaps indicates inhibition of specific Na^+-dependent transporters, such as SLC1A1, SVCT1, and B^0AT1/SLC6A19.

In the renal cortex aspartate, glutamate, and alanine showed increased concentrations. Specifically, Na^+-dependent carrier molecules for transport of the two dicarboxylic amino acids aspartate and glutamate seem to be impaired in the infection with *E. caproni*.

13.3.6.2 Holistic Influence of the Gut Microbiota

Trimethylamine concentration increased in the urine and colon of *E. caproni*-infected mice, which implies an increased synthesis in infected animals, compared to the healthy state of the rodent. As compensation to this overproduction, both colonic uptake and urinary excretion are increased.

A shift in the microbial composition toward bacterial species, which are able to degrade trimethylammonium compounds like choline, carnitine, or GPC to TMA with further oxidation to TMAO, could explain the increase of the TMA and TMAO in urine, as well as the depletion of choline and GPC in plasma. On the other hand, this microbial shift is unlikely to contribute to the changed concentrations of GPC in the kidney and phosphocholine in the liver, respectively, as the osmotic imbalance in the former and the higher rate of lipid degradation in the latter will outweigh the microbial contribution. Interestingly, the decrease of the SCFAs in the fecal samples of *E. caproni*-infected mice is not reflected in any of the assessed gut tissues, whereas particularly in the case of colon, a reduced propionate, acetate, and butyrate concentration would be expected to mirror the decrease in the feces. Acetate was the only urinary SCFA which demonstrated a subsequent depletion.

The presence of an intestinal fluke would be expected to have a considerable impact on the microbial community of the gut because of the proximity and thus direct interaction of the fluke and bacterial colonies. Yet, our integrated data suggest that the overall expression of metabolites, which are related to a changed gut microbiota, is not stronger compared to two previously assessed blood fluke-rodent models; indeed *S. mansoni*-infected mice and *S. japonicum*-infected hamsters showed the same urinary pattern of gut-microbial-modified metabolites, including changes in hippurate and trimethylamine (Wang et al. 2004, 2006).

13.3.6.3 Metabolite Flow Through Different Physiological Compartments

It is possible to monitor some of the metabolic changes across different physiological compartments. Most of these changes have their origin in the malabsorption of the ileum, such as isoleucine and valine, which are excreted in the feces in higher concentrations and thus, show lower levels in ileum and plasma. Also alanine and aspartate seem to be absorbed at lower rate in the ileum, but are increased in the kidney. Glucose, in contrast, shows increased levels in the ileum, which reflects in higher uptake from the blood circulation at the beginning of the infection. The marked decrease of plasma leucine probably induces the relatively lower levels of 2-ketoisocaproate in urine, as it is a transamination product of the former in the urine (Mitch and Chan 1979). Finally, creatine is excreted in higher amounts in urine, reflecting the depletion in plasma.

A further use of metabolic profiling is the characterization of the parasite itself. The main biochemical components of *E. caproni* consisted of creatine, taurine, and betaine, and the primary excretion products released in the intestinal are free fatty acids and sterols (Schneck et al. 2004). It is possible that the fluke-derived compounds affect the metabolic profile in the host. However, since we illustrate a protocol using an infection with the intestinal fluke with an average of 28 worms counted upon dissection of infected mice, the host weight-to-parasite mass ratio would be approximately 300:1; thus, it is unlikely to see a contribution of those excretory products in the host organism.

13.4 Concluding Remarks

The pressing need for novel trematocidal drugs has been emphasized (Keiser and Utzinger 2007b). We have summarized different studies that have demonstrated that the *E. caproni*-mouse model and *Echinostoma* cultured *in vitro* are useful tools for trematocidal drug discovery.

Metabolic profiling of an array of biological samples derived from mice infected with *E. caproni* demonstrates a novel angle for diagnosis in parasitology. A variety of disease-specific biomarkers have been demonstrated, and the approach is noninvasive and highly efficient. Furthermore, the potential of metabolic profiling to recover metabolic information in nearly every physiological compartment of a host organism allows crosslinking different pathological events and delivers a global picture of disease, which makes it a promising tool in systems biology for the purpose of infection diagnosis and eventually will provide opportunities for monitoring response to trematocidal agents.

Acknowledgments J. Keiser (project no. PPOOA–114941) and J. Utzinger (project no. PPOOB–102883 and no. PPOOB–119129) are grateful to the Swiss National Science Foundation for personal career development grants. This investigation received further financial support from Imperial College London, and Nestlé provided personal funds for Y. L. Wang. The authors thank Kirill Veselkov for giving access to his peak alignment script in Matlab. Last but not least we are grateful to Professor Bernard Fried and Professor Rafael Toledo for inviting us to write this chapter, and for their careful reading and copy-editing.

References

Akoachere, M., Buchholz, K., Fischer, E., Burhenne, J., Haefeli, W. E., Schirmer, R. H., and Becker, K. 2005. In vitro assessment of methylene blue on chloroquine-sensitive and -resistant *Plasmodium falciparum* strains reveals synergistic action with artemisinins. Antimicrobial Agents and Chemotherapy 49: 4592–4597.

Andrew, E. R., Bradbury, A., and Eades, R. G. 1959. Removal of dipolar broadening of nuclear magnetic resonance spectra of solids by specimen rotation. Nature 183: 1802–1803.

de Aquino, T. M., Liesen, A. P., da Silva, R. E. A., Lima, V. T., Carvalho, C. S., de Faria, A. R., de Araujo, J. M., de Lima, J. G., Alves, A. J., de Melo, E. J. T., and Goes, A. J. S. 2008. Synthesis, anti-*Toxoplasma gondii* and antimicrobial activities of benzaldehyde 4-phenyl-3-thiosemicarbazones and 2-[(phenylmethylene)hydrazono]-4-oxo-3-phenyl-5-thiazolidineacetic acids. Bioorganic and Medicinal Chemistry 16: 446–456.

Azzouz, S., Maache, M., Sanchez-Moreno, M., Petavy, A. F., and Osuna, A. 2007. Effect of alkyllysophospholipids on some aspects of the metabolism of *Leishmania donovani*. Journal of Parasitology 93: 1202–1207.

Balfour, C. D., Rossi, M., and Fried, B. 2001. Effects of a 100 metacercarial cyst inoculum on the host–parasite relationship of *Echinostoma caproni* and ICR mice. Journal of Helminthology 75: 321–324.

Bang, J. W., Crockford, D. J., Holmes, E., Pazos, F., Sternberg, M. J., Muggleton, S. H., and Nicholson, J. K. 2008. Integrative top-down system metabolic modeling in experimental disease states via data-driven Bayesian methods. Journal of Proteome Research 7: 497–503.

Barker, H. A., D'Ari, L., and Kahn, J. 1987. Enzymatic reactions in the degradation of 5-aminovalerate by *Clostridium aminovalericum*. Journal of Biological Chemistry 262: 8994–9003.

Beckonert, O., Keun, H. C., Ebbels, T. M., Bundy, J., Holmes, E., Lindon, J. C., and Nicholson, J. K. 2007. Metabolic profiling, metabolomic and metabonomic procedures for NMR spectroscopy of urine, plasma, serum and tissue extracts. Nature Protocols 2: 2692–2703.

Bender, D. A. 2002. Introduction to nutrition and metabolism, Boca Raton: CRC Press.

Borges, M. N., and Figueroa-Villar, J. D. 2001. NMR interaction studies of aromatic guanyl hydrazones with micelles: model for mechanism of action of cationic antibiotics. Biopolymers 62: 9–14.

Brunet, L. R., Joseph, S., Dunne, D. W., and Fried, B. 2000. Immune responses during the acute stages of infection with the intestinal trematode *Echinostoma caproni*. Parasitology 120: 565–571.

Chai, J. Y., Murrell, K. D., and Lymbery, A. J. 2005. Fish-borne parasitic zoonoses: status and issues. International Journal for Parasitology 35: 1233–1254.

Claridge, T. D. W. 1999. High-resolution NMR techniques in organic chemistry, Oxford: Elsevier.

Clayton, T. A., Lindon, J. C., Everett, J. R., Charuel, C., Hanton, G., Le Net, J. L., Provost, J. P., and Nicholson, J. K. 2003. An hypothesis for a mechanism underlying hepatotoxin-induced hypercreatinuria. Archives of Toxicology 77: 208–217.

Cook, G. C. 1971. Impairment of glycine absorption by glucose and galactose in man. Journal of Physiology 217: 61–70.

Cross, J. H., and Basaca-Sevilla, V. 1986. Studies on *Echinostoma ilocanum* in the Philippines. Southeast Asian Journal of Tropical Medicine and Public Health 17: 23–27.

Dawson, R. M. 1955. The role of glycerylphosphorylcholine and glycerylphosphorylethanolamine in liver phospholipid metabolism. Biochemical Journal 59: 5–8.

Deras, I. L., Aubin, S. M. J., Blase, A., Day, J. R., Koo, S., Partin, A. W., Ellis, W. J., Marks, L. S., Fradet, Y., Rittenhouse, H., and Groskopf, J. 2008. PCA3: a molecular urine assay for predicting prostate biopsy outcome. Journal of Urology 179: 1587–1592.

Ebbels, T. M. D., Holmes, E., Lindon, J. C., and Nicholson, J. K. 2004. Evaluation of metabolic variation in normal rat strains from a statistical analysis of ^{1}H NMR spectra of urine. Journal of Pharmaceutical and Biomedical Analysis 36: 823–833.

Fiehn, O. 2002. Metabolomics-the link between genotypes and phenotypes. Plant Molecular Biology 48: 155–171.

Fleck, S. L., Birdsall, B., Babon, J., Dluzewski, A. R., Martin, S. R., Morgan, W. D., Angov, E., Kettleborough, C. A., Feeney, J., Blackman, M. J., and Holder, A. A. 2003. Suramin and suramin analogues inhibit merozoite surface protein-1 secondary processing and erythrocyte invasion by the malaria parasite *Plasmodium falciparum*. Journal of Biological Chemistry 278: 47670–47677.

Foxall, P. J. D., Parkinson, J. A., Sadler, I. H., Lindon, J. C., and Nicholson, J. K. 1993a. Analysis of biological fluids using 600 MHz proton NMR spectroscopy: application of homonuclear two-dimensional J-resolved spectroscopy to urine and blood plasma for spectral simplification and assignment. Journal of Pharmaceutical and Biomedical Analysis 11: 21–31.

Foxall, P. J. D., Spraul, M., Farrant, R. D., Lindon, L. C., Neild, G. H., and Nicholson, J. K. 1993b. 750 MHz ^{1}H-NMR spectroscopy of human blood plasma. Journal of Pharmaceutical and Biomedical Analysis 11: 267–276.

Fried, B., and Huffman, J. E. 1996. The biology of the intestinal trematode *Echinostoma caproni*. Advances in Parasitology 38: 311–368.

Fried, B., Irwin, S. W. B., and Lowry, S. F. 1990. Scanning electron microscopy of *Echinostoma trivolis* and *E. caproni* (Trematoda) adults from experimental infections in the golden hamster. Journal of Natural History 24: 433–440.

Fried, B., Schmidt, K. A., and Sorensen, R. E. 1997. In vivo and ectopic encystment of *Echinostoma revolutum* and chemical excystation of the metacercariae. Journal of Parasitology 83: 251–254.

Halton, D. W. 2004. Microscopy and the helminth parasite. Micron 35: 361–390.

Haswell-Elkins, M., and Levri, E. 2003. Food-borne trematodes. In Manson's Tropical Diseases, eds. G. C. Cook, and A. I. Zumla, pp. 1471–1486, London: Saunders.

Horutz, K., and Fried, B. 1995. Effects of *Echinostoma caproni* infection on the neutral lipid content of the intestinal mucosa of ICR mice. International Journal for Parasitology 25: 653–655.

Huffman, J. E., and Fried, B. 1990. *Echinostoma* and echinostomiasis. Advances in Parasitology 29: 215–269.

Keiser, J., and Utzinger, J. 2004. Chemotherapy for major food-borne trematodes: a review. Expert Opinion on Phamacotherapy 5: 1711–1726.

Keiser, J., and Utzinger, J. 2005. Emerging foodborne trematodiasis. Emerging Infectious Diseases 11: 1507–1514.

Keiser, J., and Utzinger, J. 2007a. Food-borne trematodiasis: current chemotherapy and advances with artemisinins and synthetic trioxolanes. Trends in Parasitology 23: 555–562.

Keiser, J., and Utzinger, J. 2007b. Advances in the discovery and development of novel trematocidal drugs. Expert Opinion on Drug Discovery 2 (Suppl.): S9–S23.

Keiser, J., Xiao, S. H., and Utzinger, J. 2006a. Effect of tribendimidine on adult *Echinostoma caproni* harbored in mice, including scanning electron microscopic observations. Journal of Parasitology 92: 858–862.

Keiser, J., Brun, R., Fried, B., and Utzinger, J. 2006b. Trematocidal activity of praziquantel and artemisinin derivatives: in vitro and in vivo investigations on adult *Echinostoma caproni*. Antimicrobial Agents and Chemotherapy 50: 803–805.

Keiser, J., Utzinger, J., Tanner, M., Dong, Y., Vennerstrom, J. L. 2006c. The synthetic peroxide OZ78 is effective against *Echinostoma caproni* and *Fasciola hepatica*. Journal of Antimicrobial Chemotherapy 58: 1193–1197.

Keiser, J., Xiao, S. H., Tanner, M., and Utzinger, J. 2006d. Artesunate and artemether are effective fasciolicides in the rat model and in vitro. Journal of Antimicrobial Chemotherapy 57: 1139–1145.

Keun, H. C., Ebbels, T. M., Antti, H., Bollard, M. E., Beckonert, O., Schlotterbeck, G., Senn, H., Niederhauser, U., Holmes, E., Lindon, J. C., and Nicholson, J. K. 2002. Analytical reproducibility in ^{1}H NMR-based metabonomic urinalysis. Chemical Research in Toxicology 15: 1380–1386.

Krotov, A. I., and Gusel'nikova, L. M. 1971. Use of white mice infected with *Echinostoma miyagawai* Ischii, 1932, for experimental therapy of intestinal trematodiases. Meditsinskaia Parazitologiia i Parazitarnye Bolezni 40: 679–682.

Kwon, E. D., Zablocki, K., Peters, E. M., Jung, K. Y., Garcia-Perez, A., and Burg, M. B. 1996. Betaine and inositol reduce MDCK cell glycerophosphocholine by stimulating its degradation. American Journal of Physiology 270: 200–207.

Labow, B. I., and Souba, W. W. 2000. Glutamine. World Journal of Surgery 24: 1503–1513.

Lämmler, G. 1968. Chemotherapy of trematode infections. Advances in Chemotherapy 3: 153–251.

Leger, N., and Notteghem, M. J. 1975. The fasciolidal activity of a new compound, brotianide on *Echinostoma caproni* Richard, 1964. Annales Pharmaceutiques Francaises 33: 273–277.

Leger, N., Notteghem, M. J., and Cavier, R. 1973. Un test pharmacodynamique pour l'essai des douvicides. Bulletin de la Societe de Pathologie Exotique et de ses Filiales 66: 732–736.

Lenz, E. M., Bright, J., Knight, R., Westwood, F. R., Davies, D., Major, H., and Wilson, I. D. 2005. Metabonomics with ^{1}H-NMR spectroscopy and liquid chromatography-mass spectrometry applied to the investigation of metabolic changes caused by gentamicin-induced nephrotoxicity in the rat. Biomarkers 10: 173–187.

Lesaffer, G., De Smet, R., Belpaire, F. M., Van Vlem, B., Van Hulle, M., Cornelis, R., Lameire, N., and Vanholder, R. 2003. Urinary excretion of the uraemic toxin *p*-cresol in the rat: contribution of glucuronidation to its metabolization. Nephrology, Dialysis, Transplantation 18: 1299–1306.

Li, J. V., Wang, Y. L., Saric, J., Nicholson, J. K., Dirnhofer, S., Singer, B. H., Tanner, M., Wittlin, S., Holmes, E., Utzinger, J. 2008. Global metabolic responses of NMRI mice to an experimental *Plasmodium berghei* infection. Journal of Proteome Research 7: 3948–3956.

Lindon, J. C., Nicholson, J. K., and Wilson, I. D. 2000. Directly coupled HPLC-NMR and HPLC-NMR-MS in pharmaceutical research and development. Journal of Chromatography B, Biomedical Sciences and Applications 748: 233–258.

Lindon, J. C., Holmes, E., and Nicholson, J. K. 2001. Pattern recognition methods and applications in biomedical magnetic resonance. Progress in NMR Spectroscopy 39: 1–40.

Lun, Z. R., Gasser, R. B., Lai, D. H., Li, A. X., Zhu, X. Q., Yu, X. B., and Fang, Y. Y. 2005. Clonorchiasis: a key foodborne zoonosis in China. Lancet Infectious Diseases 5: 31–41.

Marchesi, J. R., Holmes, E., Khan, F., Kochhar, S., Scanlan, P., Shanahan, F., Wilson, I. D., and Wang, Y. L. 2007. Rapid and noninvasive metabonomic characterization of inflammatory bowel disease. Journal of Proteome Research 6: 546–551.

Martin, F. P. J., Verdu, E. F., Wang, Y., Dumas, M. E., Yap, I. K. S., Cloarec, O., Bergonzelli, G. E., Corthesy-Theulaz, I., Kochhar, S., Holmes, E., Lindon, J. C., Collins, S. M., and Nicholson, J. K. 2006. Transgenomic metabolic interactions in a mouse disease model: interactions of *Trichinella spiralis* infection with dietary *Lactobacillus paracasei* supplementation. Journal of Proteome Research 5: 2185–2193.

Maurer, K., Decere, M., and Fried, B. 1996. Effects of the anthelmintics clorsulon, rafoxanide, mebendazole and arprinocid on *Echinostoma caproni* in ICR mice. Journal of Helminthology 70: 95–96.

Merchant, M. L., and Klein, J. B. 2007. Proteomics and diabetic nephropathy. Seminars in Nephrology 27: 627–636.

Mitch, W. E., and Chan, W. 1979. α Ketoisocaproate stimulates branched-chain amino acid transaminase in kidney and muscle. American Journal of Physiology 236: E514–E518.

Miyakawa, H., Woo, S. K., Dahl, S. C., Handler, J. S., and Kwon, H. M. 1999. Tonicity-responsive enhancer binding protein, a rel-like protein that stimulates transcription in response to hypertonicity. Proceedings of the National Academy of Sciences USA 96: 2538–2542.

Nicholson, J. K. 2006. Global systems biology, personalized medicine and molecular epidemiology. Molecular Systems Biology 2: 52.

Nicholson, J. K., and Wilson, I. D. 1989. High resolution proton magnetic resonance spectroscopy of biological fluids. Progress in NMR Spetroscopy 21: 449–501.

Nicholson, J. K., Foxall, P. J. D., Spraul, M., Farrant, R. D., and Lindon, J. C. 1995. 750 MHz ^{1}H and ^{1}H-^{13}C NMR spectroscopy of human blood plasma. Analytical Chemistry 67: 793–811.

Nicholson, J. K., Lindon, J. C., and Holmes, E. 1999. 'Metabonomics': understanding the metabolic responses of living systems to pathophysiological stimuli via multivariate statistical analysis of biological NMR spectroscopic data. Xenobiotica 29: 1181–1189.

Nicholson, J. K., Connelly, J., Lindon, J. C., Holmes, E. 2002. Metabonomics: a platform for studying drug toxicity and gene function. Nature Reviews Drug Discovery 1: 153–161.

Notteghem, M. J., Leger, N., and Cavier, R. 1979. Study of fluke-killing activity of flubendazole on *Echinostoma caproni* Richard, 1964. Annales Pharmaceutiques Francaises 37: 153–156.

Notteghem, M. J., Leger, N., and Forget, E. 1980. Comparison of the fluke-killing activity of some compounds derivated from benzimidazole. Annales Pharmaceutiques Francaises 38: 61–63.

Ohdoi, C., Nyhan, W. L., and Kuhara, T. 2003. Chemical diagnosis of Lesch-Nyhan syndrome using gas chromatography-mass spectrometry detection. Journal of Chromatography B, Analytical Technologies in the Biomedical and Life Sciences 792: 123–130.

Ordovas, J. M., and Mooser, V. 2006. Metagenomics: the role of the microbiome in cardiovascular diseases. Current Opinion in Lipidology 17: 157–161.

Ramsay, I. R., and Pullammanappallil, P. C. 2001. Protein degradation during anaerobic wastewater treatment: derivation of stoichiometry. Biodegradation 12: 247–257.

Rim, H. J. 1984. Therapy of fluke infections in the past. Arzneimittelforschung 34: 1127–1129.

Rochfort, S. 2005. Metabolomics reviewed: a new "omics" platform technology for systems biology and implications for natural products research. Journal of Natural Products 68: 1813–1820.

Saric, J., Li, J. V., Wang, Y. L., Keiser, J., Bundy, J. G., Holmes, E., and Utzinger, J. 2008a. Global metabolic responses to *Echinostoma caproni* infection in NMRI mice. PLoS Neglected Tropical Diseases z: e754.

Saric, J., Wang, Y. L., Li, J., Coen, M., Utzinger, J., Marchesi, J. R., Keiser, J., Veselkov, K., Lindon, J. C., Nicholson, J. K., and Holmes, E. 2008b. Species variation in the fecal metabolome gives insight into differential gastrointestinal function. Journal of Proteome Research 7: 352–360.

Schlotterbeck, G., Ross, A., Dieterle, F., and Senn, H. 2006. Metabolic profiling technologies for biomarker discovery in biomedicine and drug development. Pharmacogenomics 7: 1055–1075.

Schmidt, J. 1998. Effects of benzimidazole anthelmintics as microtubule-active drugs on the synthesis and transport of surface glycoconjugates in *Hymenolepis microstoma*, *Echinostoma caproni*, and *Schistosoma mansoni*. Parasitology Research 84: 362–368.

Schneck, J. L., Fried, B., and Sherma, J. 2004. Effects of tonicity on the release of neutral lipids in *Echinostoma caproni* adults and observations on lipids in excysted metacercariae. Parasitology Research 92: 285–288.

Seibel, B. A., and Walsh, P. J. 2002. Trimethylamine oxide accumulation in marine animals: relationship to acylglycerol storage. Journal of Experimental Biology 205: 297–306.

Selmer, T., and Andrei, P. I. 2001. *p*-Hydroxyphenylacetate decarboxylase from *Clostridium difficile*. A novel glycyl radical enzyme catalysing the formation of *p*-cresol. European Journal of Biochemistry 268: 1363–1372.

Shukla-Dave, A., Roy, R., Bhaduri, A. P., and Chatterjee, R. K. 2000. Effects of 2-deoxy-D-glucose on *Acanthocheilonema viteae*: rodent filariids as studied by multinuclear NMR spectroscopy. Physiological Chemistry and Physics and Medical NMR 32: 1–12.

Smith, J. L., Wishnok, J. S., and Deen, W. M. 1994. Metabolism and excretion of methylamines in rats. Toxicology and Applied Pharmacology 125: 296–308.

Thwaites, D. T., and Anderson, C. M. 2007. H^+-coupled nutrient, micronutrient and drug transporters in the mammalian small intestine. Experimental Physiology 92: 603–619.

Tinga, N., De, N., Vien, H. V., Chau, L., Toan, N. D., Kager, P. A., and Vries, P. J. 1999. Little effect of praziquantel or artemisinin on clonorchiasis in Northern Vietnam. A pilot study. Tropical Medicine and International Health 4: 814–818.

Trygg, J., and Wold, S. 2002. Orthogonal projections to latent structures (O-PLS). Journal of Chemometrics 16: 119–128.

Trygg, J., Holmes, E., and Lundstedt, T. 2007. Chemometrics in metabonomics. Journal of Proteome Research 6: 469–479.

Utzinger, J., Xiao, S. H., Tanner, M., and Keiser, J. 2007. Artemisinins for schistosomiasis and beyond. Current Opinion in Investigational Drugs 8: 105–116.

Vasta, J. D., Fried, B., and Sherma, J. 2008. High performance thin layer chromatographic analysis of neutral lipids in the urine of BALB/c mice infected with *Echinostoma caproni*. Parasitology Research 102: 625–629.

Wang, Y. L., Holmes, E., Nicholson, J. K., Cloarec, O., Chollet, J., Tanner, M., Singer, B. H., and Utzinger, J. 2004. Metabonomic investigations in mice infected with *Schistosoma mansoni*: an approach for biomarker identification. Proceedings of the National Academy of Sciences USA 101: 12676–12681.

Wang, Y. L., Utzinger, J., Xiao, S. H., Xue, J., Nicholson, J. K., Tanner, M., Singer, B. H., and Holmes, E. 2006. System level metabolic effects of a *Schistosoma japonicum* infection in the Syrian hamster. Molecular and Biochemical Parasitology 146: 1–9.

Wang, Y., Lawler, D., Larson, B., Ramadan, Z., Kochhar, S., Holmes, E., and Nicholson, J. K. 2007. Metabonomic investigations of aging and caloric restriction in a life-long dog study. Journal of Proteome Research 6: 1846–1854.

Wang, Y. L., Utzinger, J., Saric, J., Li, J. V., Burckhardt, J., Dirnhofer, S., Nicholson, J. K., Singer, B. H., Brun, R., and Holmes, E. 2008. Global metabolic responses of mice to *Trypanosoma brucei brucei* infection. Proceedings of the National Academy of Sciences USA 105: 6127–6132.

Willson, M., Lauth, N., Perie, J., Callens, M., and Opperdoes, F. R. 1994. Inhibition of glyceraldehyde-3-phosphate dehydrogenase by phosphorylated epoxides and alpha-enones. Biochemistry 33: 214–220.

Wong, J. M., de Souza, R., Kendall, C. W., Emam, A., and Jenkins, D. J. 2006. Colonic health: fermentation and short chain fatty acids. Journal of Clinical Gastroenterology 40: 235–243.

Wright, E. M., Hirayama, B., Hazama, A., Loo, D. D., Supplisson, S., Turk, E., and Hager, K. M. 1993. The sodium/glucose cotransporter (SGLT1). Society of General Physiologists Series 48: 229–241.

Xiao, S. H., Wu, H. M., Tanner, M., Utzinger, J., and Wang, C. 2005. Tribendimidine: a promising, safe and broad-spectrum anthelmintic agent from China. Acta Tropica 94: 1–14.

Xiao, S. H., Keiser, J., Chollet, J., Utzinger, J., Dong, Y., Endriss, Y., Vennerstrom, J. L., and Tanner, M. 2007. In vitro and in vivo activities of synthetic trioxolanes against major human schistosome species. Antimicrobial Agents and Chemotherapy 51: 1440–1445.

Yancey, P. H., Clark, M. E., Hand, S. C., Bowlus, R. D., and Somero, G. N. 1982. Living with water stress: evolution of osmolyte systems. Science 217: 1214–1222.

Yoshimura, H. 1965. The life cycle of *clonorchis sinensis*: A comment on the presentation in the seventh edition of Craig and Fanst's clinical parasitology. Journal of Parasitology 6: 961–966.

Yu, L., Blaser, M., Andrei, P. I., Pierik, A. J., and Selmer, T. 2006. 4-Hydroxyphenylacetate decarboxylases: properties of a novel subclass of glycyl radical enzyme systems. Biochemistry 45: 9584–9592.

Zablocki, K., Miller, S. P., Garcia-Perez, A., and Burg, M. B. 1991. Accumulation of glycerophosphocholine (GPC) by renal cells: osmotic regulation of GPC:choline phosphodiesterase. Proceedings of the National Academy of Sciences USA 88: 7820–7824.

Zeisel, S. H., Mar, M. H., Howe, J. C., and Holden, J. M. 2003. Concentrations of choline-containing compounds and betaine in common foods. Journal of Nutrition 133: 1302–1307.

Index

A

Acanthoparyphium, 5, 8
Acanthoparyphium tyosenense, 10, 152, 153
Acris crepitans, 76
Adult echinostomes
  chemoattraction and mate finding, 99
  growth and development, 97–98
  mating behavior, 100–101
Amphibians
  echinostomes
    emergence of, 268–269
    hosts involved in, 254–257
    infection pathology and patterns, 265–268
    second intermediate hosts, 265
    vs. Ribeiroia, 264
  environmental changes, macroparasite infections
    biodiversity loss, 272–273
    climate change, 273
    importance of, 275
    land use and wetland characteristics, 270
    nutrient pollution, 270–271
    pesticide contamination, 271
  predation risk, 274, 275
  Ribeiroia
    emerging infection, 262–263
    frog deformities and trematode infection, 253, 258, 259
    hosts involved in, 254–257
    malformation associated with, 252
    mass malformations, 262
    Ribeiroia-induced pathology, 259–261
  trematodes, importance in, 250–252
Anas boschas fereae, 165
Anas platyrhynchos, 66
Anthelminthic drugs, 298–299
Antibody responses
  immune proteomic approach, 197–198
  primary echinostome infections, 196–197
Antigenic characterization, adult echinostomes, 199–200
Antioxidant factors, 42
Argentophilic papillae, 11
Artificial spring water (ASW), 119, 120
Artyfechinostomum malayanum, 154
Artyfechinostomum oraoni, 154–155
Artyfechinostomum sufrartyfex, 154
Austrovenus stutchburyi, 74

B

Babesia, 289
Batrachochytrium dendrobatidis, 251
Bellamya aeruginosa, 22
Benzimidazoles, 298
Bge cells, 40, 65
B. glabrata, 14, 137
Biomphalaria glabrata, 14, 36, 162
  embryonic (Bge), 39
  hemocytes, effect of echinostome excretory/secretory products on, 42
Bivalvia, 63
Bulinidae, 14

C

Caballerotrema, 5
Ca ions, 41
Cathaemasiidae, 7
Cercariae
  chaetotaxy and its patterns, 3, 15
  Echinochasmus spp., 9
  mortality, 63
  Parorchis acanthus, 10
  second intermediate hosts studies, 18
  survival and effects of temperature, 16
  in vitro encystment, 123

Cercariae and metacercariae, of *E. caproni* and *E. trivolvis*
  effects of copper sulfate toxicity on, 77
  metolachlor and atrazine on host–parasite system, 78
  toxicity of antifouling biocides tributyltin (TBTO) and, 77–78
Cercarial tail of *Echinostoma* species, 15
Cercaria yamagutii, 152
Cerithidea cingulata, 152
Cerithidea largillierti, 152
Chemotherapy, 296
Cipangopaludina chinensis malleata, 160
Cipangopaludina japonica, 165
Cipangopaludina malleata, 165
Cipangopaludina sp., 160
43- and 45- collar-spined *Echinoparyphium,* 5
37-collar-spined *Echinostoma,* 6, 63
  cercariae encyst and develop to metacercariae in vitro, 63
37-collar-spined echinostomes, 8
37-collar-spined revolutum group of *Echinostoma,* 6
Collar spines, importance for taxonomic study, 148
Column liquid chromatography
  methods and applications, 243–244
  principles, 243
Corbicula lindoensis, 162
Corbicula producta, 165
Corbicula sucplanta, 162
Crassostrea virginica, 73
Curtuteria australis, 11
Cytochrome c oxidase subunit 1 (CO1), 7
Cytokine responses, 198–199

D

Digenea, 36
Digestive gland–gonad complex (DGG)
  metallic ion study, 244–245
  sterol content, 242
Digoniostoma pulchella, 154
DNA genome studies, echinostomes
  EST sequencing, 213
  internal transcribed spacers (ITS) sequence, 210–211
  putative gene library, 212
Dreissena polymorpha, 64, 75
Drepanocephalus, 3
Dulcolax™, 119

E

Echinochasmus, 8, 63
Echinochasmus beleocephalus, 156
Echinochasmus fujianensis, 7, 155
Echinochasmus japonicus, 155–156
Echinochasmus jiufoensis, 156
Echinochasmus liliputanus, 22, 63, 133
Echinochasmus liliputanus cercariae, 123
Echinochasmus perfoliatus, 157
Echinoparyphium, 3, 5, 8
Echinoparyphium aconiatum, 65
Echinoparyphium elegans, 7, 63
Echinoparyphium megacirrus, 18
Echinoparyphium paraulum, 165
Echinoparyphium recurvatum, 10, 133, 134, 158–160
Echinoparyphium recurvatum metacercariae, 63
Echinoparyphium serratum, 92
Echinoparyphium sp., 10
Echinoparyphium spp., 253, 264
Echinostoma, 2, 3
  wild animals
    Asia and Europe, 132
    North America and Central and South America, 131
  zoonotic infections and main intermediate hosts, 133
Echinostoma angustitestis, 160
Echinostoma caproni
  cercariae, 15
  echinostome cysts, 65
  echinostome infections, 37
  experimental models, 134
  immune evasion, 40
  infection, metabolic profiling
    biofluids utility, 312
    dignostics metabolites, 307–309
    plasma, 310
    stool, 311–312
    urine, 310–311
  metacercariae pathology, 75
  miracidium, 11, 13
  newly hatched miracidium, 38
  in ovo cultivation, 120–121
  primary infection, 96
  species-rich genera, 6–8
  sporocysts, 42
Echinostoma caproni mouse model
  integrative metabonomics
    cross cellular transport, 317–318
    holistic influence, gut microbiota, 318
    metabolite flow, 319

intestinal profile changes
amino acids levels, 317
glucose concentration, 316
SCFAs role, 314–315
MAS NMR spectroscopy
kidney, 313
liver, 313–314
spleen, 314
metabonomic investigation
biomarker identification and diagnostic potential, 306–312
experimental protocol, 305–306
metabonomics studies, 302–305
trematocidal drug discovery and development
drug sensitivity assay, 300
scanning electron microscopic (SEM) studies, 300–301
spectroscopy, 302
in vivo evaluation, 297–299
E. cinetorchis
cercariae, 63
description, 160, 161
E. echinatum, 160–162
E. friedi, 12
E. hortense, 162, 164, 174
E. ilocanum, 164–165
E. luisreyi, 65
E. macrorchis, 165
E. paraensei, 17, 37, 65, 134
E. revolutum, 133, 137, 165–166, 254–257, 264, 265, 267
Echinostoma, sequencing ND1 and ITS genes, 7
Echinostomatidae
classification, 4–5
cysts, 64
flukes, 148
genera of, 2
major and minor species, 130–134
E. trivolvis, 63, 112, 254–257, 264, 267
Echinostome cercariae
morphology and immunocytochemistry, 15
second intermediate host, 63
survival and infectivity characteristics, 16
Echinostome coinfection
helminths
nematodes, 287
tapeworms, 286–287
trematodes, 284–286
heterologous echinostomes
concurrent infections, 282–283
primary infections, 283–284
protozoans
Babesia, 289
plasmodium, 287–289
Trypanosomes, 290
Echinostome infection
emerging disease in green frogs, 76
host resistance, 172
metacercariae excystment, 91–93
Echinostome life cycle maintenance
Dr. Bernard Fried's laboratory studies, 114–115
laboratories, 113
mammals in the wild, life cycle initiation source, 119
obtaining material from, 116
relationships between echinostomes, 117–118
Echinostome-like digeneans, 2, 5, 8
Echinostome-like metacercariae, *Ribeiroia,* 21–22
Echinostome-like *Ribeiroia,* rhythmic emergence, 17
Echinostome-like trematode group, 5
Echinostome metacercariae
cysts storage, 23
encystment, 62–63
invertebrate and vertebrate hosts
behavioral and morphological changes, 73–74
marinades effects, 78–79
pathology of, 75–77
pesticides and herbicides effects, 77–78
second intermediate hosts, 67–72
second intermediate hosts
development stages, 65–66
excystation, 67
site selection in, 63–64
Echinostome miracidia, 13
Echinostome-related genera, 8
Echinostomes
adult echinostomes
growth and development of, 97–98
mating behavior of, 98–99
column liquid chromatography
methods and applications, 243–244
principles, 243
definitive host establishment, 93–96
avian hosts, 94
concurrent infections with other helminths, 95–96
E. trivolvis, compatibility with hamsters, 95
increased dosage levels' influence, 95

Echinostomes (*cont.*)
rats as hosts, low compatibility with *E. caproni,* 94
recovery rates, *E. paraensei,* 95
use of rodents as experimental hosts, 94
description, 186
emergence of, 268–269
excystation, 93, 121–122
experimental models
life cycle of *Echinostoma* spp. in wild, 135
major echinostome species, 134–137
minor echinostome species, 137
gas chromatography
methods and applications, 241–243
principles, 241
genomics
DNA and RNA studies, 210–213
prospects and limitations, 213–214
taxonomic and phylogenetic studies, 209–210
genomics and proteomics, 208–209
host specificity and infectivity, 90–91
human infection
E. hortense infection endoscopic findings, 171
pathology and symptomatology, 167–171
immune mechanisms
antibody responses, 196–198
cytokine responses, 198–199
eosinophils, 193–194
goblet cells, 194–196
mast cells, 192–193
immunobiology
different strains and host species, 189–190
immunosuppressed mice, 189
rodent models, 188–189
immunodiagnosis
antibody detection, 200
seroantigen and coproantigen detection, 201
inductively coupled plasma-optical emission spectrometry
methods and apllications, 245
principles, 244
infection pathology and patterns
in amphibian kidney, 263, 267
geographic and frequency distribution, 266
population level impacts, 267, 268
innate immunity, 44
Biomphalaria Glabrata
antidigenean responses in, 47–49
immune system, 44–45
Echinostoma vs. Schistosoma, tools for functional validation, 51
immune-relevant genes, predicted identity of, 50
invertebrate immunity, perceived simplicity of, 45–46
invertebrates, immunological memory in, 46–47
transcriptomic and proteomic studies, 49, 51
metacercarial excystment, 91–93
molecular-based studies, 6
natural and experimental infections, 138–139
pathology, 190–192
proteomics
analytical procedures, 217–220
immunoproteomics, 220–222
prospects and limitations, 222–223
protein identification, 214–217
resistance manifestations to infections
fecundity reduction, 188
host immunosuppression, 186
primary infections, 187
secondary infections, 187–188
Riberoia ondatrae life cycles, 254–257
second intermediate hosts, 265
and snails, exploring complex interactions, 35–37
systematics, 2
thin-layer chromatography
amino acid analysis, 239–240
carbohydrates detection, 240
chromatogram quantification, 234–235
layers and mobile phases, 232–233
lipid studies, 235–239
lipophilic pigments identification, 240–241
principles, 230–231
sample application and plate development, 233
sample preparation, 231–232
statistical data analysis, 235
zone detection and identification, 234
vertebrate host fecundity, 101–104
vs. Ribeiroia, 264
in the wild
helminths co-infection, 138
major and minor species, 130–134
worm distribution and habitat location, 96–97

Echinostomes and related genera
    description of life cycles, 10–11
    free-living stages
        cercaria, 15–19
        miracidium, 11–14
    generalized life cycle of
            echinostome, 10
    life cycle pattern, 9–11
    parasitic stages
        invertebrate hosts, 19–24
        vertebrate hosts, 24–25
Echinostome trematodes, 62
E. cinetorchis, 11
Ectosymbiotic flatworm, 18
E. deserticum, 3
E. echinatum, 6, 8, 9, 65
E. friedi, 2, 6, 8, 102, 103, 134
E. hortense, 91
E. ilocanum, 78
E. japonicus, 7
E. jurini, 6, 11
E. liliputanus cercariae, 65
E. luisreyi, 3
E. malayanum, 78, 119, 154
Embden-Meyerhof-Parnas pathway, 42
E. megacirrus, 10
Emergence pattern analysis, 17
E. miyagawai, 6
Encysted metacercaria, 75–76
Eosinophils, 193–194
E. paraensei, 6, 7, 8, 9, 11, 13, 14, 138
E. paraensei and *E. caproni* miracidia hatch,
        12, 13
E. parvocirrus, 6
Episthmium caninum, 166
E. pseudorecurvatum, 7
E. recurvatum, 7
E. recurvatum metacercarial cysts, 66
E. revolutum, 6, 7, 8, 9, 11, 13, 65
ES proteomes, 42
E. trivolvis, 6, 7, 8, 9, 11, 13, 65, 91, 93, 96,
        102, 103, 134
E. trivolvis excystation, 92
Euparyphium, 3, 8
Euparyphium albuferensis, 3, 96
Excretory/secretory products (ESP)
    description, 208–209
    ex vivo studies, 41
    protein characterization, 216–217
    proteomic studies
        echinostome sporocysts, 43–44
        glycolytic enzymes, 42–43
        molecular processes, 41–42
Expressed sequence tags (EST), 208, 223

F
Fasciola revoluta, 165
Fascioloides magna, 40
Flukes, 148, 152, 166
Food-borne trematode infections, 140

G
Gas chromatography
    methods and applications
        fatty acid determination, 242–243
        sterols identification, 241–242
    principles, 241
Gastropoda, 40
GenBank, 7, 43
Genomics
    definition, 208
    echinostomes
        DNA and RNA studies, 210–213
        prospects and limitations,
            213–214
        ribosomal and mitochondrial markers,
            209–210
Glassaulax didyma, 152
Global biodiversity, of animals, 130
Glycerophosphocholine (GPC) activity
    cross cellular transport, 317–318
    kidney and liver, 313–315
Glycolytic enzymes, 42
Glycomics, 222
Gnathopogon strigatus, 155
Goblet cells
    echinostome expulsion, 195–196
    in human echinistomes, 172–173
    hyperplasia, 194–195
Gyraulus convexiusculus, 154

H
Helisoma, 14
Helisoma trivolvis, 112
Helminths, echinostome coinfection
    nematodes, 287
    tapeworms, 286–287
    trematodes
        E. caproni infections, 285–286
        Schistosoma spp, 284–285
H. elongata larvae, 63
Hemocyte-mediated cytotoxicity, 42
Heterologous echinostomes, 282–284
Himasthla, 5, 8
Himasthla elongata, 63, 121
Himasthla escamosa, 10
Himasthla leptosoma, 92

Himasthla muehlensi, 166
Himasthla quissetensis, 9, 133
Himasthla rhigedana, 63, 133
Human echinostomiasis
  diagnosis, treatment, and control, 174–175
  immunology of, 172
    cytokines and Th1/Th2 immune responses, 173–174
    effector cells and antibodies, role of, 173
    goblet cells and mast cells role, 172–173
    host resistance to, 172
  symptoms and diagnosis, 140
Human-infecting echinostomes
  egg size, 175
  geographical distribution of, 152
  human/animal infections' sources, 151
  morphological characters and life cycle, 149–150
  taxonomy, biology and epidemiology of, 150
Hypoderaeum, 3, 8
Hypoderaeum conoideum, 7, 10, 134, 166–167
Hypomesus olidus, 155
Hypoxic environment, 39

I
Idiopoma javanica, 162
Immune mechanisms, echinostome infections
  antibody responses, 196–198
  cytokine responses, 198–199
  eosinophils, 193–194
  goblet cells, 194–196
  mast cells, 192–193
Immunobiology, echinostomes
  different strains and host species, 189–190
  immunosuppressed mice, 189
  rodent models, 188–189
Immunodiagnosis
  antibody detection, 200
  seroantigen and coproantigen detection, 201
Immunome, 220, 222
Indoplanorbis exustus, 154, 166
Inductively coupled plasma-optical emission spectrometry
  methods and applications, 245
  principles, 244
Integrative metabonomics
  cross cellular transport, 317–318
  holistic influence, gut microbiota, 318
  metabolite flow, 319
Internal transcribed spacers 1 and 2 (ITS1 and ITS2), 7
Intestinal profile changes, *E.caproni* infection
  amino acids levels, 317
  glucose concentration, 316
  SCFAs role, 314–315
Intramolluskan echinostome stages, 12
Intramolluskan parasitic stages, 21
Isthmiophora, 3, 8
Isthmiophora, chemo-orientation in, 16
Isthmiophora melis, 167
ITS gene, 7

K
Karyotyping, 6

L
Larus crassiostris, 153
Larval and adult echinostomes
  maintenance of, 119
  transplantation to new sites, 121
Larval echinostomes
  immune evasion strategies
    ex vivo studies, 41
    proteomic studies, 41–44
    in vivo studies, 40
Littorina littorea, 63, 121, 166
Locke's ASW (1:1) medium, 123
Lunatia fortuni, 152
Lymnaea, 40
Lymnaea columella, 14, 137
Lymnaea corvus, 167
Lymnaea (Bullastra) cumingiana, 78, 154
Lymnaea peregra, 14, 63, 160, 167
Lymnaea pervia, 160
Lymnaea tumida, 166
Lymnaeidae, 14
Lymnaeid snails, for *H. conoideum* miracidia, 13

M
Mactra veneriformis, 152
MAPK-associated signaling pathways, 51
MAS NMR spectroscopy
  kidney, 313
  liver, 313–314
  spleen, 314
Mass spectrometry (MS), 303–304
Mast cells
  echinostome infections resistance, 192–193
  human echinostomes, 172–173

Melanitta fusca stejnegeri, 152
Melanitta nigra americana, 152
Metabolic profiling, 297
 applications, 302–303
 chemometric tools, 304–305
 E.caproni infection
  biofluids utility, 312
  diagnostics metabolites, 307–309
  plasma, 310
  stool, 311–312
  urine, 310–311
 experimental protocol, 305–306
 integrative metabonomics
  cross cellular transport, 317–318
  holistic influence, gut microbiota, 318
  metabolite flow, 319
Metabonomics. *See* Metabolic profiling
Metacercariae, echinostomes
 cysts viability and copper, 23–24
 excystment, 91–93
 site selection, second intermediate hosts, 63–64
 in situ evaluation, 79
Mical-like protein, 42–43
Miracidia
 E. caproni, 13, 39
 E. friedi, 13–14
 E. trivolvis, 13
 Euparyphium albuferensis, 13
Misgurnus anguillicaudatus, 160
Mitochondrial DNA (mtDNA), 7
Movement patterns, of cercariae, 15
Myofibril F-actin staining, 15
Mytilus edulis, 63

N
N. braziliensis, 96
Nectomys squamipes, 137, 140
Nematodes, echinostome coinfection, 287
Neocanthoparyphium echinatoides, 7
Nicotinamide adenine dinucleotide dehydrogenase subunit 1 (ND1), 7
Nippostrongylus braziliensis, 96
Nonhost snails, 13
N. squamipes, 138, 139
Nucella lapillus, 10
Nuclear magnetic resonance (NMR) spectroscopy. *See also* MAS NMR spectroscopy
 biofluids analysis, 303
 metabolic profiling, protocol, 305–306
 Trypanosoma brucei, 302
Nuclear ribosomal DNA (rDNA), 7

O
Odontobutis obcura interrupta, 163
Oncomelania, 40
Ondatra zibethica, 25
Oxidative damage, 42
Oysters, 73

P
19 papilla-like structures, 11
Parafossarulus manchouricus, 155
Parafossarulus striatulus, 157
Parasitic stages, echinostomes
 invertebrate host, 19–24
  metacercariae, 21–24
  sporocysts and rediae, 19–21
Parasitism, 140
Parorchis, 5, 8
Parorchis acanthus, 92
Paryphostomum, 3
P. combesi, 10
P. distomatosa, 10
Petasiger, 8
Petasiger encyst, 63
Petasiger variospinosus, 10
Philophthalmidae, 5
Philophthalmus lucipetus, 10
Physa acuta, 160
Physa alexandrina, 160
Physa fontinalis, 160
Physa marmorata, 14, 137
Physastra hungerfordiana, 78
Physidae, 14
Pila conica, 165
Pila luzonica, 78
Pila scutata, 154
Pisidium coreanum, 73
P. islandicus, 10
Planorbella, 14
Planorbella trivolvis, 264
Planorbidae, 14
Planorbis compressus japonicus, 165
Planorbis corneus, 166
Planorbis planorbis, 160
Plasmodium, 287–289
Plasmodium falciparum, 112
Polymerase chain reactions (PCR), 6
Praziquantel, 119
Principal component analysis (PCA), 304
Prosobranchia, 40, 63
Proteomics
 analytical procedures
  imaging and comparison, 219

Proteomics (*cont.*)
protein identification, mass spectroscopy, 219–220
protein separation and staining, 218–219
sample preparation, 217–218
definition, 208–209
echinostomes
antigens identification, 220–222
prospects and limitations, 222–223
protein characterization, 214
protein identification
enzymes and ligand-binding molecules, 214–215
ESP characterization, 216–217
Protozoans, echinostome coinfection, 287–290
Babesia, 289
plasmodium, 287–289
Trypanosomes, 290
Pseudechinoparyphium echinatum, 7
Pseudomyicola spinosus, 74
Psilostomidae, 7

R
Radix auricularia coreana, 160
Radix peregra, 14
Radix quadrasi, 78
Rana pipiens, 64
Rana temporaria, 160
Random amplification of polymorphic DNA (RAPD), 6
Rattus rattus, 25
R. clamitans, 64, 76
R. congolensis, 7
Reactive oxygen species (ROS), 43
Retractile apical papillae, 11
Revolutum group, 5
Ribeiroia
cercariae, 258, 267, 271
emerging infection, 262–263
first intermediate hosts, 5, 9, 14
frog deformities and trematode infection
frequency distribution, 261
geographic distribution, 259, 260
limb malformations, 258
hosts involved in, 254–257
malformation associated with
emerging infection, 262
representative malformation, 252
Ribeiroia-induced pathology, 259–261
Ribeiroia, ITS2 sequences, 7
Ribeiroia ondatrae, 251, 253–257, 264
R. marini, 7
RNA genome studies. *See* DNA genome studies, echinostomes
R. ondatrae, 7
ROS-scavenging molecules, 43
R. peregra, 14
Ruditapes philippinarum, 152

S
Scanning electron microscopic (SEM) studies, 300–301
Schistosoma haematobium, 37
Schistosoma intercalatum, 40
Schistosoma japonicum, 40
Schistosoma mansoni, 36, 37, 40, 138
Schistosoma mansoni sporocysts, 42
Schistosoma mattheei, 40
Second intermediate hosts, echinostome metacercariae
developmental stages
cysts wall and shape, 65
surface topography, echinostome digenean, 66
excystation, 67
invertebrate and vertebrate hosts, 67–73
metacercarial site selection in, 63–64
Segmentina hemisphaerula, 73, 160
Segmentina nitiella, 165
Sensilla in cercariae, 6
Singhia, 2, 5
Snail intermediate hosts, echinostomes
host exploitation *vs.* host immune response, 39–40
miracidia host-finding behavior, 37–39
S. nitiella, 165
Solen grandis, 152
Solen strictus, 152
Spectroscopy, 297, 302
Sporocysts, 40
Stephanoprora, 3, 8
Stephanoprora aylacostoma, 10
Stephanoprora uruguayense, 10
Strigeidida, 36

T
Tapeworms, 286–287
Taxidea taxus, 25
Tegument, 11
Temnocephala chilensis, 18
Thin-layer chromatography
applications in echinostomes
amino acids, 239–240

carbohydrates, 240
lipids, 235–239
lipophilic pigments, 240–241
chromatogram quantification, 234–235
data analysis, 235
layers and mobile phases, 232–233
principles, 230–231
sample application and plate development, 233
sample preparation, 231–232
zone detection and identification, 234
Toxic pollutants, 17
Trematocidal drug discovery and development
drug sensitivity assay, 300
echinostomicidal activity evaluation
anthelminthic studies, 297–298
drug screening studies, 298–299
scanning electron microscopic (SEM) studies, 300–301
spectroscopy, 302
Trematode infections, 296
Trematodes, 250, 251, 284–286. *See also Echinostoma; Ribeiroia*
Trematodes, exposed to ROS, 43
Trypanosomes, 290
Trypsin-bile salts (TB) medium, 79
Tympanotonus microptera, 152

U
Uroproctepisthmium, 3

V
Valvata piscinalis, 160
Venus mercenaria, 166
Vertebrate host, echinostomes fecundity
factors affecting fecundity, 101–103
methods for measuring fecundity, 103–104
Viviparus javanicus, 165
Viviparus malleatus, 165

W
Wild echinostome zoonoses, potential of, 139–140
Wildlife conservation, impact of parasitism, 140–141

Z
Zebra mussels, 64, 75
Zoonotic disease, 134

MIX
Papier aus verantwortungsvollen Quellen
Paper from responsible sources
FSC® C105338

If you have any concerns about our products,
you can contact us on
ProductSafety@springernature.com

In case Publisher is established outside the EU,
the EU authorized representative is:
Springer Nature Customer Service Center GmbH
Europaplatz 3, 69115 Heidelberg, Germany

Printed by Libri Plureos GmbH
in Hamburg, Germany